BASICS OF
MECHANICAL ENGINEERING

BASICS OF
MECHANICAL ENGINEERING

R.K. Singal
Professor & Consultant
IILM Academy of Higher Learning, Greater Noida.

Mridul Singal
B.Tech. (IIT, Kanpur), M.Tech. (IISC, Bangalore),
IBM India Pvt. Ltd.

Rishi Singal
B.Tech. (PEC Chandigarh)
Accenture India Pvt. Ltd.

I.K. International Publishing House Pvt. Ltd.
NEW DELHI • BANGALORE

Published by

I.K. International Pvt. Ltd.
S-25, Green Park Extension
Uphaar Cinema Market
New Delhi–110 016 (India)
E-mail:info@ikinternational.com
Website: www.ikbooks.com

ISBN 978-81-89866-20-4

© 2007 I.K. International Publishing House Pvt. Ltd.

Reprint 2014

All rights reserved. No part of this publication may be reproduced, stored
in a retrieval system, or transmitted in any form or any means: electronic,
mechanical, photocopying, recording, or otherwise, without the prior written
permission from the publisher.

Published by Krishan Makhijani for I.K. International Publishing House
Pvt. Ltd., S-25, Green Park Extension, Uphaar Cinema Market, New Delhi–110
016 and Printed by Rekha Printers Pvt. Ltd., Okhla Industrial Area, Phase II,
New Delhi–110 020.

Preface

"Basics of Mechanical Engineering" is meant for the students of all branches of engineering and subject teachers. The book will also be helpful to engineers preparing for various competitive examinations and memberships.

The primary attractions of the book are the presentation style of the text, its simple and clear language, and use of comparison tables and illustrative drawings. The difficulties usually faced by new engineering students have been carefully dealt with while preparing the book.

In addition to exhaustive theory, a large number of numerical problems have been selected from university examination papers and question banks, graded properly, solved and arranged in various chapters. Exhaustive question banks on theory problems have also been added at the end of every chapter. The students need supplementary material for the subject, the same has been provided as annexures in the book.

The authors are grateful to the publisher for bringing out this book in a very short time. The authors are specially thankful to Mr. Anand Singh Aswal, Sales Executive of I.K. International Publishing House Pvt. Ltd. Although the authors are constantly in touch with the students while teaching and understanding their requirements and difficulties, any constructive criticism will be welcome with gratitude.

Authors

Contents

Part II: Force and Structure Analysis (Engineering Mechanics)

Part III: Stress and Strain Analysis (Strength of Materials)

PART I: THERMODYNAMICS

Fundamental Concepts and Definitions

1.1 THERMODYNAMICS

Thermodynamics is the basic science that deals with the conversion of heat into mechanical work. The application of basic thermodynamics for the solution of engineering problems is called applied thermodynamics or engineering thermodynamics. Applied thermodynamics plays a vital role in the design and development, analysis and optimisation of various thermal machines, plants and systems connected with power production and power consumption.

The whole of heat supplied (Q_1) to a heat engine (Fig. 1.1) from a heat source cannot be converted into mechanical energy or work (W). Some portion of heat (Q_2) at temperature T_2 has to be degraded and rejected to the environment or heat sink. The portion of heat energy which is not available for conversion into work is measured by a thermodynamic property called *entropy*. The part of heat which is available for conversion into maximum work is called *exergy*. Thus, thermodynamics is a science which deals with the computation of energy, exergy and entropy.

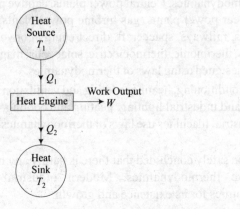

Fig. 1.1 Heat engine

Thermodynamics is based on four laws:

1. Zeroth law deals with *equality of temperature* and helps in *measurement of temperature.*

2. First law is the law of conservation of energy and introduces the *concept of internal energy*.

3. Second law puts limit on the amount of conversion *of heat into work* and introduces the *principle of increase of entropy*.

4. Third law defines the *absolute zero of entropy*.

These laws are based on experimental observations and have no mathematical proof. Their validity stems from the fact that these laws have not been violated.

1.2 SCOPE OF THERMODYNAMICS

Thermodynamics has very wide applications:

1. Thermal engineering is a very important associate branch of mechanical, chemical, metallurgical, automobile, aerospace, marine, textile, energy technology and environmental engineering. Thermodynamics is the basic science of thermal engineering.

2. Wherever, there is combustion, heating or cooling, exchange of heat for carrying out chemical reactions, conversion of heat into work for producing mechanical or electrical power, propulsion of rockets, railway engines, ships, etc. application of thermal engineering is required.

3. Process engineering of pharmaceutical, refinery, fertilizer, food processing, sugar, paper and pulp, organic and inorganic chemical plants is based on thermodynamics.

4. All process and engineering industries, agriculture, transport, commercial and domestic activities use thermal engineering for their design and operation.

5. Energy technology and power sectors are purely dependent on the laws of thermodynamics. Central power plants, captive power plants based on coal, nuclear power plants, gas turbine power plants, engines for automobiles, ships, railways, spacecraft, direct energy conversion devices such as fuel cells, thermionic, thermoelectric, solar cells, magneto-hydrodynamic plants are designed using laws of thermodynamics.

6. Airconditioning, heating, cooling and ventilation plants, domestic, commercial and industrial lighting, heating and cooling, agricultural, transport and industrial machines use laws of thermodynamics for their design and operation.

It can be safely concluded that there is hardly any human activity which does not involve thermodynamics. Modern industrial society depends upon thermodynamics for its existence and growth.

1.3 STUDY OF THERMODYNAMICS

Special techniques are used for the study of thermodynamics. Highly simplified substitutes and idealizations are made to replace actual physical materials and actions.

1. A region or a quantity of matter is identified as *system* for the study and analysis of a problem. The system is isolated from the surrounding through a real or imaginary boundary.
2. Thermodynamics makes no hypothesis about the structure of matter of the system. The system is regarded as a *continuum* i.e., continuous distribution of matter.
3. The matter is described *macroscopically* based on time-averaged measured values of the properties of the system.
4. Thermodynamics deals with heat and work and also properties of substances that transport heat and work between system and surrounding. The properties are grouped as measurable *properties* in the laboratory with the help of an instrument such as pressure, temperature, volume, mass, etc. and properties which can be calculated from experimental data using thermodynamic relations such as internal energy, enthalpy, specific heat, etc.
5. The concept of *quasi-static process* is used and it is assumed that the system is very near to equilibrium state during a process.

1.4 TYPES OF SYSTEMS

1.4.1 Thermodynamic System

Whenever a change is to be studied or a problem is to be analysed, it is necessary to identify a region or a quantity of matter. In thermodynamics, a real or imaginary boundary is drawn around the region of consideration. The shape and size of the boundary may be fixed or may change. The boundary may be in motion or at rest. The composition of the matter inside the system may be fixed or may change because of chemical or nuclear reactions. Everything within the boundary is called the system, while external to the boundary is called the surrounding. The system and surrounding together constitute the universe. All transfer of mass and energy (heat and work) between the system and the surrounding are evaluated at the boundary.

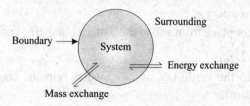

Fig. 1.2 Thermodynamic system.

The first step in the thermodynamic analysis is the selection of a system, its boundary and its surrounding. A careful and judicious selection of the system and its boundary can simplify the analysis very much. Thermodynamic analysis includes the study of transfer of mass and energy across the boundary of the system.

1.4.2 Types of Boundaries

(i) *Adiabatic boundary*: The walls do not allow heat transfer to take place across them. For example, insulated walls of a geyser.

(ii) *Diathermic boundary*: The walls allow heat interaction between the system and the surrounding. For example, hot plate, heat exchanger tubes.

(iii) *Control surface*: A control surface is a boundary which encloses a constant or control volume of the system. The matter can continuously flow in and out of the control volume and heat and work can cross the control surface. The control surface may be real or imaginary which may be fixed in shape, position and orientation. For example, steam generator whose volume is fixed.

(a) Closed system or control mass

In a closed system, the mass within the boundary remains constant. A closed system can permit exchange of heat and work with the surrounding but does not permit any mass transfer across the boundary.

The physical nature and chemical composition of the mass of the system may change. For example, water may evaporate into steam or steam may condense into water. A chemical reaction may occur between two or more components of the closed system.

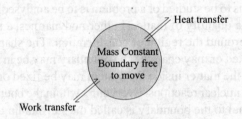

Fig. 1.3 Closed system

Example 1: *Car battery*

Electric supply takes place from and to the battery but there is no material transfer.

Example 2: *Tea kettle*

Heat is supplied to the kettle but mass of water remains constant (neglecting escape of some steam).

(b) Open system or control volume

In an open system, both mass and energy (heat and work) cross the boundary. The mass within the system may not be constant.

If the size of the boundary is fixed it is called control surface and the system of constant volume is called control volume. It is an open system.

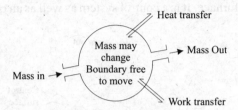

Fig. 1.4 Open system

Example 1: *Water pump*

Water enters at low level and pumped to a higher level, pump being driven by an electric motor. The mass (water) and energy (electricity) cross the boundary of the system (pump and motor).

Example 2: *Scooter engine*

Air and petrol enter and burnt gases leave the engine. The engine delivers mechanical energy to the wheels.

Example 3: *Boilers, turbines, heat exchangers*

Fluid flows through them and heat or work is taken out or supplied to them.

Most of the engineering machines and equipment are open systems or control volume systems.

(c) Isolated system

In an isolated system, neither energy nor mass are allowed to cross the boundary. The system has fixed mass and energy. It is a special type of closed system that does not interact with the surrounding.

Example 1: *Thermoflask* containing a fluid.

Example 2: *Universe* constituting a system and its surrounding.

(d) Special systems

   (i) *Adiabatic system*: A system with adiabatic boundary can only exchange work and no heat with the surrounding. All adiabatic systems are thermally insulated from their surroundings.

   (ii) *System with control surface*: A control system has control volume which is separated from its surrounding by a real or imaginary control surface which is fixed in shape, position and orientation, flow in and out of the control system (volume) and heat and work can cross the control surface. This is also an open system.

Example: *Steam generator*

The volume of the generator is fixed. Water is pumped in and steam comes out. Heat is supplied from the furnace. It is a control system as well as an open system.

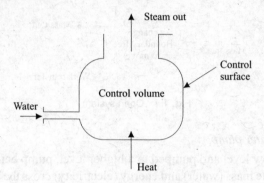

Fig. 1.5 Control system

(iii) *Homogeneous and Heterogeneous systems*:
A homogeneous system has single phase.

Example: Ice, water, dry saturated steam, wet air, mixture of ammonia in water, water plus nitric acid.

A heterogeneous system has more than one phase.

Example: Mixture of ice and water, mixture of water and mercury, wet steam, mixture of oil and water.

(i) Classification of Thermodynamic Systems

The different types of thermodynamic systems can be represented by a tree diagram, Fig. 1.6. (see p. 9)

(ii) Comparison of Systems

Two basic types of systems are closed system and open system. The comparison is given in Table 1.1. (see p. 9)

(iii) Specifications of Thermodynamic Systems

The common systems are classified. The mass and energy flow are also given in Table 1.2. (see p. 10)

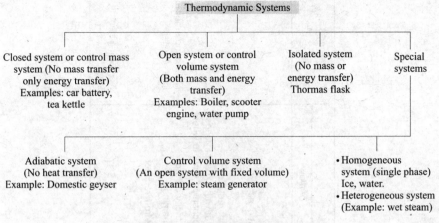

Fig. 1.6 Classification of systems

Table 1.1 Comparison of open and closed systems

Open system	Closed system
1. Both mass and energy can cross the boundary.	1. Only energy can cross the boundary.
2. Mass of the system can be constant or vary.	2. Mass of the system is constant.
3. Flow processes take place in open system.	3. Non-flow processes take place in closed system.
4. Special case: Control volume system with control surface.	4. Special case: Isolated system with no energy interaction.
5. Examples: Turbines, pumps, compressors.	5. Example: Piston-cylinder engines, bomb calorimetre, car battery, tea kettle.

1.5 STUDY OF SYSTEM MATTER

Fluids in the form of gas, vapour, liquid or mixtures are used in thermodynamic devices to transport energy between the system and its surrounding. The following concepts are used for thermodynamic substances in the study of thermodynamics.

1.5.1 Phase

A quantity of matter homogeneous (uniform) in physical structure and chemical composition throughout is called a phase. If the matter is all gas, all liquid or all solid, it has physical uniformity.

Similarly, if chemical composition does not vary from one part of the system to another, it has chemical uniformity.

Examples of one phase system are a single gas, a single liquid, a mixture of gases or solution of liquids contained in a vessel.

A system consisting of liquid and gas is a two-phase system.

Table 1.2 Specifications of thermodynamic systems

S. No.	System	Type of system	Mass		Energy		Remarks
			Inflow	outflow	Inflow	outflow	
1.	Carburretor	Open	Fuel + air	Fuel-air mixture	–	–	–
2.	Radiator of car	Open	Hot water & cold air	Cooled water and hot air	Heat of water	Heat of air	
3.	Scooter engine	Open	Air + petrol	Burnt gases	Chemical energy of fuel	- Heat radiation - Heat with exhaust gases - Mechanical work to wheel	
4.	Car battery	Closed	–	–	Electricity during charging	Electricity during discharging	Neglecting small loss of water
5.	Kitchen refrigerator	Closed	–	–	Electricity	Heat from radiator	
6.	Fan	Open	Air	Air	Electricity	–	–
7.	Pressure cooker	Closed	–	–	Heat	–	Neglecting steam loss

Water at triple point exists as water, ice and steam simultaneously forms a three-phase system.

(a) Pure substance

A pure substance has homogeneous and invariable chemical composition even though there is a change of phase.

Examples: Liquid water, mixture of liquid water and steam, mixture of ice and water. The mixture of liquid air and gaseous air is not a pure substance.

1.5.2 Concept of Continuum

Matter consists of large number of molecules which are in a random state of motion. They experience inter-molecular forces of attraction and repulsion. They are separated from each other by a distance of the order of mean-free path. Thus, from a microscopic point of view, matter is neither continuous nor homogeneous. However, the number of molecules is extremely large. Matter appears to be continuous on a macroscopic scale because of such a large number of molecules in a small volume and because properties of matter are continuously subjected to an averaging effect to the contributions of the large number of molecules. Thus, the matter can be treated as a continuum and physical properties, i.e., mass, density, etc. associated with matter contained within a small volume will be regarded as distributed uniformly over that volume.

For thermodynamic analysis, when the linear dimensions of the system under consideration is quite large as compared to the inter-molecular distance, this continuum hypothesis is fully justified. However, when the linear dimension of the system is comparable to the inter-molecular distance (such as rarefied gas flow), this continuum assumption fails and microscopic and statistical method must be used.

As a consequence of continuum hypothesis, each intensive property is assumed to have a definite value at a point. Therefore, classical thermodynamics makes no hypothesis about the structure of the matter of the system. The volumes of the system considered are very large compared to molecular dimensions. The system is considered a continuum. It is assumed to contain continuous distribution of matter. There are no voids or cavities. The pressure, temperature, density and other properties are the average values of action of many molecules and atoms. Such idealization is a must for solving most problems. The laws and concepts of thermodynamics are independent of structure of matter.

1.5.3 Macroscopic and Microscopic Approach

Thermodynamic system can be studied in two ways:
  (i) from a macroscopic point of view
  (ii) from a microscopic point of view.

In a macroscopic approach, the overall behaviour of the system is studied without any consideration of atomic structure of the matter. Such studies come under the domain of classical thermodynamics. The state or condition of the system can be completely described by measured values of pressure, temperature and volume which are called macroscopic or time-averaged variables. The results obtained are of sufficient accuracy and validity. Engineering thermodynamic analysis is macroscopic and most of the analysis is made by it.

In the microscopic approach, the focus is on the statistical behaviour of the matter. It is based on kinetic theory. The matter consists of a large number of molecules which move randomly in chaotic fashion. At a particular moment, each molecule has a definite position, velocity and energy. The characteristics change very frequently due to collision between molecules. The overall behaviour of the matter is predicted by statistically averaging the behaviour of individual molecules. It helps to give deeper understanding of laws of thermodynamics. However, it is rather complex, cumbersome and time-consuming.

The comparison of the two approaches can be summarised as follows:

Table 1.3 Comparison of macroscopic and microscopic approaches

Macroscopic approach	Microscopic approach
1. The system is considered a continuum i.e., continuous distribution of matter without voids or cavities.	1. The system is considered to contain a large number of molecules moving randomly in chaotic fashion.
2. The thermodynamic properties are measured values which are time-averaged variables.	2. The thermodynamic properties are found out by statistically averaging the behaviour of individual molecules.
3. It is used in classical thermodynamics.	3. It is used in statistical thermodynamics.
4. The results obtained are of sufficient accuracy and validity.	4. The results obtained are very accurate.
5. Engineering devices are designed and analysed macroscopically.	5. For research and study of laws of thermodynamics deeply, microscopic study is used.
6. The approach is simple and quick.	6. It is rather complex, cumbersome and time consuming.

1.6 THERMODYNAMIC PROPERTIES

Property is a quantity to describe the state of a system. The number of properties required to describe a system depends upon the nature of the systems. However, each property has a single value at each state. Each state can be represented by a point on a graph with any two properties as coordinates. All properties are point functions.

Special *characteristics* of thermodynamic properties, therefore, are:

1. Property is a quantity to describe the state of a system.
2. The number of properties required to describe a system depends upon the nature of the system.

3. Each property has a single value at each state.
4. Each state can be represented by a point on a graph with any two properties as coordinates.
5. All properties are point functions.
6. Any operation in which one or more properties of a system change is called a change of state which may result in a process or a cycle.
7. Thermodynamic properties are exact differentials.
8. Heat and work are path functions and not properties of the system.
9. A property is a macroscopic characteristic of a system to which numerical value can be assigned.
10. The value of a property is unique and depends on the condition of the system.
11. Thermodynamic properties can be classified as extensive or intensive. Extensive properties depend on the extent or size of the system. Intensive properties do not depend on the size of the system and have same value throughout a system.
12. There are some properties which can be measured directly in the laboratory with the help of instruments. For example, pressure, temperature, volume, mass, etc. There are some properties which are defined by combining mathematically other properties. A third class of properties are defined by laws of thermodynamics. For example, internal energy enthalpy, entropy, specific heats. Helmholz function, Gibb's function, etc.
13. Properties are defined only when a system is in equilibrium.

1.6.1 Extensive and Intensive Properties

There are two types of properties:

(a) Extensive property

It depends on the extent or size of the system, and is the sum of the properties of various parts of the system. Volume, energy, enthalpy, entropy are examples of extensive properties.

$$V = V_1 + V_2 + V_3 + \ldots$$
$$= \Sigma V_i$$

where
$$V = \text{Total volume of the system } [\text{m}^3]$$
$$V_i = \text{Volume of the sub-systems } [\text{m}^3]$$

(b) Intensive property

It does not depend on the extent or size of the system. Examples are pressure, temperature, density, composition, viscosity, etc.

$$p = p_1 = p_2 = p_3 = \dots$$

where
$$p = \text{pressure of the total system } [P_a]$$

$$p_1, p_2, p_3 \dots = \text{pressure of sub-systems } [P_a]$$

(c) Specific property

It is the extensive property per unit mass of the system. This is an intensive property and is expressed by a lower case letter.

$$\text{Specific volume, } v = \frac{V}{m} \; [\text{m}^3/\text{kg}]$$

$$\text{Specific energy, } e = \frac{E}{m} \; [\text{kJ/kg}]$$

Like extensive properties, specific properties are also additive.

$$v = v_1 + v_2 + v_3 + \dots$$

(d) Molar property

It is the extensive property per unit mole.

$$\text{Molar volume} = \frac{\text{Total volume}}{\text{Mole number}}$$

$$\bar{v} = \frac{v}{n} \; [\text{m}^3/\text{mol}]$$

1.6.2 Specific Volume and Density

Volume (V) is the space occupied by a substance and is measured in m^3.
Mass is an indication of the quantity of matter present in the system and is measured in kg.

Specific volume (v) is the volume for unit mass and is measured in m^3/kg. It is an intensive property and may vary from point and point.

$$v = \frac{V}{m} \; [\text{m}^3/\text{kg}]$$

The specific volume may be expressed either on mass basis or mole basis.

A *mole* of a substance has mass numerically equal to the molecular weight of the substance. One mol of oxygen has a mass of 32 g and 1 kg mol (or kmol) of nitrogen has a mass of 28 kg.

A bar over the symbol is used to designate the property on a mole basis. The molar specific volume ($\bar{v}$) is the volume per mole and is given as m^3/mol (or m^3/k mol).

Density (ρ) is the mass for unit volume and is given in kg/m³.

$$\rho = \frac{m}{V} = \frac{1}{v} \ [\text{kg/m}^3]$$

Density is an intensive property and may vary from point to point within a system and may also vary with time. Therefore, average density of a system is defined as mass per unit volume of the system.

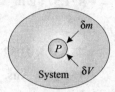

Fig. 1.7 Continuum

Consider mass δm of small volume δV of a system. The ratio $\dfrac{\delta m}{\delta V}$ is the average mass density of the system within the volume δV. If the volume is shrunk to $\delta V'$ about a point P which is regarded as a continuum, the density at a point is defined as:

$$\rho = \underset{\delta V \to \delta V'}{\text{Lt}} \frac{\delta m}{\delta V}$$

Specific weight (w): Weight is a force exerted by the system due to gravitational acceleration. The weight (W) of the system is the product of mass (m) and local gravitational acceleration (g) and is measured in newton (N).

$$W = mg \ [N]$$

Specific weight (w) is the weight per unit volume.

$$w = \frac{W}{V} \ [\text{N/m}^3]$$

Example 1.1: A pump discharges a liquid into a drum at the rate of 0.032 m³/s. The drum, 1.5 m in diameter and 4.20 m in length, can hold 3500 kg of liquid. Find the density of the liquid and mass flow rate of the liquid handled by the pump.

[U.P.T.U II Sem. 2004-2005]

Solution: The volumetric flow rate $\dot{V} = 0.032 \text{ m}^3/\text{s}$

Drum diameter, $d = 1.5 \text{ m}$

Drum height, $h = 4.20 \text{ m}$

Mass of liquid, $m = 3500 \text{ kg}$

The drum volume, $\dot{V} = \dfrac{\pi}{4}d^2 h$

$$V = \frac{\pi}{4}(1.5)^2 \times 4.20$$

$$= 7.422 \text{ m}^3.$$

The liquid density, $\rho = \dfrac{m}{V} = \dfrac{3500}{7.422}$

$$= 471.57 \text{ kg/m}^3 \quad \textbf{Ans.}$$

The mass flow rate, $\dot{m} = \rho \dot{V}$

$$= 471.57 \times 0.032$$

$$= 15 \text{ kg/s} \quad \textbf{Ans.}$$

Example 1.2: A fan delivers 4 m³ of air having a mass of 5 kg. Calculate the weight, specific weight, density and specific volume of air being delivered.

Solution: Volume, $\qquad V = 4 \text{ m}^3$

Mass, $\qquad\qquad m = 5 \text{ kg}$

Weight, $\qquad\quad W = mg$

$$= 5 \times 9.81$$

$$= 49.05 \text{ N} \quad \textbf{Ans.}$$

Specific weight, $\quad w = \dfrac{W}{V} = \dfrac{49.05}{4}$

$$= 12.26 \text{ N/m}^3 \quad \textbf{Ans.}$$

Mass density $\qquad \rho = \dfrac{m}{V} = \dfrac{5}{4}$

$$= 1.25 \text{ kg/m}^3 \quad \textbf{Ans.}$$

Specific volume, $\quad v = \dfrac{V}{m} = \dfrac{4}{5}$

$$= 0.80 \text{ m}^3/\text{kg} \quad \textbf{Ans.}$$

1.6.3 Pressure

Pressure is the normal force exerted by a system against unit area of boundary surface. The pressure (p) at a point is the same in all directions.

$$p = \frac{F}{A} \text{ [N/m}^2]$$

If δA is a small area and $\delta A'$ is the smallest area from continuum consideration and δF_n is the component of force normal to δA, then the pressure p at a point is defined as:

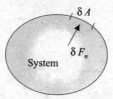

Fig. 1.8   Pressure of continuum

$$p = \mathop{Lt}_{\delta A \to \delta A'} \frac{\delta Fn}{\delta A}$$

The limit of δA approaches $\delta A'$ which is the smallest value from continuum consideration. The value does not approach zero.

The pressure is expressed in N/m^2 or pascal (Pa)

$$1 \text{ Pa} = 1 \text{ N/m}^2$$

Pascal is the force of one newton acting on an area of 1 m^2. The unit of pascal is very small. Engineers express pressure in kilopascal (*kPa*) and Megapascal (*Mpa*)

$$1 \text{ kPa} = 10^3 \text{ Pa}$$

$$1 \text{ MPa} = 10^6 \text{ Pa}$$

Two other units, not within SI system are widely used. These are bar and standard atmosphere.

$$1 \text{ bar} = 10^5 \text{ N/m}^2 = 10^5 \text{ Pa}$$

$$1 \text{ atm} = 101.325 \text{ kPa} = 1.01325 \text{ bar}$$

$$= 760 \text{ mm of mercury}$$

$$= 760 \text{ Torr.}$$

The pressure exerted by 1 mm column of mercury is called Torr.

$$1 \text{ bar} = 750 \text{ mm of mercury}$$

$$= 750 \text{ Torr}$$

$$= 100 \text{ kPa}$$

(a) Absolute and Gauge Pressure

The pressure at a given point can be specified relative to zero pressure or to atmospheric pressure.

The pressure relative to absolute zero pressure is called absolute pressure. This is the pressure of the system. In thermodynamic calculations, the value of absolute pressure is considered.

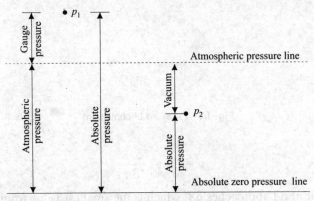

Fig. 1.9   Absolute and Gauge pressure

The pressure relative to the atmosphere is called gauge pressure. This is the pressure indicated by the instrument.

Absolute pressure = Gauge pressure + Atmospheric pressure

$$p_1 = p_{abs} = p_g + p_{at}.$$

If the pressure of the system is less than atmosphere, it can be designated as negative gauge pressure or positive pressure called vacuum.

$$p_2 = p_{at} - p_{vac}.$$

(b) Measurement of pressure

There are a number of pressure measuring devices.

1. *Bourdon gauge*: This measures the difference of the system pressure inside a tube and atmospheric pressure acting at the outside of the tube.

The pressure gauge consists of a metallic bent hollow tube which tries to un-bend when there is a change between the inside (system) pressure and outside (atmospheric) pressure. This moves a pointer through a suitable gear-and-lever mechanism against a calibrated scale.

2. *U-tube manometer*: It is the simplest pressure measuring device consisting of a glass U-tube filled with a suitable liquid (mercury, coloured water) whose density is known. The height of the balanced column of the liquid is measured and converted into desired pressure units. Manometers are used to measure gauge, atmospheric and absolute pressure.

$$p_{abs} = p_g + p_{at}.$$
$$p_g = \rho g h \text{ [N/m}^2]$$

where ρ = Density of manometer fluid

= 13616 kg/m^3 for mercury.

= 1000 kg/m^3 for water.

g = Acceleration due to gravity

$\quad = 9.81 \text{ m/s}^2$

h = Height of liquid column in manometer [m].

$p_{at} = \rho g h_b \, [\text{N/m}^2]$

h_b = Height of liquid column in barometer [m].

Example 1.3: A manometer measures the pressure of a tank as 250 cm of mercury. For the density of mercury $13.6 \times 10^3 \text{ kg/m}^3$ and atmospheric pressure 101 kPa, calculate the tank pressure in MPa.

[U.P.T.U., II Sem., 2000-01]

Solution: The height of manometer liquid, $h = 250$ cm

$$= 250 \times 10^{-2} \text{ m}.$$

Density of mercury, $\qquad \rho = 13.6 \times 10^3 \text{ kg/m}^3$.

Atmospheric pressure, $\quad p_{at} = 101$ kPa

$$= 101 \times 10^{-3} \text{ MPa}.$$

The absolute pressure in the tank,

$$p_{ab} = p_{at} + p_g = p_{at} + \rho g h$$
$$= 101 \times 10^{-3} + [13.6 \times 10^3 \times 9.81 \times 250 \times 10^{-2}] \times 10^{-6}$$

= 0.435 MPa. Ans.

Example 1.4: A U-tube manometer using mercury shows that the gas pressure inside a tank is 30 cm. Calculate the gauge pressure of the gas inside the vessel. ($g = 9.78 \text{ m/s}^2$, density of mercury = 13,550 kg/m³).

[U.P.T.U. I Sem, 2003]

Solution: The manometer height, $h = 30$ cm $= 30 \times 10^{-2}$ m

$$g = 9.78 \text{ m/s}^2$$
$$\rho = 13550 \text{ kg/m}^3$$

The gas pressure, $\qquad p = \rho g h$

$$= 13550 \times 9.78 \times 30 \times 10^{-2}$$
$$= 39755.7 \text{ Pa}$$

= 39.756 kPa Ans.

Example 1.5: A vacuum gauge measures vacuum of 57 kN/m² while the barometer reading is 765 mm of Hg. Find the absolute pressure in mm Hg.

[U.P.T.U. II Sem., 2003-2004]

Solution: $\quad p_{vac} = 57 \text{ kN/m}^2$

$$h_b = 765 \text{ mm}$$

Assume
$$\rho = 13.6 \times 10^3 \text{ kg/m}^3$$
$$g = 9.81 \text{ m/s}^2$$
$$p_{abs} = p_{at} - p_{vac}$$
$$= \rho g h_b - \rho g h_g$$
$$p_{vac} = \rho g h_g = 13.6 \times 10^3 \times 9.81 \times h_g$$
$$= 57 \times 10^3$$

$\therefore$
$$h_g = \frac{57 \times 10^3}{13.6 \times 10^3 \times 9.81} = 0.427 \text{ m}$$

$\therefore$
$$p_{abs} = 13.6 \times 10^3 \times 9.81 \ (0.765 - 0.427)$$
$$= 13.6 \times 10^3 \times 9.81 \times 0.338 \text{ N/m}^2$$

$\therefore$
$$p_{abs} = 0.338 \text{ m of Hg}$$
$$= \textbf{338 mm of Hg} \quad \textbf{Ans.}$$

Example 1.6: A barometer reads 760 mm of mercury. What would be the absolute pressure, if

(a) a pressure gauge connected to a steam turbine at inlet registers 28 bar

(b) a vacuum gauge connected in the exhaust line of steam turbine reads 760 mm of Hg.

Solution: (a) Barometer reading, $h_b = 760$ mm of Hg

Pressure gauge reading, $\quad p_g = 28$ bar
$$= 28 \times 10^5 \text{ N/m}^2$$

Absolute pressure in the turbine inlet,
$$p_{abs} = p_{at} + p_g$$

Assume density of mercury, $\quad \rho = 13.6 \times 10^3 \text{ kg/m}^3$

Acceleration due to gravity, $\quad g = 9.81 \text{ m/s}^2$
$$p_{abs} = \rho g h_b + p_g$$

$$= 13.6 \times 10^3 \times 9.81 \times \frac{760}{1000} + 28 \times 10^5$$

$$= 1.01396 \times 10^5 + 28 \times 10^5$$

$$= \textbf{29.014 bar} \quad \textbf{Ans.}$$

(b) $\qquad\qquad\qquad p_{abs} = p_{at} - p_{vac} = \textbf{0 N/m}^2 \quad \textbf{Ans.}$

Example 1.7: The gas used in the gas engine trial was tested. The pressure of the gas supply was 10 cm of water column. Find the absolute pressure of the gas if the barometer reading is 760 mm of Hg.

Solution: The gauge pressure of gas, $h_g = 10$ cm of water

$$= \frac{10}{100} \text{ m of water}$$

Barometer reading, $h_b = 760$ mm of Hg

$$= \frac{760}{1000} \text{ m of Hg.}$$

$$p_{abs} = p_{at} + p_g$$
$$= \rho p h_b + \rho g h_g$$

$$= \left[13.6 \times 10^3 \times 9.81 \times \frac{760}{1000} + 1000 \times 9.81 \times \frac{10}{1000} \right] \times 10^{-3}$$

$$= 102.377 \text{ kPa}$$

$$= \mathbf{1.0238 \ bar} \quad \textbf{Ans.}$$

Example 1.8:   A vessel of cylindrical shape of 50 cm diameter and 75 cm height, contains 4 kg of gas. The pressure gauge mounted on the vessel indicates 620 mm of Hg above atmosphere. If the barometer reading is 760 mm of Hg, calculate the absolute pressure of gas in bar. Also determine the density and specific volume of the gas.

Solution:

Vessel diameter, $d = 50 \text{ cm} = \dfrac{50}{100} \text{ m}$

Vessel height, $h = 75 \text{ cm} = \dfrac{75}{100} \text{ m.}$

The gauge pressure of gas supply,

$$h_g = 620 \text{ mm of Hg}$$

$$= \frac{620}{1000} \text{ m of Hg.}$$

The barometer reading, $h_b = 760$ mm of Hg

$$= \frac{760}{1000} \text{ m of Hg}$$

The mass of the gas, $m = 4$ kg.

The volume of vessel, $V = \dfrac{\pi}{4} d^2 h$

$$V = \frac{\pi}{4} \times \left(\frac{50}{100}\right)^2 \times \frac{75}{100} = 0.147 \, \text{m}^3$$

Density of gas, $\qquad \rho = \dfrac{m}{V} = \dfrac{4}{0.147} = \textbf{27.21 kg/m}^3 \quad \textbf{Ans.}$

Specific volume, $\qquad v = \dfrac{V}{m} = \dfrac{0.147}{4} = \textbf{0.03675 m}^3\textbf{/kg} \quad \textbf{Ans.}$

$$p_{abs} = p_{at} + p_g$$
$$= \rho g h_b + \rho g h_g$$

$$= \left[13.6 \times 10^3 \times 9.81 \times \frac{760}{1000} + 13.6 \times 10^3 \times 9.81 \times \frac{620}{1000}\right] \times 10^{-3}$$

$$= 184 \, \text{kPa}$$
$$= \textbf{1.84 bar} \quad \textbf{Ans.}$$

Example 1.9: A turbine is supplied with steam at a pressure of 20 bar gauge. After expansion in the turbine, the steam passes to condenser which is maintained at a vacuum of 250 mm of mercury by means of pumps. Express the inlet and exhaust steam pressure is N/m^2 and kPa.

Solution:

Steam pressure, $\qquad p_g = 20 \, \text{bar} = 20 \times 10^5 \, \text{N/m}^2$

Condenser vacuum, $\qquad p_{vac} = 250 \, \text{mm Hg} = \dfrac{250}{1000} \, \text{m Hg.}$

Assume density of mercury, $\quad \rho = 13.6 \times 10^3 \, \text{kg/m}^3$

Accelerator due to gravity, $\quad g = 9.81 \, \text{m/s}^2$

Barometer reading, $\qquad h_b = 760 \, \text{mm of Hg}$

(i) $\qquad\qquad\qquad p_{abs} = p_{at} + p_g$
$$= \rho g h_b + p_g$$
$$= 13.6 \times 10^3 \times \frac{760 \times 9.81}{1000} + 20 \times 10^5$$
$$= 101396.16 + 20,00000 = 2101396.2 \, \text{N/m}^2$$
$$= \textbf{2101.4 kPa.} \quad \textbf{Ans.}$$

(ii) $\qquad\qquad\qquad p_{abs} = p_{at} - p_{vac}$
$$= \rho g h_b - \rho g h_v$$
$$= 13.6 \times 10^3 \times 9.81 \frac{(760 - 250)}{1000} = 68042 \, \text{N/m}^2$$
$$= \textbf{68 kPa} \quad \textbf{Ans.}$$

Example 1.10: Convert the following readings of pressure into kPa assuming that the barometer reads 760 mm Hg.

(a) 90 cm Hg gauge

(b) 40 cm Hg vacuum

(c) 1.2 m of H_2O g.

Solution:

Barometer reading, $\quad h_b = 760$ mm Hg

$$= \frac{760}{1000} \text{ m Hg}$$

(a) Pressure gauge reading, $\quad h_g = 90$ cm Hg

$$= \frac{90}{100} \text{ m Hg.}$$

$$p_{ab} = p_{at} + p_g$$

$$= \rho g h_b + \rho g h_g$$

$$= \rho g (h_b + h_g)$$

$$= \left[13.6 \times 10^3 \times 9.81 \left(\frac{760}{1000} + \frac{90}{100} \right) \right] \times 10^{-3}$$

$$= \textbf{221.47 kPa.} \quad \textbf{Ans.}$$

(b) Vacuum gauge reading $\quad = 40$ cm of Hg

$$= \frac{40}{100} \text{ m of Hg}$$

$$p_{ab} = p_{at} - p_{vac}$$

$$= \rho g h_b - \rho g h_{vac}$$

$$= \rho g (h_b - h_{vac})$$

$$= \left[13.6 \times 10^3 \times 9.81 \left(\frac{760}{1000} - \frac{40}{100} \right) \right] \times 10^{-3}$$

$$= \textbf{48 kPa} \quad \textbf{Ans.}$$

(c) Pressure gauge reading, $\quad h_g = 1.2$ m H_2O

$$p_{ab} = p_{at} + p_g$$

$$= \rho g h_b + \rho g h_g$$

$$= \left[13.6 \times 10^3 \times 9.81 \times \frac{760}{1000} + 1000 \times 9.81 \times 1.2 \right] \times 10^{-3}$$

$$= \textbf{113.17 kPa} \quad \textbf{Ans.}$$

1.6.4 Temperature

Temperature may be defined as degree of heat or coldness of a body or environment. If two bodies are brought in contact, heat will flow from hot body at a high temperature to a cold body at a lower temperature. Temperature is the thermal potential causing the flow of heat energy. It is an intensive thermodynamic property independent of extent or size of the system. The temperature of a body is measured with the help of an instrument called thermometer. Thermometers work on different principles using different working substances. These are graduated with different temperature scales.

(a) Equality of temperature

When two bodies at different temperatures are brought into contact, after sometime, they attain a common temperature and are then said to exist in thermal equilibrium.

"Two systems have equal temperature if there are no changes in their properties when they are brought in thermal contact with each other".

(b) Temperature scales

1. *Centigrade and Fahrenheit temperature scale*: These scales are based on two fixed points: the ice point and steam point.

Scale	Ice point	Steam point	Interval	Unit
Centigrade	0	100	100	°C
Fahrenheit	32	212	180	°F

Considering linear correlation between temperature t and thermodynamic property x.

$$t_i = a \times i + b$$
$$t_s = a \times s + b$$

where i and s stand for ice and steam

$$a = \frac{t_s - t_i}{x_s - x_i} \text{ and } b = t_i - \frac{t_s - t_i}{x_s - x_i}.x_i$$

∴ unknown temperature,

$$t = \left[\frac{t_s - t_i}{x_s - x_i}\right] x + \left[t_i - \frac{t_s - t_i}{x_s - x_i}.x_i\right]$$

$$= t_i + (t_s - t_i) \times \frac{x - x_i}{x_s - x_i}$$

Centigrade scale

$$t°, C = 0 + 100 \; \frac{x - x_i}{x_s - x_i}$$

$$= 100 \frac{x - x_i}{x_s - x_i}$$

(c) Fahrenheit scale

$$t°, F = 32 + 180 \; \frac{x - x_i}{x_s - x_i}$$

The correlation between temperatures measured on Centigrade and Fahrenheit scales can be established as:

$$\frac{t°,C}{100} = \frac{(t°,F - 32)}{180}$$

or

$$t°, C = \frac{5}{9}(t°, F - 32)$$

2. *Celsius scale*: The Celsius temperature scale employs a degree of the same magnitude as that of ideal gas scale, but its zero point is shifted.

The Celsius temperature of the triple point of water is 0.01°C.

The steam point and ice point on Celsius scale are:

$$t_s = 100.00°C, \quad t_i = 0.00°C.$$

The only Celsius temperature fixed by definition is that of triple point.

3. *Kelvin and Rankine absolute scales*: The absolute Kelvin temperature scale has fixed point at triple point of water equal to 273.16.

$$T, K = t°C + 273.16$$

The triple point of water on Rankine scale is $\left(\frac{9}{5} \times 273.16 \right) 491.688$.

$$T, R = t°, F + 459.67$$

In engineering applications, the above correlations are approximated as:

$$T, K = t°, C + 273$$
$$T, R = t°, F + 460.$$

4. *Thermodynamic temperature scale*: The efficiency of an ideal engine operating on the Carnot cycle between any two temperatures is given by:

$$\eta = \frac{T_1 - T_2}{T_1} = \frac{Q_1 - Q_2}{Q_1}$$

where Q_1 is the heat absorbed at temperature T_1 and Q_2 is the heat rejected at temperature T_2

$$\therefore \qquad \frac{T_1}{T_2} = \frac{Q_1}{Q_2}$$

This expression forms the basis of thermodynamic temperature scale devised by Lord Kelvin.

"The ratio of any two temperatures on the thermodynamic scale (Kelvin scale) is equal to the ratio of the heat absorbed to the heat rejected by a Carnot engine operating between these temperatures."

   5. *Electrical resistance thermometer*: The change of resistance of platinum wire with temperature is calibrated as temperature scale.

$$R = R_i(1 + At + Bt^2)$$

where $\qquad R_i$ = Resistance at ice point

A and B are constants.

   6. *Thermocouple*: An e.m.f. is generated in a circuit due to Seeback effect which can be calibrated into a temperature scale.

$$E = a + bt + ct^2 + dt^3$$

$$E = \text{e.m.f. generated}$$

$$a, b, c, d \text{ are constants.}$$

Example 1.11: The temperature of human body is 99.2°F. Determine the temperature in degree Celsius (°C), Rankine (R) and Kelvin (K) scales.

Solution:

$$t°, C = \frac{5}{9}(t° F - 32)$$

$$= \frac{5}{9}(99.2 - 32) = \textbf{37.33°C} \quad \textbf{Ans.}$$

$$T, K = t°, C + 273.15 = 37.33 + 273.15$$

$$= \textbf{310.48 K} \quad \textbf{Ans.}$$

$$T, R, = t°, F + 459.67 = 99.2 + 459.67 = \textbf{558.87 R} \quad \textbf{Ans.}$$

Example 1.12: Estimate the triple point of water in Fahrenheit, Rankine and Kelvin scales. [U.P.T.U. July, 2002]

Solution:

Triple point of water, $\qquad t°C = 0.01°C$

$$t°, F = \frac{9}{5} \times t°C + 32 = \frac{9}{5} \times 0.01 + 32 = \textbf{22.018°F}$$

$$T, K = t°, C + 273.15 = 0.01 + 273.16 = \textbf{273.16 K}$$
$$T, R = t°, F + 459.67 = 32.018 + 459.67$$
$$= \textbf{491.688 R} \quad \textbf{Ans.}$$

Example 1.13: During temperature measurement, it is found that a thermometer gives the same temperature in °C and °F. Express the temperature value in R.

[U.P.T.U. Dec. 2002]

Solution:

$$t°, C = \frac{5}{9} (t°, F - 32)$$

$$t°, C = t°, F = x \quad \text{(say)}$$

$$\therefore \quad x = \frac{5}{9} (x - 32) = -40$$

$$\therefore \quad t°, C = -40°C$$

$$t°, F = -40°F$$

$$T, R = t°F + 459.67 = -40 + 459.6$$

$$= \textbf{419.67 R} \quad \textbf{Ans.}$$

Example 1.14: A platinum resistance thermometer has resistance of 2.8 ohms at 0°C and 3.8 ohms at 100°C. Calculate the temperature when the resistance indicated is 5.8 ohm. [U.P.T.U. II Sem. 2003-2004]

Solution: Let the temperature resistance correlation be:

$$R_t = R_0 (1 + \alpha t)$$

At 0°C, $\quad R_t = 2.8 \text{ ohm}$

$\therefore \quad R_t = R_0 = 2.8 \text{ ohm}$

At $t = 100°C$, $\quad R_t = R_0 (1 + \alpha t)$

$$3.8 = 2.8(1 + 100 \alpha)$$

$\therefore \quad \alpha = 0.00357$

when $\quad R_t = 5.8 \text{ ohm}$

$$5.8 = 2.8 (1 + 0.00357t)$$

$\therefore \quad t = \textbf{300°C} \quad \textbf{Ans.}$

Example 1.15: The e.m.f. in a thermocouple with the test junction at $t°C$ on gas thermometer scale and reference junction at ice point is given by

$$E = 0.20t - 5 \times 10^{-4} t^2 \text{ [mV]}$$

The multivoltmeter is calibrated at ice point and steam point. What will this thermocouple read in a place where the gas thermometer reads 50°C?

Solution:

At ice point, $t = 0°C$ and $E = 0$ mV

At steam point, $t = 100°C$

$$E = 0.2 \times 100 - 5 \times 10^{-4}(100)^2$$

$$= 15 \text{ mV}$$

At 50°C, $E = 0.20 \times 50 - 5 \times 10^{-4}(50)^2$

$$= 8.75 \text{ mV}.$$

When gas thermometer reads 50°C, the thermocouple will read $= \dfrac{100}{15} \times 8.75$

$$= \textbf{58.33°C} \quad \textbf{Ans.}$$

Example 1.16: A certain thermometer using pressure as the thermometric property gives the value of pressure as 1.86 and 6.81 at ice point and steam point respectively. The temperatures of ice point and steam point are assigned the number 32 and 212 respectively. Determine the temperature corresponding to $p = 2.5$ if the temperature t is defined in term of pressure p as $t = a \ln p + b$, where a and b are constants. [U.P.T.U. 2005-2006]

Solution:

$$t = a \ln p + b$$

At ice point,

$$32 = a \ln 1.86 + b = 0.620\, a + b$$

At steam point,

$$212 = a \ln 6.81 + b = 1.918\, a + b$$

Subtracting,

$$180 = 1.298a$$

$\therefore$ $\quad a = 138.7$

$$b = -54.07$$

$\therefore$ $\quad t = 138.7 \ln p - 54.07.$

For $\quad p = 2.5$

$$t = 138.7 \ln 2.5 - 54.07$$

$$= \textbf{73°C} \quad \textbf{Ans.}$$

1.6.5 Equation of State

The functional relationship among the properties, pressure, molar or specific volume and temperature is known as equation of state or characteristic equation of a gas. This can be developed by combining Boyle's Law and Charle's law for a perfect gas:

$$pv = RT$$
$$pV = m.RT$$
$$p\bar{v} = MRT$$
$$pV = n(MR)T$$

These equations are called characteristic gas equations.

Where
$$p = \text{pressure of gas } [\text{N/m}^2]$$
$$V = \text{volume of gas } [\text{m}^3]$$
$$m = \text{mass of gas } [\text{kg}]$$

$$v = \frac{V}{m} \ [\text{m}^3\text{/kg}]$$

$$\bar{v} = \text{molar volume } [\text{m}^3\text{/mol}]$$
$$T = \text{absolute temperature of gas } [\text{K}]$$
$$R = \text{characteristic gas constant } [\text{N} - \text{m/kg-K}]$$
$$M = \text{molecular mass of gas}$$
$$n = \text{number of moles of the gas.}$$

(a) Universal Gas Constant

The universal gas constant or molar constant $\left(\bar{R}\right)$ of a gas is the product of the characteristic gas constant (R) and the molecular mass (M) of the gas.

$$\bar{R} = MR$$

The value of $\bar{R}$ is the same for all gases. Its value is 8314 J/kgmol- K or 8.314 kJ/kgmol-K.

From Avogadro's law, when $p = 760$ mm Hg $(1.013 \times 10^5 \text{ N/m}^2)$

$$T = 273.15 \text{ K and } \bar{v} = 22.4 \text{ m}^3\text{/kgmol,}$$

$$\bar{R} = \frac{1.013 \times 10^5 \times 22.4}{273.15} = 8.3143 \text{ kJ/kgmol-K.}$$

Example 1.17: Determine the mass of nitrogen in a vessel of 2.5 m³ capacity. The pressure and temperature are 80 bar and 25°C respectively. Take $R = 297$ J/kg-K.

Solution:

Volume, $\quad V = 2.5 \text{ m}^3$

Pressure, $\quad p = 80 \text{ bar} = 80 \times 10^5 \text{ N/m}^2$

$$\text{Temperature, } T = 25 + 273 = 298 \text{ K}$$

$$R = 297 \text{ J/kg-K.}$$

Using equation of state,

$$pV. = mRT$$

$$\therefore \quad m = \frac{pV}{RT} = \frac{80 \times 10^5 \times 2.5}{297 \times 298}$$

$$= \textbf{225.97 kg} \quad \textbf{Ans.}$$

Example 1.18: A pressure bottle stores 5 m³ of an inert gas at pressure of 10 MPa and temperature 27°C. Calculate mass, density, specific volume and molar volume of gas. Take molar mass of gas as 28.

Solution:

Volume of gas, $\quad V = 5 \text{ m}^3$

Pressure, $\quad p = 10 \text{ MPa} = 10 \times 10^6 \text{ N/m}^2$

Temperature, $\quad T = 27 + 273 = 300 \text{ K}$

Molar mass, $\quad M = 28$

$$\overline{R} = 8314.3 \text{ J/kgmol-K.}$$

Gas constant, $\quad R = \dfrac{\overline{R}}{M} = \dfrac{8314.3}{28} = 297 \text{ J/kgK.}$

Applying equation of state,

$$pV = mRT$$

$$\therefore \quad m = \frac{pV}{RT} = \frac{10 \times 10^6 \times 5}{297 \times 300} = \textbf{561.28 kg} \quad \textbf{Ans.}$$

Density, $\quad \rho = \dfrac{m}{V} = \dfrac{561.28}{5} = \textbf{112.26 kg/m}^3 \quad \textbf{Ans.}$

Specific volume, $\quad v = \dfrac{1}{\rho} = \dfrac{1}{112.26} = \textbf{0.0089 m}^3\textbf{/kg} \quad \textbf{Ans.}$

Molar volume, $\quad \overline{v} = \dfrac{V \times M}{m} = \dfrac{5 \times 28}{561.28} = \textbf{0.2494 m}^3\textbf{/kgmol.}$

Example 1.19: 10 kgmol of a gas occupies a volume of 603.1 m³ at temperature of 140°C while its density is 0.464 kg/m³. Find its molecular weight and gas constant and its pressure. [U.P.T.U. I Sem., 2003-2004]

Solution:

Number of moles $n = 10$ kgmol

Volume, $V = 603.1$ m^3

Temperature, $T = 140 + 273 = 413$ K.

Density, $\rho = 0.464$ kg/m^3.

Specific volume $v = \dfrac{1}{\rho} = \dfrac{1}{0.464} = 2.155$ m^3/kg

Characteristic gas equation

$$pV = n\overline{R}\,T$$

$$p = \frac{n\overline{R}\,T}{V} = \frac{10 \times 8314 \times 413}{603.1} = \textbf{0.569 bar}$$

Again, $pv = RT$

$\therefore$

$$R = \frac{pv}{T} = \frac{0.569 \times 10^5 \times 2.155}{413} = \textbf{296.9 J/kg-K.}$$

Molecular weight, $M = \dfrac{\overline{R}}{R} = \dfrac{8314}{296.9} = \textbf{28}\quad\textbf{Ans.}$

Example 1.20: A tank of 0.35 m^3 capacity contains H_2S gas at 300 K. When 2.5 kg of gas is withdrawn, the temperature in the tank becomes 288 K and pressure 10.5 bar. Calculate the mass of gas initially present in the tank. Also determine the initial pressure of gas. [U.P.T.U. 2005-06]

Solution:

Volume, $V_1 = V_2 = 0.35$ m^3

$T_1 = 300$ K

$T_2 = 288$ K

$p_2 = 10.5$ bar $= 10.5 \times 10^5$ N/m^2

The molecular mass of H_2S gas

$$= 2 + 32 = 34$$

$\therefore$

$$R = \frac{\overline{R}}{M} = \frac{8314}{34} = 244.53 \text{ J/kg-K.}$$

Applying equation of state to final condition,

$$p_2 V_2 = m_2 R T_2$$

$$m_2 = \frac{p_2 V_2}{R T_2} = \frac{10.5 \times 10^5 \times 0.35}{244.53 \times 288} = \textbf{5.2 kg}$$

$$m_1 = 5.2 + 2.5 = 7.7 \text{ kg} \quad \textbf{Ans.}$$

Applying equation of state to initial conditions,

$$p_1 V_1 = m_1 R T_1$$

$$\therefore \qquad p_1 = \frac{m_1 R T_1}{V_1} = \frac{7.7 \times 244.53 \times 300}{0.35}$$

$$= \textbf{16.14 bar} \quad \textbf{Ans.}$$

Example 1.21: A high-altitude test chamber occupies a volume of 40 m³. During operation, its pressure is reduced to 0.45 bar and a temperature of 5°C from initial conditions of 1 bar and 35°C. How many kg of air have been removed from the chamber during the process?

Solution: Initial conditions of test chamber are:

Volume, $\qquad V_1 = 40 \text{ m}^3$

Pressure, $\qquad P_1 = 1 \text{ bar} = 10^5 \text{ N/m}^2$

Temperature, $\qquad T_1 = 35 + 273 = 308 \text{ K}$

For air, $\qquad R = 297 \text{ J/kg-K}$

Applying equation of state,

$$p_1 V_1 = m_1 R T_1$$

$$\therefore \qquad m_1 = \frac{p_1 V_1}{R T_1} = \frac{10^5 \times 40}{297 \times 308} = 43.73 \text{ kg.}$$

Final conditions of the test chamber are:

$$V_2 = 40 \text{ m}^2$$
$$P_2 = 0.45 \text{ bar} = 0.45 \times 10^5 \text{ N/m}^2.$$
$$T_2 = 5 + 273 = 278 \text{ K}$$

Applying equation of state,

$$p_2 V_2 = m_2 R T_2$$

$$\therefore \qquad m_2 = \frac{p_2 V_2}{R T_2} = \frac{0.45 \times 10^5 \times 40}{297 \times 278} = 21.8 \text{ kg}$$

Amount of air removed,

$$m_1 - m_2 = 43.73 - 21.80$$

$$= \textbf{21.93 kg} \quad \textbf{Ans.}$$

Example 1.22: Determine the molecular weight of a gas, if its two specific heats are: $C_p = 2.286$ kJ/kg-K and $C_v = 1.768$ kJ/kg-K. Universal gas constant $= 8.3143$ kJ/kg-mol.

Solution:

$$C_p = 2.286 \text{ kJ/kg-K}$$
$$C_v = 1.768 \text{ kJ/kg-K}$$

∴
$$R = C_p - C_v = 2.286 - 1.768$$
$$= 0.518 \text{ kJ/kg-K}$$
$$\overline{R} = 8.3143 \text{ kJ/kg-mol.}$$

Molecular weight,

$$M = \frac{\overline{R}}{R} = \frac{8.3143}{0.518} = 16 \quad \textbf{Ans.}$$

Example 1.23: Determine the size of a spherical balloon filled with hydrogen at 30°C and atmospheric pressure for lifting 400 kg payload. Atmospheric air is at temperature of 27°C and barometer reading is 75 cm of mercury.

[U.P.T.U. II Sem. 2001-2002]

Solution:

Assume size of balloon $= V m^3$

Pressure, $\quad p_1 = \rho gh = 13.6 \times 10^3 \times 9.81 \times \dfrac{75}{100} \times 10^{-5}$

$$= 1 \text{ bar}$$

Gas constant for hydrogen,

$$R = \frac{\overline{R}}{m} = \frac{8314}{2} = 4157 \text{ J/kg-K}$$

Temperature $\quad T = 27 + 273 = 300 \text{ K}$

Applying equation of state,

$$p_1 V_1 = m_1 R T_1$$

Mass of hydrogen filling the balloon,

$$m_1 = \frac{p_1 V}{R T_1} = \frac{1 \times 10^5 \times V}{4158 \times (273 + 30)} = 0.079392 \ V \text{ kg}$$

Mass of atmospheric air displaced,

$$m_2 = \frac{p_1 V}{R T_1} = \frac{1 \times 10^5 \times V}{287 \times 300} = 1.16144 \ V \text{ kg.}$$

Payload $= m_2 - m_1$

$$= (1.16144 - 0.079392)V = 400$$

∴
$$V = 369.669 \text{ m}^3$$

Size of balloon,

$$\frac{4}{3}\pi R^3 = 369.669$$

$$R = \textbf{4.55 m} \quad \textbf{Ans.}$$

1.6.6 Derived Thermodynamic Properties

These properties are calculated from experimental data by the use of generalized thermodynamic relations. Examples are:

1. Internal energy
2. Enthalpy
3. Entropy
4. Specific heats
5. Joule-Thompson coefficient
6. Helmholz function
7. Gibb's function

These properties will be discussed in subsequent sections.

1.7 THERMODYNAMIC EQUILIBRIUM

The thermodynamic properties are measured when the system is in thermodynamic equilibrium. Similarly, processes are studied for systems in thermodynamic equilibrium.

Equilibrium is the state of an isolated system where there is no change at anytime. A system having same values of properties (pressure, temperature, velocity, elevation, composition, etc.) throughout is in thermodynamic equilibrium. The system is simultaneously in the state of mechanical, thermal, electrical and chemical equilibrium.

1. *Thermal equilibrium* denotes uniformity of temperature or absence of temperature gradient and heat flow.
2. *Mechanical equilibrium* denotes uniformity of pressure or absence of unbalanced forces.
3. *Electrical equilibrium* denotes equality of electrical potential and absence of current flow.
4. *Chemical equilibrium* denotes absence of phase change or chemical reaction and there is no mass diffusion.

A system in thermodynamic equilibrium, when isolated, is incapable of spontaneous change. The major part of thermodynamic deals with systems which are in thermodynamic equilibrium, or idealized to be in equilibrium.

The conditions of equilibrium can be illustrated with the help of the following table.

Table 1.4 Thermodynamic equilibrium

S. No.	Equilibrium state	Conditions of equilibrium
1.	Thermal equilibrium	Absence of temperature gradient and heat flow
2.	Mechanical equilibrium	Absence of pressure difference and unbalanced forces
3.	Electrical equilibrium	Absence of electrical potential difference and current flow
4.	Chemical equilibrium	Absence of phase change, mass diffusion and chemical reaction
5.	Thermodynamic equilibrium	The system should be simultaneously in the state of thermal, mechanical, electrical and chemical equilibrium

1.8   THERMODYNAMIC PROCESSES

1. *State* of a system is the condition of the system identified or described by its properties such as pressure, temperature, volume, etc. As different properties are related to each other, the state of the system can be described by specifying a limited number of properties. The values of other properties can be determined from the values of the few used to specify the state.

The number of properties required to describe a system depends upon the nature of the system. Each state can be represented by a point on a graph with any two properties as coordinates. However, each property has a single value at each state. All properties are point functions.

Example: When the piston is at position 1, the gas inside the cylinder has pressure p_1, volume V_1 and temperature t_1. Therefore, the state of the system is p_1, V_1 and T_1 and is represented by point 1 on $p - V$ diagram.

2. *Change of state*: The state of a system changes when any of the properties of a system changes. Any operation in which one or more of properties of a system change is called change of state.

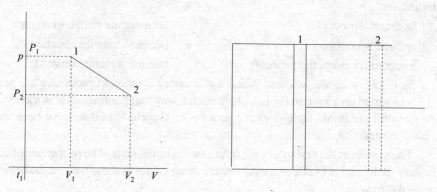

Fig. 1.10   State of a system

Example: The piston moves outwards to position 2. The volume increases from V_1 to V_2, pressure decreases from p_1 to p_2 and temperature decreases form t_1 to t_2.

Any such operation in which properties of the system change is called change of state. The new state is described by point 2.

    3. *Path*: The successive states passed during a change of state is called path of change of state.

In the above example, the line joining states 1 and 2 is called path of change of state.

    4. *Process*: When the path of change of state is completely specified, it is called a process. It is the change of the system from one equilibrium state to another equilibrium state.

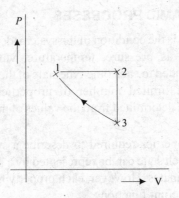

Fig. 1.11 A process and cycle

Example: Process 1-2 is a constant pressure process. Process 2-3 is a constant volume process.

The processes during which any one property remains constant are specified by using prefix *iso* before the property.

Example:

Isothermal process	:	temperature remains constant
Isobaric process	:	pressure remains constant
Isochonic or isometric process	:	volume remains constant

    5. *Cycle or cyclic process*: When a process or a series of processes are performed on a system (Fig. 1.11) in such a way that the final state is identical with the initial state, a cyclic process or a cycle is said to have been executed.

The system at the end of a cycle returns to its initial state. The properties of the system, at the end of cyclic process have the same values as these were in the beginning of the cycle.

Example: Process 1-2 is isobaric, process 2-3 is isochonic and process 3-1 is isothermal (say), then 1-2-3-1 is a thermodynamic cycle.

1.8.1 Point and Path Functions

A property has a unique value at any state. The change in property between two equilibrium states is independent of the history of the process. Mathematically speaking, a property is an exact differential.

$$\int_1^2 dp = p_2 - p_1$$

Since the change in the value of properties depends only on the initial and final states and is independent of the path of the process, they are also called *state functions*. The change in the value of any property of the system for a cyclic process is zero.

$$\oint dp = 0$$

When two properties locate a *point* on the graph (co-ordinate axes) then these properties are called *point functions*.

Example: Pressure, temperature, volume, etc. are properties of the system and are exact differential.

$$\int_1^2 dV = V_2 - V_1$$

There are certain quantities which cannot be located on a graph by a *point* but are given by the area on the graph. The area for the graph for a process is a function of the path of the process. Such quantities are called *path functions*.

Example: Heat and work, etc. These are inexact differentials.

$$\int_1^2 \delta Q \neq Q_2 - Q_1 \text{ and is shown as } {}_1Q_2 \text{ or } Q_{1-2}$$

$$\int_1^2 \delta W \neq W_2 - W_1 \text{ and is shown as } {}_1W_2 \text{ or } W_{1-2}$$

Note: The operator d is used to denote exact differential and δ as inexact differential.

Non-properties, the change in the value of which depends on the path of the process between any two equilibrium states are called path functions.

Example:

(i) $pdv + vdp = d(pv)$

Its integration $\int d(pv) = pv$

Therefore, $pdv + vdp$ is an exact differential and hence a property of the system.

(ii) In pdv, p is a function of v and are connected by a path line on $p-v$ diagram.

$\int pdv$ can be determined only if a relationship between variation of p and v is known. Thus, pdv is not an exact differential and, hence, not a property.

1.8.2 Quasi-Static Process

Quasi-static process is an important concept in thermodynamics. Properties are defined only when there is an equilibrium. However, all real processes occur when there is no equilibrium. Some potential difference must exist within the system between a system and its surroundings to promote a change of state during a thermodynamic process.

In the analysis of real processes, it is assumed that the system deviates from the equilibrium state by a very small margin. All the intermediate states during a process are assumed as equilibrium states. Such a process in which the system is always very near to the equilibrium state throughout the process is called quasi-equilibrium or *quasi-static process.*

A quasi-static process is an idealized process to which many real processes closely approach. The process can be treated without much error.

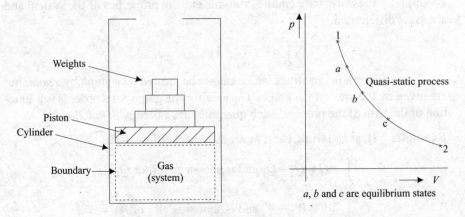

Fig. 1.12 Quasi-static process

Consider a system (gas) enclosed in a piston cylinder arrangement. Initially the system is in equilibrium and is represented by point 1 on $p-V$ diagram. If a small weight placed on the top of the piston is removed, there will be unbalanced force between the system and surrounding. The piston will move upwards, increasing the volume and decreasing both pressure and temperature of the system. The system comes to rest and is represented by point "a". Further removal of small weights from the top of the piston results in intermediate equilibrium states 'b', 'c' till final state 2 is reached. The locus of points 1, a, b, c ... 2 is called a quasi-static or quasi-equilibrium process and can be represented graphically by a continuous line on a state diagram.

The departure of the state of the system will be negligibly small. A quasi-static process can be viewed as sufficiently slow. In thermodynamics, quasi-static processes are considered extensively.

1.8.3 Reversible and Irreversible Processes

(a) Reversible process

Reversible process is another concept for the ideal process. A process is said to be reversible if at any point during the course of the process, it can be reversed and both the system and surroundings can be restored to their original states. If a system passes through a continuous series of equilibrium states during a process, these states can be located on a property diagram.

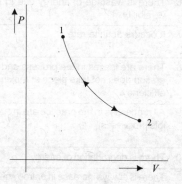

Fig. 1.13 Reversible process

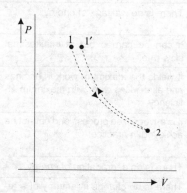

Fig. 1.14 Irreversible process

Line 1-2 has been drawn by joining points of equilibrium states of a process and represent a reversible process. If the process is repeated in the reverse direction, the initial state 1 is restored and the same line 2-1 is traced.

(b) Irreversible process

If a system passes through a sequence of non-equilibrium states during a process, these states cannot be located on any property diagram as each property does not have a unique value in the entire system. Such a process is called irreversible process and is represented by a broken line on a property diagram joining initial and final state points which are in equilibrium. If irreversible process 1-2 is repeated in the reverse direction, the initial state 1 is not restored.

The real processes in engineering are irreversible processes due to presence of the fluid friction, mechanical friction, heat transfer due to finite temperature difference, etc. These are the main causes of irreversibility.

The main features and examples of reversible and irreversible processes as can be differentiated by the following comparison table.

Table 1.5 Comparison of reversible and irreversible process

Reversible Process	Irreversible Process
1. The system passes through a continuous series of equilibrium states.	1. The system passes through a sequence of non-equilibrium states.
2. Reversible process can be plotted by a line on any property diagram as the system has unique values of properties.	2. Irreversible process cannot be plotted on a property diagram as the system does not have unique values of properties.
3. The process can be reversed back to its initial state.	3. The process cannot be reversed back to its initial state.
4. It can be carried out in both the directions without violating second law of thermodynamics.	4. It can be carried out in one direction only.
5. It leaves no after effects or changes in the system or surrounding.	5. There are permanent changes in the surroundings.
6. There is no wastage of energy.	6. There is wastage of energy due to irreversibility.
7. It can be carried out at infinitesimal slow rate.	7. It occurs at finite rate.
8. It yields the maximum work in engines and all devices work with maximum efficiency.	8. There are losses in the process and operation does not take place at maximum efficiency.
9. It is an idealized process and cannot be achieved in practice.	9. All processes in the practice are irreversible processes.
10. There is no degradation of energy.	10. There is degradation of energy.
11. There is no change in entropy of a reversible cyclic process.	11. There is always increase in entropy of irreversible cyclic process.
12. The process takes place with the absence of fluid friction, mechanical friction and infinitely small difference of temperature and pressure between system and surrounding.	12. The process takes place with viscous flow, fluid friction, mechanical friction and energy transfer as heat with finite temperature difference.
13. Examples: Motion without friction, frictionless adiabatic process, controlled expansion and compression, Carnot cycle.	13. Examples: Free expansion and throttling process, fluid flow with friction, mixing of dissimilar gases, Rankine cycle.

(c) Causes of irreversibility

(i) Mechanical irreversibility It is associated with fluid friction between the molecules due to turbulence at high speeds as well as friction due to viscocity. The frictional resistance heats the fluid and there is loss of work. The heat cannot be completely recovered as work.

The mechanical irreversibility is also due to friction in bearings, piston and cylinder walls and other contact surfaces with relative motion. There is also windage losses due to friction between rotating parts and atmospheric air.

(ii) Thermal irreversibility It is due to energy transfer as heat with finite temperature difference between system and surrounding. The heat transfer takes place due to temperature gradients. The heat transfer from low temperature to higher temperature is not possible without the aid of external work.

(iii) Free expansion No work is done in free expansion and process cannot be reversed back, i.e., restoring initial parameters without doing work of compression.

(iv) Mixing of fluids No work is done during mixing of fluids. The fluids cannot be separated into original condition without the help of work.

1.9 FORMS OF ENERGY

Energy is a fundamental concept like God, and it is very difficult to define energy. However, energy can be defined as the capability to produce an effect or capacity to do work. Energy has certain characteristics:

1. Energy is available in different forms.
2. Energy can neither be created nor destroyed.
3. Energy can be transformed from one form to another form.
4. Energy can be stored.
5. Energy can be transported from one system to another system or from one place to another.
6. An energy is associated with a potential. For example, thermal potential (temperature difference) for heat energy or pressure difference for mechanical energy or electrical potential for electrical energy. Free flow of energy takes place only from a higher potential to a lower potential.
7. The energy is measured as N-m or in joules. The energy density is expressed as J/kg. Different sources and forms of energy have different densities.
8. Energy does not have absolute value. It is measured relative to a datum. The common data are ground level, atmospheric pressure and a temperature of 0 K. Heat energy is assumed to have zero value at 0 K. Similarly, potential energy is zero at ground level.
9. The transfer of heat energy takes place due to temperature gradient. All other forms of energy exchange may take place due to difference in other intensive properties. These forms of energy can also be called work.
10. The forms of energy are graded as per their availability or exergy content. A high grade energy like electrical energy can be converted into mechanical work without significant loss. On the other hand, whole of low grade energy like heat cannot be converted into mechanical work. Some part of heat is further degraded (into a very low temperature heat) and rejected to environment.

There are many *forms* of energy.

1. Electrical energy.
2. Mechanical energy including shaft energy, kinetic energy, potential energy, flow energy, etc.
3. Chemical energy expressed as calorific value of fuel, heat of combustion, Gibb's free energy.
4. Heat energy including enthalpy and internal energy, solar energy, geothermal energy.
5. Nuclear energy of fission and fusion.

However, in thermodynamics, a system can process two types of energy.

Stored Energy

The energy possessed by a system within its boundaries is called stored energy. For example, potential energy (PE), kinetic energy (KE), internal energy (U).

Transit Energy

The energy, which is not possessed by the system but can cross its boundaries is called transit energy. For example, heat, work and electrical energy.

The stored energy is a thermodynamic property whereas the transit energy is not a thermodynamic property as it depends upon the path.

1.9.1 Potential Energy (PE)

The energy possessed by a body or system by virtue of its position or elevation above the datum (ground) level. The work is done due to its falling on earth's surface.

Potential energy,

$$PE = WZ = mgZ$$

where W = weight of body, N

   m = mass of body, kg

   Z = distance of fall of body, m

   g = acceleration due to gravity = 9.81 m/s^2

The units of potential energy,

$$PE = WZ$$

$$= Nm$$

or $PE = mgZ$

$$= \text{kg} \times \frac{\text{m}}{\text{s}^2} \times \text{m} = \text{Nm} \qquad \left[\because 1\,\text{N} = 1\,\frac{\text{kg-m}}{\text{s}^2} \right]$$

·The total work done by the body in falling from initial position Z_1 to final position Z_2,

$$W_{1-2} = \int_{Z_1}^{Z_2} mg\,dz = mg(Z_2 - Z_1)$$

Therefore, work done by the body represents the change in its potential energy. The change in potential energy in a gravitation field can be determined from the values of Z_1 and Z_2 and does not depend how change has occurred. Therefore, potential energy is a property of the system as it does not depend upon the path of the fall.

The magnitude of potential energy depends upon the mass of the system and hence it is an extensive property.

1.9.2 Kinetic Energy (KE)

The energy possessed by a body or a system by virtue of its mass and velocity of motion is called kinetic energy.

$$\text{KE} = \frac{mV^2}{2}$$

where $\qquad m = \text{mass of the body, kg}$

$\qquad\qquad V = \text{velocity of the body, m/s}$

$$\text{KE} = \text{kg} \times (\text{m/s}^2) = \frac{\text{kgm}}{\text{s}^2}.\text{m} = \text{N-m.}$$

According to Newton's law of motion, the force

$$F = \text{mass} \times \text{acceleration}$$

$$= m \times a = m.\frac{dV}{dt}$$

$$= m.\frac{dV}{ds}.\frac{ds}{dt} = mV.\frac{dV}{ds} \qquad \left[\because V = \frac{ds}{dt} \right]$$

$$W_{1-2} = \int_{S_1}^{S_2} F\,ds = \int_{V_1}^{V_2} mV\,dV$$

$$W_{1-2} = \frac{1}{2}m\left(V_2^2 - V_1^2\right)$$

The work done by a body is the change of its kinetic energy.

Kinetic energy can be computed by knowing the mass of the body and velocity relative to a specific coordinate frame at any instant without the consideration how this velocity was attained. Therefore, kinetic energy is a state function and hence it is the property of the system. It depends upon the mass of the system, hence, it is an extensive property.

1.9.3 Internal Energy (U)

The energy can be stored within a system and can be transferred from one system to another. The energy can be stored at microscopic and macroscopic level of the system. The kinetic energy and potential energy of a system are associated with the coordinate frames selected and can be specified by the macroscopic parameters of the mass, velocity and elevation.

The energy possessed by a body or a system by virtue of its molecular arrangement and motion of molecules is called internal energy or intrinsic energy or microscopic energy. This energy is associated with thermodynamic state of the system. For perfect gases, internal energy depends only on the temperature of the system as it is assumed that the molecular forces of attraction and repulsion are absent. For non-permanent gas, vapour or liquid, the internal energy strongly depends on temperature and weakly on pressure.

For all gases,

$$U = U(T, p)$$

For ideal gases,

$$U = U(T)$$

The main characteristics of internal energy are:

1. Internal energy is associated with the thermodynamic state of the system, therefore, it is a point function and thermodynamic property.

2. Internal energy depends upon the mass of the system, hence it is an extensive property.

3. As per Joule's law,

$$dU = m\, C_v\, dT$$

$\therefore$
$$U_2 - U_1 = m\int_{T_1}^{T_2} C_v\, dT$$

$$= m\, C_v(T_2 - T_1)$$

$$C_v = \left(\frac{d\mathrm{U}}{dT}\right) \text{ for ideal gas.}$$

4. When a system undergoes a thermodynamic process of change of state, both heat and work transfer take place. The net energy transfer is stored within the system and is called stored energy.

$$dQ - dW = dE$$

$$Q_{1-2} - W_{1-2} = E_2 - E_1$$

The total energy of the system,

$$E = PE + KE + U$$

For a closed or non-flow thermodynamic system, when there is no change of PE and also there is no flow of mass into or out of the system, there is no kinetic energy present.

$$\therefore \quad Q_{1-2} - W_{1-2} = U_2 - U_1$$

5. When a closed system undergoes a thermodynamic cycle, the net heat transfer is equal to the net work transfer.

$$\sum_i Q_i = \sum_i W_i$$

The cyclic integral of heat transfer is equal to cyclic integral of work transfer.

$$\oint dQ = \oint dW$$

$$\therefore \qquad \oint d\mathrm{U} = 0$$

6. For an isolated system, when

$$Q_{1-2} = 0; \quad W_{1-2} = 0$$

$$\therefore \qquad U_2 = U_1$$

1.9.4 Total Stored Energy (E)

Total energy of the system is the sum of three types of energy.

$$E = PE + KE + U$$

$$= mgh + \frac{1}{2}mV^2 + U$$

For unit mass,

$$e = pe + ke + u$$

∴ $$e = gh + \frac{V^2}{2} + u$$

In special case when PE = 0 and KE = 0

$$E = U$$

or $$e = u$$

Example 1.24: A cage of 80 kg of mass is travelling at a speed of 10 m/s on a ropeway. If the cage is 40 m above ground level, estimate the kinetic energy and potential energy of the cage.

Solution: Mass, $m = 80$ kg

$$V = 10 \text{ m/s}$$

$$h = 40 \text{ m}$$

Potential energy,

$$\text{PE} = mgh$$

$$= 80 \times 9.81 \times 40 \times 10^{-3}$$

$$= 41.39 \text{ kJ}$$

Kinetic energy,

$$\text{KE} = \frac{mV^2}{2} = \frac{1}{2} \times 80 \times (10)^2 \times 10^{-3}$$

$$= 4 \text{ kJ}$$

Example 1.25: An artificial satellite of 500 kg mass is travelling towards the moon. Calculate the kinetic energy and potential energy (i) relative to the earth when 50 km above launching pad and with a speed of 50 km/h. Take earth's gravitational field equal to 7.9 m/s^2 (ii) relative to moon when travelling at the same speed and 50 km from its destination when 1 kg of mass has a weight of 3 N.

Solution: (i) *With respect to earth*

$$m = 500 \text{ kg}$$

$$v = 2500 \text{ km/h} = \frac{2500 \times 10^3}{3600} = 694.45 \text{ m/s}$$

$$g = 7.9 \text{ m/s}^2$$

$$h = 50 \text{ km} = 50 \times 10^3 \text{ m}$$

Kinetic energy,

$$\text{KE} = \frac{mV^2}{2} = \frac{500(694.45)^2}{2} = 120.\, 6 \times 10^6 \text{ J}$$

$$= 120.\, 6 \text{ MJ}$$

Potential energy,

$$PE = mgh = 500 \times 7.9 \times (50 \times 10^3)$$
$$= 197.5 \times 10^6 \text{ J}$$
$$= 197.5 \text{ MJ}$$

(ii) *With respect to moon*

$$KE = \frac{1}{2}mV^2 = \frac{1}{2} \times 500 \times (694.45)^2$$
$$= 120.6 \times 10^6 \text{ J} = 120.6 \text{ MJ}$$
$$W = mg$$
$$g = \frac{W}{m} = \frac{3}{1} = 3 \text{ m/s}^2$$
$$PE = mgh$$
$$= 500 \times 3 \times (50 \times 10^3)$$
$$= 75 \times 10^6 \text{ J} = 75 \text{ MJ}$$

Example 1.26: A system undergoes a cyclic process through four states 1-2, 2-3, 3-4, 4-1. Find the values of X_1, X_2, Y_1, Y_2, Y_3 in the following table.

[U.P.T.U., I Sem., 2005-06]

Process	Heat transfer	Work transfer		Change of internal energy
	kJ/min	kW	kJ/min	kJ/min
1-2	800	5.0	300	Y_1
2-3	400	X_1		600
3-4	–400	X_2		Y_2
4-1	0	3.0	180	Y_3

Solution:

Process 1-2:
$$Q_{1-2} - W_{1-2} = U_2 - U_1$$
$$800 - 300 = Y_1$$
∴
$$Y_1 = 500 \text{ kJ/min}$$

Process 2-3:
$$Q_{2-3} - W_{2-3} = U_3 - U_2$$
$$400 - X_1 = 600$$
∴
$$X_1 = 400 - 600 = -200 \text{ kJ/min}$$

Process 4-1:
$$Q_{4-1} - W_{4-1} = U_1 - U_4$$
$$0 - 180 = Y_3$$
∴
$$Y_3 = -180 \text{ kJ/min}$$

For the cycle: $\oint dQ - \oint dW = \oint dU$

$\therefore$ $\qquad 500 + 600 + Y_2 - 180 = 0$

$\therefore$ $\qquad\qquad\qquad Y_2 = -920 \text{ kJ/min}$

Example 1.27: The specific heat at constant pressure of a gas is given by the following relation:

$$C_p = 0.85 + 0.00004\,T + 5 \times 10\,T^2$$

where T is in Kelvin. Calculate the change in internal energy of 10 kg of gas when its temperature is raised from 300 K to 2300 K. Take that the ratio of specific heats to be 1.5. [U.P.T.U., I Sem., 2003-04]

Solution: $\qquad m = 10 \text{ kg}$

$\qquad\qquad T_1 = 300 \text{ K}$

$\qquad\qquad T_2 = 2300 \text{ K}$

$$\frac{C_p}{C_V} = 1.5$$

$$C_p = 0.85 + 0.00004\,T + 5 \times 10\,T^2$$

$$C_v = \frac{C_p}{1.5}$$

$$\Delta U = \int_1^2 m C_v dt$$

$$= \frac{m}{1.5} \int_{300}^{2300} 0.85 + 0.00004\,T + 5 \times 10\,T^2 dt$$

$$= \frac{10}{1.5} \left| 0.85T + \frac{0.00004}{2} T^2 + \frac{5 \times 10 T^3}{3} \right|_{300}^{2300}$$

$$= \frac{10}{1.5}[(1955 + 105.8 + 2 \times 10^{11}) - 255 + 1.8 + 4.5 \times 10^8]$$

$$= 134.887 \times 10^{11} \text{ kJ}$$

Example 1.28: A closed system undergoes a thermodynamic cycle consisting of four separate and distinct processes. The heat and work transfer in each process are tabulated below:

Process	Heat transfer (kJ/min)	Work transfer (kJ/min)
1-2	0	−10,500
2-3	30,000	0
3-4	−3,000	30,000
4-1	−9000	−1,500

Evaluate the net work output in kW, thermal efficiency of the cycle and change in internal energy.

Solution:
$$\oint dQ = 0 + 30,000 - 3,000 - 9,000 = 18,000 \text{ kJ/min}$$

$$\oint dW = -10,500 + 0 + 30,000 - 1500 = 18,000 \text{ kJ/min}$$

(a) Net work output $= \oint dW = \dfrac{18,000}{60} = 300 \text{ kW}$

(b) Thermal efficiency,

$$\eta_{th} = \frac{\text{Work output}}{\text{Heat supplied}}$$

$$= \frac{\oint dW}{Q_{2\text{-}3}}$$

$$= \frac{18,000}{30,000} \times 100 = 6\%$$

(c) Change in internal energy

$$\Delta U = \oint Q - \oint dW = 0$$

1.9.5 Heat

Heat is energy transferred across the boundary of a system due to temperature difference between the system and the surrounding. The heat can be transferred by conduction, convection and radiation.

The main characteristics of heat are:

1. Heat flows from a system at a higher temperature to a system at a lower temperature.
2. Heat is energy in transit and exists only during transfer into or out of a system. Once energy transfer is over there will be no heat. It is energy and not heat that is stored in the system.
3. *Sign convection.* Heat is positive when it flows into the system and negative when it flows out of the system. This sign convection is adopted as the

main scope of thermodynamics is to generate work by the system when heat is supplied to it.

4. *Adiabatic process*. A process in which there is no heat transfer is called an adiabatic process.

5. Heat is a path function. The amount of heat transferred as a system undergoes a change of state depends on the path of the process. Its differential is not exact,

$$\int_1^2 \delta Q \neq Q_2 - Q_1$$

$$\int_1^2 \delta Q = {}_1Q_2 \text{ or } Q_{1\text{-}2}$$

6. The rate of heat transfer is denoted as:

$$\frac{\delta Q}{dt} = \dot{Q}$$

7. *Units*. In SI system, the unit of heat is joule (J) or kilojoule (kJ). The rate of heat transfer is kJ/s or kW.

8. Heat required to raise the temperature of a body or a system is:

$$Q_{1\text{-}2} = mC(T_2 - T_1) \quad [\text{kJ}]$$

where m = mass, kg

C = Specific heat, kJ/kg-K

T_1 and T_2 = Temperatures in °C or K.

Specific heat for gases can be specific heat at constant pressure (C_p) and constant volume (C_v).

The product mC = thermal or heat capacity, kJ

mC = water equivalent, kg

1.9.6 Specific Heats of Ideal Gas

The specific heat for liquids and solids varies primarily with temperature and does not vary much with the process. For gases, specific heat also depends on the path of the thermodynamic process which may be carried out isochorically ($V = C$) isobarically ($p = C$) or polytropically ($pv^u = C$). The following two types of specific heats are important:

1. Specific heat at constant volume, C_v.

2. Specific heat at constant pressure, C_p.

The total heat supplied to a gas at constant volume,

$$Q_{1\text{-}2} = mC_v(T_2 - T_1)$$

There is no work done during heating of a gas at constant volume. The heat supplied is stored as internal energy or as increase in internal energy of gas.

$$C_v = \left(\frac{dU}{dT}\right)$$

Total heat supplied at constant pressure,

$$Q_{1\text{-}2} = mC_p(T_2 - T_1)$$

The heat supplied is used to increase the temperature of gas and to do some external work of expansion.

$$\begin{aligned} Q_{1\text{-}2} &= dU + W_{1\text{-}2} \\ &= mC_v(T_2 - T_1) + p(V_2 - V_1) \\ &= mC_v(T_2 - T_1) + mR(T_2 - T_1) \qquad [\because pV = mRT] \end{aligned}$$

$$C_p = \left(\frac{dh}{dT}\right)_{\text{ideal gas}}$$

where h is specific enthalpy,

$$h = u + pv$$

$$C \text{ for water} = 4.186 \text{ kJ/kg-K}$$

$$C_p \text{ for air} = 1.005 \text{ kJ/kg-K}$$

$$C_v \text{ for air} = 0.718 \text{ kJ/kg-K}$$

$$C_p - C_v = R$$

$$C_p/C_v = \gamma$$

Example 1.29: A metal block of 1 kg mass is heated upto 80°C in open atmosphere and subsequently submerged in 10 kg of water so as to raise its temperature by 5°C. The initial temperature of water is 25°C. Determine the specific heat of metal considering no loss to surroundings.   [U.P.T.U., I Sem., 2002-2003]

Solution: For metal,

$$m = 1 \text{ kg}$$

$$t_1 = 80°C$$

$$t_2 = \text{final temperature of water}$$

$$= 25 + 5 = 30°C$$

For water,

$$m = 10 \text{ kg}$$

$$t_1 = 25°C$$

$$t_2 = 30°C$$

Assume $\qquad C = 4.186$ kJ/kg-K for water

Heat lost by metal = Heat gained by water

$$[mC(t_1 - t_2)]_{\text{metal}} = [mC(t_2 - t_1)]_{\text{water}}$$

$$1 \times C(80 - 30) = 10 \times 4.186(30 - 25)$$

$$\therefore \qquad C = \textbf{4.186 kJ/kg-K Ans.}$$

Example 1.30: In a cylinder-piston arrangement, 2 kg of an ideal gas is expanded adiabatically from a temperature of 125°C to 30°C and it is found to perform 152 kJ of work during the process while its enthalpy change is 212.8 kJ. Find its specific heats at constant volume and constant pressure and characteristic gas constant.

[U.P.T.U., II Sem., 2002-03]

Solution: $\qquad m = 2$ kg

$$t_1 = 125°C$$

$$t_2 = 30°C$$

$$W_{1\text{-}2} = 152 \text{ kJ}$$

$$\Delta H = 212.8 \text{ kJ}$$

$$Q_{1\text{-}2} = mC_p(t_2 - t_1) = W_{1\text{-}2} + \Delta U$$

$$\Delta H = mC_p(t_2 - t_1)$$

$$\therefore \qquad C_p = \frac{212.8}{2 \times (125 - 30)} = \textbf{1.12 kJ/kg-K}$$

$$W_{1\text{-}2} = mR(t_2 - t_1)$$

$$\therefore \qquad R = \frac{W_{1\text{-}2}}{m(t_2 - t_1)} = \frac{152}{2 \times 95} = \textbf{0.8 kJ/kg-K}$$

$$\therefore \qquad C_v = C_p - R = 1.12 - 0.8 = \textbf{0.32 kJ/kg-K}$$

Example 1.31: Determine the molecular weight of a gas if its two specific heats are: $C_p = 2.286$ kJ/kg-K and $C_v = 1.768$ kJ/kg-K. Universal gas constant = 8.3143 kJ/kg-K.

[U.P.T.U., II Sem., 2002-03]

Solution: $\qquad C_p = 2.286$ kJ/kg-K

$$C_v = 1.768 \text{ kJ/kg-K}$$

$$\overline{R} = 8.3143 \text{ kJ/kg mol-K}$$

$$R = C_p - C_v = 2.286 - 1.768 = 0.518 \text{ kJ/kg-K}$$

Molecular weight, $\qquad M = \dfrac{\overline{R}}{R} = \dfrac{8.3143}{0.518} = \textbf{16} \quad \textbf{Ans.}$

Example 1.32: Estimate the heat required to raise the temperature of fluid from 300 K to 400 K if its specific heat is given by

$$C = (0.2 + 0.002\,T),\ \text{kJ/kg-K}$$

The mass of fluid is 2 kg. What will be the mean specific heat?

Solution:
$$Q_{1\text{-}2} = m \int_{T_1}^{T_2} C\,dT$$

$$= 2 \int_{300}^{400} (0.0 + 0.002\,T)\,dT$$

$$= 2 \left| 0.2T + \frac{0.002\,T^2}{2} \right|_{300}^{400}$$

$$= 2\{0.2(400 - 300) + 0.001[(400)^2 - (300)^2]\}$$

$$= \textbf{180 kJ}$$

Also
$$Q_{1\text{-}2} = mC_m(T_2 - T_1)$$

$$C_m = \frac{Q_{1\text{-}2}}{m(T_2 - T_1)} = \frac{180}{2(400 - 300)} = \textbf{0.9 kJ/kg-K}$$

Example 1.33: It is proposed to heat 2 litres of water initially at 15°C to a temperature of 80°C. What time will be required with an electric heater of 0.5 kW capacity. There is no heat loss to the surrounding.

Solution:
$$Q_{1\text{-}2} = mC\Delta T$$
$$m = 2 \times 10^{-3} \times 1000 = 2\ \text{kg}$$
$$Q_{1\text{-}2} = 2 \times 4.186 \times (80 - 15) = 544.18\ \text{kJ}$$

Power of heater, $\quad \dot{Q} = 0.5\ \text{kW} = 0.5\ \text{kJ/s}$

∴ Time required, $\quad t = \dfrac{Q_{1\text{-}2}}{\dot{Q}} = \dfrac{544.18}{0.5} = 1088.62\ \text{s}$

$$= \textbf{18.14 minutes}\quad \textbf{Ans.}$$

1.9.7 Work

Work is done when the point of application of a force moves in the direction of force.

$$W = \text{Force} \times \text{displacement}$$

$$= F \times x$$

If the force changes with displacement, then

$$W = \int_{1}^{2} F(x)\, dx$$

The work is the scalar product of force and displacement. The thermodynamic definition of work is related with the concept of system and surrounding. The work may be defined as follows:

(i) *Obert:* "Work is defined as the energy transferred (without transfer of mass) across the boundary of a system because of an intensive property difference other than temperature that exists between the system and surrounding".

Pressure difference results in mechanical work and electrical potential difference results in electrical work.

(ii) *Keenan:* "Work is said to be done by a system during a given operation if the sole effect of the system on things external to the system (surroundings) can be reduced to the raising of a weight".

Example 1: If a piston is pushed by a gas pressure in a cylinder, the expansion of gas can be reduced to the raising of a weight W against gravity by manipulating surroundings external to system as shown in Fig. 1.15.

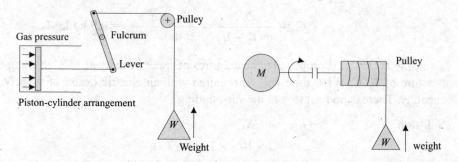

Fig. 1.15 Work done by gas expansion   **Fig. 1.16** Work done by electric motor

Example 2: When a motor rotates, it can be reduced to raising of a weight W through a rope and pulley arrangement as shown in Fig. 1.16.

The gas engine and motor have been manipulated to raise a weight. Hence these devices are performing work as per definition of Keenan.

The main *characteristics* of work are:

1. Work is a means of transferring energy and exists only so long as this transfer takes place.

2. Work is energy in transit.

3. *Unit of work:* In SI system, unit of work is joule (J)

$$1 \text{ J} = 1 \text{ N-m}$$

Power is the time rate of doing work and is designated as W. The unit of power is watt (W) or (kW)

$$1 \text{ W} = 1 \text{ J/s}$$

4. *Sign Convection:* The work is positive when done by system and negative when done on the system.

$$W_{system} + W_{surroundings} = 0$$

When a system does positive work, its surroundings do an equal amount of negative work.

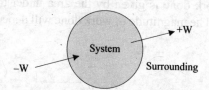

Fig. 1.17 Sign convection of work

5. *Expansion or compression work:* In a cylinder-piston arrangement, when the gas expands, it pushes the piston out. Positive work is done by the system during expansion of gas.

When the piston is pushed inside the cylinder due to the pressure of the surrounding or any external force, the gas is compressed. Negative work is done by the system (as it receives work from the surrounding) in compression of gas.

6. *Work a path function:* Work is evaluated in terms of macroscopic force and displacement

$$W_{1-2} = \int_{1}^{2} F dX$$

where $F = $ Force, N

$X = $ Displacement, m

W depends upon the details of interaction taking place between the system and surroundings and not just the initial and final state of the system. Therefore work is a path function and, therefore, not a property of the system. The differential of work is not an exact differential.

$$\therefore \qquad \int_{1}^{2} \delta W \neq W_2 - W_1$$

$$\int_{1}^{2} \delta W = {}_{1}W_{2} \text{ or } W_{1\text{-}2}$$

1.9.8 Displacement Work

When a piston moves in a cylinder from position 1 to position 2 with volume changing from V_1 to V_2, the amount of work done by the system is given by

$$W_{1\text{-}2} = \int_{V_1}^{V_2} p\,dV$$

The value of work done is given by the area under the process 1-2 on p-V diagram (Fig. 1.18). The magnitude of work done will depend upon the thermodynamic process.

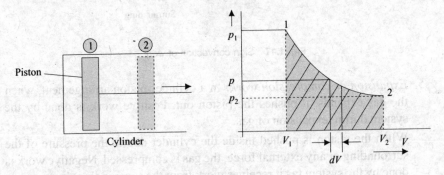

Fig. 1.18 Displacement work

1. *Isobaric process* (p = constant)

$$W_{1\text{-}2} = \int_{V_1}^{V_2} p\,dV$$

$$= p(V_2 - V_1)$$

2. *Isochoric process* (V = constant)

$$W_{1\text{-}2} = \int_{V_1}^{V_2} p\,dV$$

$$= p(V_2 - V_1) = 0$$

$$\therefore \quad V_1 = V_2$$

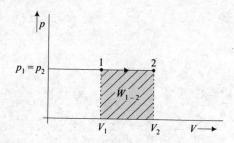

Fig. 1.19 Isobaric process

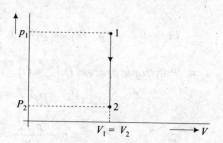

Fig. 1.20 Isochoric process

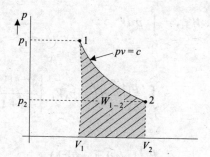

Fig. 1.21 Isothermal process

3. *Isothermal process* (pV = constant)

$$W_{1\text{-}2} = \int_{V_1}^{V_2} pdV$$

$$p_1V_1 = p_2V_2 = pV = C$$

$\therefore \qquad \dfrac{V_1}{V_2} = \dfrac{p_2}{p_1}$

$\therefore \qquad p = \dfrac{p_1V_1}{V}$

$\therefore \qquad W_{1\text{-}2} = \int_{V_1}^{V_2} \dfrac{p_1V_1}{V} dV$

$$= p_1V_1 \int_{V_1}^{V_2} \dfrac{dV}{V}$$

$$= p_1V_1 \ln \dfrac{V_2}{V_1}$$

or
$$W_{1\text{-}2} = p_1 V_1 \ln \frac{p_1}{p_2}$$

4. *Polytropic process* $(pV^n = C)$
$$p_1 V_1{}^n = p_2 V_2{}^n = pV^n = C$$

$\therefore$
$$p = \frac{p_1 V_1^n}{V^n}$$

$$W_{1\text{-}2} = \int_{V_1}^{V_2} p\, dV$$

$$= \int_{V_1}^{V_2} \frac{p_1 V_1^n}{V^n} dV = p_1 V_1^n \left| \frac{V^{-n+1}}{-n+1} \right|_{V_1}^{V_2}$$

$$= \frac{p_1 V_1^n}{1-n} \left(V_2^{1-n} - V_1^{1-n} \right)$$

$\therefore$
$$p_1 V_1{}^n = p_2 V_2{}^n$$

$\therefore$
$$W_{1\text{-}2} = \frac{p_2 V_2^n \times V_2^{1-n} - p_1 V_1^n \times V_1^{1-n}}{1-n}$$

$$= \frac{p_1 V_1 - p_2 V_2}{n-1} = \frac{p_1 V_1}{n-1}\left[1 - \left(\frac{p_2}{p_1} \right)^{\frac{n-1}{n}} \right]$$

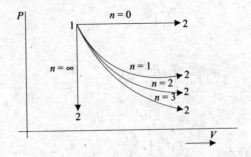

Fig. 1.22 Polytropic process

5. *Adiabatic process:* In adiabatic process, $n = \gamma$, where $\gamma = \dfrac{C_p}{C_v}$.

$$\therefore \qquad W_{1\text{-}2} = \frac{p_1 V_1 - p_2 V_2}{\gamma - 1}$$

1.9.9 Flow Work

Flow work or flow energy is required to move a certain quantity of fluid into and out of the control volume against the existing pressure.

If a working substance with pressure $p_1(\text{N/m}^2)$ flows through an area $A_1(\text{m}^2)$ and moves through a distance $x_1(\text{m})$, the energy or work required to move the working substance,

$$FE = \text{Force} \times \text{Distance}$$
$$= (p_1 A_1) \times x_1$$
$$= p_1 V_1 \text{ (J)} \qquad\qquad [\because A_1 x_1 = V_1]$$

Flow work per unit mass

$$fe = pv \text{ [J/kg]}$$

The rate of work done on the fluid is the product of normal force and fluid velocity.

$$\dot{W}_i = (pA)_i \times V_i \text{ [W]}$$

where
$$p_i = \text{Fluid pressure at inlet [N/m}^2]$$
$$A_i = \text{Inlet area normal to flow [m}^2]$$
$$V_i = \text{Normal velocity of fluid at inlet [m/s]}$$

Similarly, time rate of work done at the exit of control volume,

$$\dot{W}_e = (pA)_e \times V_e \text{ [W]}$$

Net flow work rate,

$$\dot{W}_{net} = \dot{W}_e - \dot{W}_i$$
$$= (pA)_e \times V_e - (pA)_i \times V_i$$

From equation of continuity

$$AV = \dot{m}v$$

$$\therefore \qquad \dot{W}_{net} = (\dot{m}pv)_e - (\dot{m}pv)_i$$

The flow work is supplied to the system when fluid enters, and it is supplied to the surrounding when it leaves the system.

1.9.10 Summary of Forms of Energy

The different forms of energy discussed above can be represented pictorialy in a tree diagram as shown in Fig. 1.23.

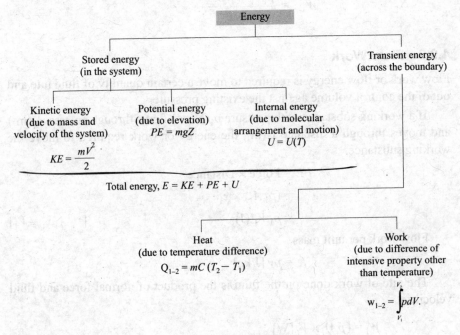

Fig. 1.23 Forms of energy

Example 1.34: An engine cylinder has a piston area of 0.12 m² and contains gas at a pressure of 1.5 MPa. The gas expands according to a process which is represented by a straight line on a pressure-volume diagram. The final pressure is 0.15 MPa. Calculate the work done by the gas on the piston if the stroke is 0.3 m.

[U.P.T.U., I Sem., 2005-06]

Solution:

Piston area, $\qquad A = 0.12 \text{ m}^2$

Stroke, $\qquad L = 0.3 \text{ m}$

$$p_1 = 1.5 \text{ MPa} = 1.5 \times 10^3 \text{ kPa}$$

$$p_2 = 0.15 \text{ MPa} = 0.15 \times 10^3 \text{ kPa}$$

Change of volumes $\quad \Delta V = AL = 0.12 \times 0.30 = 0.036 \text{ m}^3$

The work done = Area under the curve 1-2

= Area (1) + Area (2)

$\therefore \qquad\qquad W_{1-2} = \dfrac{p_1 - p_2}{2} \times \Delta V + p_2 \Delta V$

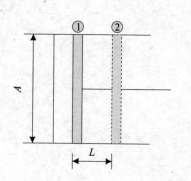
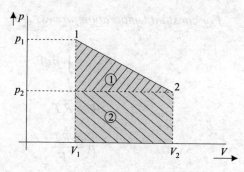

$$W_{1\text{-}2} = \left[\frac{1.5 - 0.15}{2} \times 0.036 + 0.15 \times 0.036 \right] \times 10^3$$

$$= \textbf{23.4 kJ} \quad \textbf{Ans.}$$

Example 1.35: One mole of an ideal gas at 0.1 MPa and 300 K is heated at constant pressure till volume is doubled and then it is allowed to expand at constant temperature till the volume is doubled again. Calculate the work done by the gas.

Solution: $p_1 = 0.1$ MPa

$T_1 = 300$ K

$V_2 = 2V_1$

$V_3 = 2V_2$

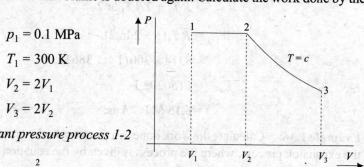

For constant pressure process 1-2

$$W_{1\text{-}2} = \int_1^2 p\,dV = p_1(V_2 - V_1)$$

$$= p_1(2V_1 - V_1) = p_1 \overline{V}_1 = \overline{R}\,T_1$$

Also $$\frac{p_1 V_1}{T_1} = \frac{p_2 V_2}{T_2}$$

∴ $$\frac{V_1}{T_1} = \frac{V_2}{T_2} \quad [\because p_1 = p_2]$$

∴ $$T_2 = T_1 \frac{V_2}{V_1} = 2T_1$$

For constant temperature process 2-3

$$W_{2-3} = \int_2^3 p \, dV$$

$$pV = \overline{R} \, T$$

$$p = \frac{\overline{R} T}{V}$$

$$\therefore \quad W_{2-3} = \int_2^3 \frac{\overline{R} T}{V} dV = \overline{R} \, T_2 \ln \frac{V_3}{V_2} = \overline{R} \, T_2 \ln 2$$

$$= 2\overline{R} \, T_1 \ln 2 \; [\because \; T_3 = T_2 \text{ and } T_2 = 2T_1]$$

Total work done

$$W_{1-3} = W_{1-2} + W_{2-3}$$

$$= \overline{R} \, T_1 + 2\overline{R} \, T_1 \ln 2$$

$$W_{1-3} = \overline{R} \, T_1(1 + 2 \ln 2)$$

$$= 8314 \times 300 (1 + 1.3863)$$

$$= 6150292 \text{ J}$$

$$= \textbf{6.15 MJ} \quad \textbf{Ans.}$$

Example 1.36: Calculate the work done in a piston cylinder arrangement during the expansion process, where the process is given by the equation:

$$p = (V^2 + 6V), \text{ bar}$$

the volume changes from 1 m^3 to 4 m^3 during expansion.

[U.P.T.U., I Sem., 2004-05]

Solution:
$$V_1 = 1 \text{ m}^3$$
$$V_2 = 4 \text{ m}^3$$

$$W_{1-2} = \int_1^2 p \, dV$$

$$= \int_{V_1}^{V_2} (V^2 + 6V) \times 10^5 \, dV$$

$$= 10^5 \left| \frac{V^3}{3} + \frac{6V^2}{2} \right|^4$$

$$= 10^5 \left[\frac{(4^3 - 1^3)}{3} + 3(4^2 - 1^2) \right]$$

$$= 10^5 = \left[\frac{63}{3} + 15 \times 3 \right] = 66 \times 10^5 \text{ J}$$

$$= \textbf{6.6 MJ} \quad \textbf{Ans.}$$

Example 1.37: In a piston-cylinder arrangement, the gas is initially at 150 kPa and occupies a volume of 0.03 m^3. The gas is heated so that volume of gas increase to 0.1 m^3. Calculate the work done by the gas if volume is inversely proportional to the pressure.

Solution:
$$p_1 = 150 \text{ kPa}$$
$$V_1 = 0.03 \text{ m}^3$$
$$V_2 = 0.1 \text{ m}^3$$
$$V \propto \frac{1}{p}$$

$\therefore$
$$pV = C = p_1 V_1 = p_2 V_2$$

$\therefore$
$$p = \frac{C}{V}$$

$$W_{1\text{-}2} = \int p \, dV = C \int_{V_1}^{V_2} \frac{dV}{V} = C \ln \frac{V_2}{V_1} = p_1 V_1 \ln p_1 V_1$$

$$= 150 \times 0.03 \ln \frac{0.1}{0.03} = 5.41 \text{ kJ} \quad \textbf{Ans.}$$

Example 1.38: Compute the work which a pump shall have to do upon water in an hour to just force the water into a tank (closed) having pressure of 1.0 MPa at the rate of 1.5 m^3/min horizontally from an open well.

Solution:
$$p = 1 \text{ MPa} = 1 \times 10^3 \text{ kPa}$$
$$V = 1.5 \text{ m}^3/\text{min} = 1.5 \times 60 = 90 \text{ m}^3/\text{hr}$$
$$W = pV = 1 \times 10^3 \times 90 = \textbf{90} \times \textbf{10}^3 \textbf{ kJ} \quad \textbf{Ans.}$$

Example 1.39: The flow energy of 0.124 m^3/min of a fluid crossing a boundary to a system is 18 kW. Find the pressure at this point.

Solution: Flow energy,

$$FE = 18 \text{ kW} = 18 \times 10^3 \text{ W}$$

$$V = 0.124 \text{ m}^3/\text{min} = \frac{0.124}{60} \text{ m}^3/\text{s}$$

Now, $$FE = pV$$

∴

$$p = \frac{FE}{V} = \frac{18 \times 10^3}{0.124/60} = 8.7 \times 10^6 \text{ N/m}^2$$

$$= 87 \text{ bar} \quad \textbf{Ans.}$$

1.9.11 Comparison between Work and Heat Transfer

There are many similarities and dissimilarities between heat and work.

Table 1.6 Comparison between work and heat

Heat	Work
I. Similarities	
1. Heat is a transient phenomenon. The systems do not process heat. When a system undergoes a change, heat transfer takes place.	1. Work is a transient phenomenon. The systems do not process work. When a system undergoes a change, work done may occur.
2. Heat is a boundary phenomenon. It is observed at the boundary of the system.	2. Work is a boundary phenomenon. It is observed at the boundary of the system.
3. Heat is the energy crossing the boundary of the system.	3. Work is the energy crossing the boundary of the system.
4. Heat is a path function and hence it is an inexact differential.	4. Work is a path function and hence it is an inexact differential.
5. Heat is not a property of the system.	5. Work is not a property of the system.
6. The magnitude of heat transfer depends upon the path followed by the system during change of state.	6. The magnitude of work done depends upon the path followed by the system during change of state.
II. Dissimilarities	
7. Heat transfer can take place in a stable system.	7. Work transfer cannot take place in a stable system.
8. Heat transfer is the energy interaction due to temperature difference only.	8. Work transfer is the energy interaction due to other reasons.
9. Heat is a low grade energy.	9. Work is high grade energy.
10. The quantity of heat transfer can be found out by area under the process curve on a T-S diagram.	10. The quantity of work done can be found out by the area under the process curve on p-V diagram.

1.10   ENTHALPY OF AN IDEAL GAS

The sum of internal energy (U) and the product of pressure and volume (pV) is called enthalpy.

$$\text{Enthalpy} = \text{Internal energy} + \text{Flow energy}$$

$$H = U + pV$$

For a unit mass,

$$h = u + pv$$

For an ideal gas,

$$pv = \bar{R}\,T \text{ and } pv = RT$$

$$\therefore \qquad v = u + (RT)$$

But for an ideal gas,

$$u = u(T)$$

$$\therefore \qquad h = h(T)$$

The enthalpy of an ideal gas is a function of temperature only.

As the enthalpy is a sum of different variables which are thermodynamic properties, therefore, enthalpy is a thermodynamic property. Its value depends upon the mass of the system, therefore, it is an extensive property. Its units are kJ. For all processes,

$$dh = C_p dT$$

$$h_2 - h_1 = C_p(T_2 - T_1)$$

1.10.1   Relation between Enthalpy and Specific Heats

$$h = u + RT \qquad \qquad \dots(1)$$

$$C_v = \left(\frac{du}{dT}\right)_{\text{ideal gas}}$$

$$C_p = \left(\frac{dh}{dT}\right)_{\text{ideal gas}}$$

$$\therefore \quad \text{From (1)}$$

$$\frac{dh}{dT} = \frac{du}{dT} + R$$

$$C_p = C_v + R$$

$$(C_p - C_v) = R$$

For molar quantities

$$\left(\overline{C}_p - \overline{C}_v\right) = \overline{R}$$

where $\overline{C}_p$ and $\overline{C}_v$ are molar specific heats.

Internal energy,

$$du = C_v\, dT$$

$$(\Delta u)_{\text{ideal gas}} = \int_{T_1}^{T_2} C_v dT$$

Enthalpy,

$$dh = Cp\, dT$$

$$(\Delta h)_{\text{ideal gas}} = \int_{T_1}^{T_2} C_p dT$$

Heat supplied at constant pressure

$$Q_{1\text{-}2} = m\,C_p(T_2 - T_1)$$

$$= W_{1\text{-}2} + \Delta U$$

$$mC_p(T_2 - T_1) = p(V_2 - V_1) + mC_v(T_2 - T_1) \qquad \text{...(2)}$$

But
$$pV = mRT$$

$\therefore$
$$p(V_2 - V_1) = mR(T_2 - T_1)$$

$\therefore$ From (2)

$$mC_p(T_2 - T_1) = mR(T_2 - T_1) + mC_v(T_2 - T_1)$$

$\therefore$
$$C_p = R + C_v$$

But
$$(C_p - C_v) = R$$

$$C_v\left(\frac{C_p}{C_v} - 1\right) = R$$

$$C_v(\gamma - 1) = R$$

$$\frac{C_p}{C_v} = \gamma \text{ by definition}$$

$\therefore$
$$C_v = \frac{R}{\gamma - 1}$$

Example 1.40: Calculate the change of enthalpy as 1 kg of oxygen is heated from 500 K to 2000 K. The value of specific heat at constant pressure is given:

$$C_p = 11.515 - \frac{172}{\sqrt{T}} - \frac{1530}{T} \text{ kJ/k mol-K}$$

Solution:

$$dh = \int_1^2 C_p dT$$

$$= \int_1^2 \left(11.515 - \frac{172}{\sqrt{T}} - \frac{1530}{T} \right) dT$$

$$= \int_{500}^{2000} \left(11.515 - 172 T^{-\frac{1}{2}} - 1530 T^{-1} \right) dT$$

$$= \left| 11.515 T + \frac{172 T^{\frac{1}{2}}}{\frac{1}{2}} + 1530 \right|_{500}^{2000}$$

$$= 11.515(2000 - 500) + 344(2000 - 500)^{\frac{1}{2}} + 1530(2000 - 500)$$

$$= 17272.5 + 13323 + 2295000$$

$$= \mathbf{232559.55 \ kJ/kg \quad Ans.}$$

QUESTION BANK NO. 1

1. Define "Thermodynamics". Discuss its scope.
2. Define and explain a thermodynamic system. Differentiate between various types of thermodynamic systems and give example of each of them.
3. Compare the important features of reversible and irreversible processes. Give examples of each.
4. Differentiate between thermal equilibrium and thermodynamic equilibrium.
5. Explain:
   (i) Thermodynamic equilibrium
   (ii) Quasi-static process
   (iii) Concept of continuum
6. Differentiate between macroscopic and microscopic point of view, giving some examples.

7. For an ideal gas, derive the relation $C_p - C_v = R$.

8. What do you understand by flow work? Is it different from displacement work? How?

9. Define and explain the following:
   (i) Path
   (ii) Process
   (iii) Thermodynamic cycle

10. What is "Thermodynamic Property"? Is it a point or path function? Giving example, describe thermodynamic variables that are a property and that are not a property.

11. Discuss the importance and relevance of following concepts in thermodynamics:
    (i) Continuum
    (ii) Quasi-static process

12. What are the reasons for irreversibility of actual process? How can we reach near reversibility condition? What is its importance in engineering thermodynamics?

13. Compare 'heat' and 'work'.

14. Explain the different forms of energy.

15. Differentiate between Universal and Characteristic gas constants.

16. State thermodynamic definition of work. Also differentiate between heat and work.

17. Differentiate amongst gauge pressure, atmospheric pressure and absolute pressure. Also give the value of atmospheric pressure in bar and mm of Hg.

18. What is energy? What are its different forms?

Laws of Thermodynamics

2.1 SIGNIFICANCE AND SCOPE OF LAWS

Thermodynamics is based on the following four laws. These laws cannot be proved mathematically. Their validity stems from the fact that these laws have not been violated. These laws are based on experimental results and observations of common experience.

2.1.1 Zeroth Law

When a body A is in thermal equilibrium with body B, and also separately with body C, then bodies B and C will be in thermal equilibrium with each other. This is known as the zeroth law of thermodynamics.

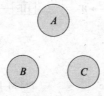

Fig. 2.1 Zeroth law

If, $T_A = T_B$ and also

$T_A = T_C$, then

$T_B = T_C$

where, T = Temperature.

This law deals with thermal equilibrium, concept of equality of temperature and forms the basis for all temperature measurements.

Example: If body C is the thermometer, body B is the reference temperature (i.e., triple point of water), body A is the unknown temperature, the thermometer compares the unknown temperature with the known reference temperature.

2.1.2 First Law

Energy can neither be created nor destroyed; it is always conserved. However, it can change from one form to another.

The first law deals with the conservation of energy and introduces the concept of internal energy. In general, all thermal machines are designed from first law of thermodynamics,

$$Q_{1-2} - W_{1-2} = dU = U_2 - U_1$$

For a closed system where changes in kinetic energy and potential are negligible, the difference in heat supplied and work done is stored as internal energy.

2.1.3 Second Law

It is the directional law of energy and also the law of degradation of energy. It is based on the following statements.

(a) Kelvin-Planck statement

"It is impossible to construct an engine working on a cyclic process, whose sole purpose is to convert heat energy from a single reservoir into an equivalent amount of work." No cyclic engine can convert whole of heat into equivalent work. Second law dictates limits on the conversion of heat into work.

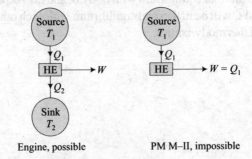

Engine, possible   PM M–II, impossible

Fig. 2.2 Kelvin-Planck statement

(b) Clausius statement

"It is impossible for a self-acting machine, working in a cyclic process, to transfer heat from a body at a lower temperature to a body at a higher temperature without the aid of an external energy".

Heat cannot flow itself from a cold body to a hot body without expenditure of mechanical work. Second law deals with the direction of flow of heat energy.

(c) Carnot statement

"No heat engine operating in a cycle between two given thermal reservoirs, with fixed temperatures, can be more efficient than a reversible engine operating between the same thermal reservoirs".

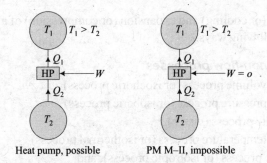

Fig. 2.3 Clausius statement

Second law provides a yardstick to evaluate the performance of an engine or a device. The portion of heat energy which is not available for conversion into work is measured by entropy.

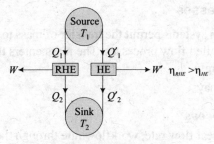

Fig. 2.4 Carnot statement

2.1.4 Third Law

Entropy has zero value at absolute zero temperature. The third law defines the absolute zero of entropy.

"At absolute zero temperature, the entropy of all homogeneous crystalline (condensed) substances in a state of equilibrium becomes zero".

This law is used in chemical engineering for measurement of chemical affinity, analysis of chemical equilibrium and study of behaviour of solids at very low temperatures.

As $T \to 0$, $S \to 0$.

2.2 CLASSIFICATION OF THERMODYNAMIC PROCESSES

The thermodynamic processes are grouped into:

2.2.1 Non-flow Processes

The processes occurring in a closed system where there is no transfer of mass across the boundary are called non-flow processes. In such processes, the energy in the form of heat and work crosses the boundary of the system.

The heating (or cooling) and expansion (or compression) of a gas may be performed in the following ways:

(a) Reversible non-flow processes

(i) Constant volume process (or isochoric process)

(ii) Constant pressure process (or isoberic process)

(iii) Hyperbolic process ($pv = c$)

(iv) Constant temperature process (or isothermal process)

(v) Adiabatic process (or isotropic process), and

(vi) Polytropic process.

(b) Irreversible non-flow process

Free expansion process (or unrestricted process)

2.2.2 Flow Processes

The processes in open systems permit the transfer of mass to and from the system. Such processes are called flow processes. The mass enters the system and leaves after exchanging energy.

The flow processes may:

(a) Steady flow process

The mass flow rate, heat flow rate, workflow rate through the system remain constant and there is no change in any properties or chemical composition of the working fluid at any given point within the system. Nozzles, turbines, compressors and other thermal machines operate as steady flow process.

(b) Unsteady flow process

The filling and evacuation of vessels are unsteady flow processes.

The complete classification of thermodynamic processes is illustrated in Fig. 2.4. The governing equations are also given in the diagram.

2.3 FIRST LAW OF THERMODYNAMICS

The first law of thermodynamics was formulated on the basis of Paddle Wheel Experiment conducted by Joule. A number of experiments ware conducted by him wherein a paddle wheel was rotated by different forms of inputs. His findings were that the work expended was proportional to increase in thermal energy.

$$Q \propto W$$

or
$$Q = \frac{W}{J}$$

where, J = proportionality constant called mechanical equivalent.

In SI system,
$$J = 1$$

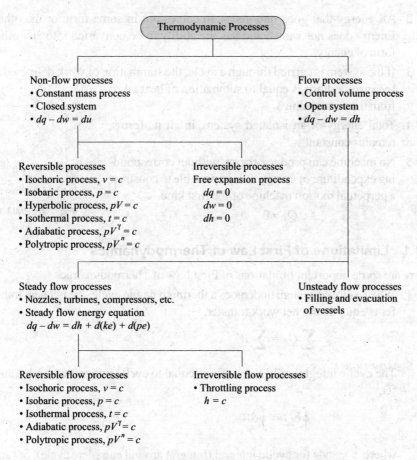

Fig. 2.5 Classification of thermodynamic processes

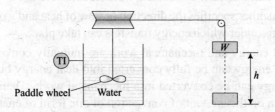

Fig. 2.6 Joule's paddle wheel experiment

The first law of thermodynamics states that work and heat are mutually convertible. The present tendency is to include all forms of energy.

The first law can be stated in many ways:

1. Energy can neither be created nor destroyed; it is always conserved. However, it can change from one form to another.

2. All energy that goes into a system comes out in some form or the other. Energy does not vanish and has the ability to be converted into any other form of energy.

3. If the system is carried through a cycle, the summation of work delivered to the surroundings is equal to summation of heat taken from the surrounding.

4. Total energy of an isolated system, in all its forms, remains constant.

5. No machine can produce energy without corresponding expenditure of energy. It is impossible to construct a perpetual motion machine of the first kind.

$$Q_1 \neq 0$$

PM M–I, impossible

Fig. 2.7   PMM-1

2.3.1   Limitations of First Law of Thermodynamics

There are some important limitations of First Law of Thermodynamics.

1. When a closed system undergoes a thermodynamic cycle, the net heat transfer is equal to the net work transfer.

$$\sum_i Q_i = \sum_i W_i$$

The cyclic integral of heat transfer is equal to cyclic integral of work transfer.

$$\oint \delta Q = \oint \delta W$$

where $\oint$ stands for cyclic integral (integral around complete cycle), δQ and δW are small elements of heat and work transfer and have same units.

The law neither specifies the direction of flow of heat and work nor gives any conditions under which energy transfers can take place.

2. The heat energy and mechanical work are mutually convertible. The mechanical energy can be fully converted into heat energy but only a part of heat energy can be converted into mechanical work. Therefore, there is a limitation on the amount of conversion of one form of energy into another form.

2.3.2   Applications of First Law to Thermodynamic Processes

According to First law, when a system undergoes a thermodynamic process (change of state) both heat and work transfer takes place. The net (difference) energy transfer is stored within the system and is called stored energy or total energy of he system.

$$\delta Q - \delta W = dE.$$

where, E is the extensive property and represents the total energy of the system at a given state.

$$Q_{1-2} - W_{1-2} = E_2 - E_1$$

Q, W and E have same units.

For a unit mass,

$$q_{1-2} - w_{1-2} = e_2 - e_1$$

where, Q_{1-2} = Heat transferred to the system during the process 1-2. It is positive when supplied to the system and negative when rejected by the system.

$\quad W_{1-2}$ = Work done by the system on the surrounding during the process 1-2.

$\quad E_1$ = Total energy of the system at state 1.

$$= PE_1 + KE_1 + U_1 + FE_1$$

$$= mgz_1 + \frac{m_1 V_1^2}{2} + U_1 + p_1 V_1$$

$\quad E_2$ = total energy of the system at state 2.

$$= PE_2 + KE_2 + U_2 + FE_2$$

$$= mgz_2 + KE_2 + U_2 + p_2 V_2$$

$\therefore \qquad Q_{1-2} - W_{1-2} = E_2 - E_1$

$$= (PE_2 + KE_2 + U_2 + FE_2) - (PE_1 + KE_1 + U_1 + FE_1)$$

$$= (PE_2 - PE_1) + (KE_2 - KE_1) - (H_2 - H_1) \ [\because H = U + FE]$$

$$= mg\,(Z_2 - Z_1) + \frac{m}{2}\,(V_2^2 - V_1^2) + (H_2 - H_1)$$

For unit mass,

$$q_{1-2} - w_{1-2} = g(Z_2 - Z_1) + \frac{V_2^2 - V_1^2}{2} + (h_2 - h_1)$$

Case I For a closed system, there is no flow energy (FE)

$\therefore \qquad q_{1-2} - w_{1-2} = g(Z_2 - Z_1) + \dfrac{V_2^2 - V_1^2}{2} + (u_2 - u_1)$

This is called non-flow energy equation.

Case II When there is no change in the potential energy of a closed system, i.e., $PE_2 = PE_1$

$\therefore \qquad q_{1-2} - w_{1-2} = (KE_2 - KE_1) + (u_2 - u_1) = \dfrac{V_2^2 - V_1^2}{2} + (u_2 - u_1)$

Case III Closed or non-flow thermodynamic system, when there is no change of *PE* and also there is no flow of mass into or out of a system, i.e.,

$$PE_2 = PE_1 \text{ and } KE_2 = KE_1$$

$$\therefore \qquad q_{1-2} - w_{1-2} = u_2 - u_1$$

Case IV Isolated system,

$$q_{1-2} = 0 \;;\quad w_{1-2} = 0$$

$$e_2 = e_1 \quad \text{and} \quad u_2 = u_1$$

This shows that the first law of thermodynamics is law of conservation of energy.

Case V Cyclic process

There is no change in the internal energy and stored energy is zero.

$$\oint \delta q = \oint \delta w.$$

2.4 FIRST LAW ANALYSIS OF PROCESSES FOR IDEAL GAS

For an ideal gas,

$$du = Cv \, dT$$
$$dh = Cp \, dT$$
$$pv = RT.$$

These equations are valid for all processes.

2.4.1 Constant Volume Process

The first law equation for a non-flow process when kinetic energy and potential energy are negligible:

$$\delta q - \delta w = du.$$

For a reversible process,

$$\delta w = p \, dv \text{ and } du = Cv \, dT$$

$$\therefore \qquad \delta q - p \, dv = Cv \, dT$$

For constant volume process

$$\delta w = p \, dv = 0$$

$$\therefore \qquad \delta q = Cv \, dT = du$$

or

$$q_{1-2} = u_2 - u_1 = \int_1^2 Cv \, dT$$

If the value of *Cv* as a function of *T* is known, integral can be evaluated.

2.4.2 Constant Pressure Process

The first law equation is:

$$\delta q - \delta w = du$$

or
$$\delta q = du + p \, dv$$
$$= d(u + pv) \qquad [\because \quad p = \text{constant}]$$
$$= dh = Cp \, dT$$

$\therefore \qquad q_{1-2} = \int_{1}^{2} Cp \, dT = h_2 - h_1$

If the value of Cp as a function of temperature is known, integral can be evaluated.

2.4.3 Constant Temperature Process

The first law equation is:
$$\delta q - \delta w = du$$
$$du = Cv \, dT = 0 \qquad (\because \quad T \text{ is constant})$$
$\therefore \qquad \delta q = p \, dv$

$$q_{1-2} = \int_{1}^{2} p \, dv$$

For an ideal gas,
$$pv = RT$$
$\therefore \qquad p = \dfrac{RT}{v}$

$\therefore \qquad q_{1-2} = w_{1-2} = \int_{1}^{2} \dfrac{RT}{v} dv$

$$= RT \ln \dfrac{v_2}{v_1}$$

$$= p_1 v_1 \ln \dfrac{v_2}{v_1}$$

Also, $\qquad w_{1-2} = p_1 v_1 \ln \dfrac{p_1}{p_2} \qquad (\because \quad p_1 v_1 = p_2 v_2)$

2.4.4 Adiabatic Process

The first law equation is,
$$\delta q - \delta q = du.$$

No heat leaves or enters the system.

$$\therefore \qquad 0 - \delta w = du$$

$$0 - p\,dv = Cv\,dT$$

$$\therefore \qquad dT = \frac{-p\,dv}{Cv} \qquad\qquad\qquad ...(a)$$

But, $\qquad pv = RT$

Differentiating,

$$pdv + vdp = RdT$$

$$\therefore \qquad dT = \frac{pdv + vdp}{R} = \frac{pdv + vdp}{(Cp - Cv)} \quad [\because R = Cp - Cv] \ \ ...(b)$$

Equating (a) and (b) for dT

$$\frac{-p\,dv}{Cv} = \frac{pdv + vdp}{(Cp - Cv)}$$

$$\therefore \qquad \frac{(Cp - Cv)}{Cv} = \frac{pdv + vdp}{-p\,dv} = -1 - \frac{vdp}{pdv}$$

$$\therefore \qquad \frac{Cp}{Cv} - 1 = -1 - \left[\frac{v}{dv} \times \frac{dp}{p} \right]$$

or $\qquad \gamma = -\left[\dfrac{v}{dv} \times \dfrac{dp}{p} \right] \qquad\qquad \left[\because \dfrac{Cp}{Cv} = \gamma \right]$

$$\therefore \qquad \gamma \left[\frac{dv}{v} \right] = -\frac{dp}{p}$$

or $\qquad \gamma \dfrac{dv}{v} + \dfrac{dp}{p} = 0$

Integrating,

$$\gamma \ln v + \ln p = \text{constant}$$

$$\ln pv^{\gamma} = \ln (\text{constant})$$

$$p_1 v_1^{\gamma} = p_2 v_2^{\gamma} = pv^{\gamma} = \text{constant}$$

$$\frac{p_1}{p_2} = \left(\frac{v_2}{v_1} \right)^{\gamma}$$

$$\frac{v_1}{v_2} = \left(\frac{p_2}{p_1}\right)^{\frac{1}{\gamma}}$$

But

$$\frac{p_1 v_1}{T_1} = \frac{p_2 v_2}{T_2}$$

or

$$\frac{v_1}{v_2} = \frac{T_1}{T_2} \times \frac{p_2}{p_1}$$

$$\therefore \quad \left(\frac{p_2}{p_1}\right)^{\frac{1}{\gamma}} = \frac{T_1}{T_2} \times \frac{p_2}{p_1}$$

or

$$\frac{T_1}{T_2} = \left(\frac{p_2}{p_1}\right)^{\frac{1}{\gamma}} \times \frac{p_2}{p_1}$$

$$= \left(\frac{p_1}{p_2}\right)^{-\frac{1}{\gamma}+1}$$

$$\therefore \quad \frac{T_1}{T_2} = \left(\frac{p_1}{p_2}\right)^{\frac{\gamma-1}{\gamma}}$$

2.4.5 Polytropic Process

The first law equation is

$$\delta q - \delta w = du$$

$$q_{1-2} - w_{1-2} = (u_2 - u_1)$$

$$\delta q - pdv = du$$

For polytropic process,

$$w_{1-2} = \frac{p_1 v_1 - p_2 v_2}{n-1} = \frac{R(T_1 - T_2)}{n-1}$$

$$\therefore \quad q_{1-2} = \frac{R(T_1 - T_2)}{n-1} + Cv(T_2 - T_1)$$

$$= \frac{R(T_1 - T_2)}{n-1} + \frac{R}{\gamma-1}(T_2 - T_1) \qquad \left[\because \ Cv = \frac{R}{\gamma-1}\right]$$

$$= R(T_1 - T_2) \left[\frac{1}{n-1} - \frac{1}{\gamma-1} \right]$$

$$= R(T_1 - T_2) \left[\frac{(\gamma-1) - (n-1)}{(n-1)(\gamma-1)} \right]$$

$$= R(T_1 - T_2) \left[\frac{\gamma-n}{(n-1)(\gamma-1)} \right]$$

$$= \frac{\gamma-n}{\gamma-1} \times \frac{R(T_1 - T_2)}{n-1}$$

Also,
$$q_{1-2} = \frac{\gamma-n}{\gamma-1} \times \frac{p_1 v_1 - p_2 v_2}{n-1}$$

$$q_{1-2} = \frac{\gamma-n}{\gamma-1} \times w_{1-2}$$

2.4.6 Free Expansion Process

Free expansion process is unrestricted expansion of a gas without any work output. As the gas cannot be compressed back to its initial state without the use of external work, the process of free expansion is highly irreversible process. The process is also called constant internal energy process as there is no exchange of heat and work between the system and the surrounding. The process can be explained by a simple example.

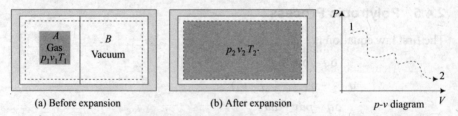

| (a) Before expansion | (b) After expansion | *p-v* diagram |

Fig. 2.8 Free expansion process

There are two chambers A and B completely insulated and separated by a membrane. Chamber A has gas at parameters $p_1 v_1 T_1$ and chamber B is completely evacuated. The state is plotted as point 1 on *p-v* diagram. The partition is removed and the gas occupies the total volume of chambers A and B. The final conditions of the system are $p_2 v_2 T_2$ are shown by point 2 on the *p-v* diagram. The free expansion is shown by a dotted line as the process is uncontrolled, irreversible and unspecified.

1. The work done is zero as there is no expansion of boundary
$$W_{1-2} = 0$$

2. The system is insulated and there is no exchange of heat with the surrounding.
$$Q_{1-2} = 0$$

3. Applying first law of thermodynamics to the closed system,
$$Q_{1-2} - W_{1-2} = U_2 - U_1$$
$$0 - 0 = U_2 - U_1$$
$$\therefore \qquad U_2 = U_1$$

The internal energy of the system during free expansion remains constant.

4. $$U_2 - U_1 = m \, Cv \, (T_2 - T_1) = 0$$
$$\therefore \qquad T_2 = T_1.$$

The process of free expansion is isothermal process.

5. $$H_2 - H_1 = m \, Cp \, (T_2 - T_1) = 0$$

The enthalpy of the system remains constant.

2.5 SUMMARY OF THERMODYNAMIC RELATIONS FOR NON-FLOW PROCESSES

The thermodynamic relations for ideal gas in different non-flow processes are summarised in Table 2.1. The following relationships have been tabulated:

1. Governing equation or p-v-T relationship
2. Work done
3. Change of internal energy
4. Change of enthalpy
5. Heat exchanged.

The values given are for unit mass.

Example 2.1: 5 m^3 of air at 2 bar and 27°C is compressed upto 6 bar pressure following $pv^{1.3}$ = constant. It is subsequently expanded adiabatically to 2 bar. Considering the two processes to be revesrsible, determine the net work. Also plot the process on T-s diagram. [U.P.T.U. I sem, 2002-03].

Solution:

1. Draw T-s diagram to show the following process:
   (i) Process 1–2: Reversible polytropic.
   (ii) Process 2–3: Reversible adiabatic

Table 2.1 Thermodynamic relations (Unit mass)

Process	Equation	Index n	$p\text{-}v\text{-}T$ relation	Specific heat C	Heat added	Work done $\int_1^2 pdv$	Change of enthalpy	Change of internal energy	Remark
Units	—	—	—	kJ/kg-K	kJ/kg	kJ/kg	kJ/kg	kJ/kg	—
Constant volume (Isochoric)	$v = c$	$n = \infty$	$\frac{p_1}{p_2} = \frac{T_1}{T_2}$	Cv	$Cv(T_2 - T_1)$	0	$Cp(T_2 - T_1)$	$Cv(T_2 - T_1)$	—
Constant pressure (Isobaric)	$p = c$	$n = 0$	$\frac{v_1}{v_2} = \frac{T_1}{T_2}$	Cp	$Cp(T_2 - T_1)$	$p(v_2 - v_1)$	$Cp(T_2 - T_1)$	$Cv(T_2 - T_1)$	—
Constant temperature (Isothermal)	$t = c$	$n = 1$	$p_1 v_1 = p_2 v_2$	∞	$p_1 v_1 \ln \frac{v_2}{v_1}$	$p_1 v_1 \ln \frac{v_2}{v_1}$	0	0	—
Reversible Adiabatic (Isentropic)	$pv^\gamma = c$	$n = \infty$	$p_1 v_1^\gamma = p_2 v_2^\gamma$	0	0	$\dfrac{p_1 v_1 - p_2 v_2}{\gamma - 1}$	$Cp(T_2 - T_1)$	$\dfrac{R(T_1 - T_2)}{\gamma - 1}$	$\dfrac{T_2}{T_1} = \left(\dfrac{v_1}{v_2}\right)^{\gamma-1} = \left(\dfrac{p_2}{p_1}\right)^{\frac{\gamma-1}{\gamma}}$ $u_2 - u_1 = \dfrac{p_1 v_1 - p_2 v_2}{\gamma - 1}$
Polytropic	$pv^n = c$	$n = n$	$p_1 v_1^n = p_2 v_2^n$	Cn	$Cn(T_2 - T_1)$	$\dfrac{p_1 v_1 - p_2 v_2}{n - 1}$	$Cp(T_2 - T_1)$	$Cv(T_2 - T_1)$	$\dfrac{T_2}{T_1} = \left(\dfrac{v_1}{v_2}\right)^{n-1} = \left(\dfrac{p_2}{p_1}\right)^{\frac{n-1}{n}}$ $C_n = Cv\left(\dfrac{\gamma - n}{1 - n}\right);\ Q_{1\text{-}2} = \dfrac{\gamma - n}{r - 1}\,W.D$ $W.D = \dfrac{R(T_2 - T_1)}{n - 1}$
Free expansion	$u_1 = u_2$ $T_1 = T_2$	—	$T_1 = T_2$	—	—	0	0	0	—

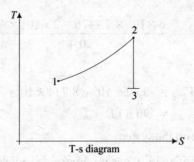

T-s diagram

2. Data given,

$$V_1 = 5 \text{ m}^3$$
$$p_1 = 2 \text{ bar} = 2 \times 10^5 \text{ N/m}^2$$
$$T_1 = 27°C + 273 = 300 \text{ K}$$
$$n = 1.3$$
$$p_2 = 6 \text{ bar} = 6 \times 10^5 \text{ N/m}^2$$
$$p_3 = 2 \text{ bar} = 2 \times 10^5 \text{ N/m}^2$$

Assume $\gamma = 1.4$ for adiabatic process 2–3,

$$m = \frac{p_1 V_1}{RT_1} = \frac{2 \times 10^5 \times 5}{287 \times 300} = 11.6 \text{ kg (take } R = 287 \text{ J/kg-K for air)}$$

3. Work done during polytropic process 1-2,

$$p_1 v_1^n = p_2 v_2^n$$

$$V_2 = V_1 \left(\frac{p_1}{p_2}\right)^{\frac{1}{n}} = 5 \times \left(\frac{2}{6}\right)^{\frac{1}{1.3}} = 2.1476 \text{ m}^3$$

$$W_{1-2} = \frac{p_1 V_1 - p_2 V_2}{n-1} = \frac{2 \times 10^5 \times 5 - 6 \times 10^5 \times 2.1476}{1.3 - 1} = -9.62 \times 10^5 \text{ N-m}$$

The negative sign shows that work is done on the system during polytropic compression process.

4. Work done during reversible adiabatic process 2-3,

$$p_3 V_1^\gamma = p_2 V_2^\gamma$$

$$V_3 = V_2 \left(\frac{p_2}{p_3}\right)^{\frac{1}{\gamma}} = 2.1476 \left(\frac{6}{2}\right)^{\frac{1}{1.4}} = 4.7 \text{ m}^3$$

$$W_{2-3} = \frac{p_2 V_2 - p_3 V_3}{\gamma - 1} = \frac{6 \times 10^5 \times 2.1476 - 2 \times 10^5 \times 4.7}{0.4} = 8.714 \times 10^5 \text{ N/m}$$

5. Total work done

$$W_{1-3} = W_{1-2} + W_{2-3} = -9.62 \times 10^5 + 8.714 \times 10^5 = -0.906 \times 10^5 \text{ N/m}$$
$$= -90.6 \text{ kJ}$$

The work is done on the system.

Example 2.2: Two kg of an ideal gas is compressed adiabatically from pressure 100 kPa and temperature 220 K to a final pressure of 400 kPa. Calculate:

(i) Initial volume

(ii) Final volume and temperature

(iii) Heat exchanged

(iv) Change in internal energy.

Take $Cp = 1$ kJ/kg-K and $Cv = 0.707$ kJ/kg-K.

Solution:

(i) *Initial volume*

$$p_1 = 100 \text{ kPa} = 100 \times 10^3 \text{ N/m}^2$$
$$T_1 = 220 \text{ K}$$
$$m = 2 \text{ kg}$$
$$R = Cp - Cv = 1 - 0.707 = 0.293 \text{ kJ/kg-K}$$
$$= 293 \text{ J/kg-K}.$$

Now, $\qquad p_1 V_1 = m RT_1$

$\therefore \qquad\qquad V_1 = \dfrac{m RT_1}{p_1} = \dfrac{2 \times 293 \times 220}{100 \times 10^3} = 1.29 \text{ m}^3$

(ii) *Final volume and temperature*

$$\gamma = \frac{Cp}{Cv} = \frac{1}{0.707} = 1.414$$

$$p_2 = 400 \times 10^3 \text{ N/m}^2$$

Now, $\qquad p_1 V_1^\gamma = p_2 V_2^\gamma$

$\therefore \qquad\qquad V_2 = V_1 \left(\dfrac{p_1}{p_2}\right)^{\frac{1}{\gamma}} = 1.29 \left(\dfrac{100}{400}\right)^{\frac{1}{1.414}}$

$$= 0.484 \text{ m}^3$$

$$p_2 V_2 = m R T_2$$

$$\therefore \qquad T_2 = \frac{p_2 V_2}{m R} = \frac{400 \times 10^3 \times 0.484}{2 \times 293} = 330.4 \text{ K}$$

(iii) *Work performed*

$$W_{1-2} = \frac{p_1 V_1 - p_2 V_2}{\gamma - 1} = \frac{100 \times 10^3 \times 1.29 - 400 \times 10^3 \times 0.484}{1.414 - 1}$$

$$= -156 \times 10^3 \text{ J}$$

$$= 156 \text{ kJ}.$$

Work is done on the system during compression process.

(iv) *Heat exchanged*

During adiabatic process, no heat is exchanged

$$\therefore \qquad Q_{1-2} = 0$$

(v) *Change in internal energy*

From first law of thermodynamics

$$dU = dQ - dW = 0 - (-156)$$

$$= \mathbf{156 \text{ kJ}} \text{ (increase)}$$

Example 2.3: The internal energy of a certain substance is expressed by the equation:

$$u = 3.62 \, pv + 86$$

where, u is given in kJ/kg

p is given in kPa and

v is in m³/kg.

A system composed of 5 kg of this substance expands from an initial pressure of 550 kPa and a volume of 0.25 m³/kg to a final pressure of 125 kPa, in a process in which pressure and volume are related by $pv^{1.2} = \text{constant}$. If the expansion process is quasi-static, determine Q, Δu and W for this process.

[U.P.T.U. II Sem., 2005-06]

Solution:

$$m = 5 \text{ kg}$$

$$p_1 = 550 \text{ kPa}$$

$$v_1 = 0.25 \text{ m}^3/\text{kg}$$

$$p_2 = 125 \text{ kPa}$$

$$n = 1.2$$

$$V_1 = mv_1 = 5 \times 0.25 = 1.25 \text{ m}^3$$
$$p_1 v_1^n = p_2 v_2^n$$

$$\therefore \qquad v_2 = v_1 \left(\frac{p_1}{p_2}\right)^{\frac{1}{n}} = 0.25 \left(\frac{550}{125}\right)^{\frac{1}{1.2}} = 0.859 \text{ m}^3/\text{kg}.$$

$$V_2 = mv_2 = 5 \times 0.859 = 4.295 \text{ m}^3$$

(i) *Work done*

$$W_{1-2} = \frac{p_1 V_1 - p_2 V_2}{n-1} = \frac{550 \times 1.25 - 125 \times 4.295}{1.2 - 1}$$

$$= \frac{687.5 - 536.875}{0.2} = \textbf{753.125 kJ} \quad \textbf{Ans.}$$

(ii) *Change of internal energy*

$$u_1 = 3.62 \, p_1 v_1 + 86$$
$$u_2 = 362 \, p_2 v_2 + 86$$
$$\therefore \qquad u_2 - u_1 = 3.62 \, (p_2 v_2 - p_1 v_1)$$
$$= 3.62 \, (125 \times 0.859 - 550 \times 0.25)$$
$$= 3.62 \, (107.375 - 137.5) = -109 \text{ kJ/kg}$$
$$\Delta U = m(u_2 - u_1) = 5 \times 109 = \textbf{545 kJ (decrease) Ans.}$$

(iii) *Heat exchange*

$$Q_{1-2} = W_{1-2} + (U_2 - U_1)$$
$$= 753.125 - 545 = \textbf{208.125 kJ} \quad \textbf{Ans.}$$

Example 2.4: A cylinder contains 0.5 m³ of air at 1.5 bar and 100°C. It is compressed polytropically to a volume of 0.125 m³ and final pressure is 9.0 bar. Determine:

  (i) The mass of air

  (ii) The value of index 'n' for pv^n = constant

  (iii) Work done, and

  (iv) Heat supplied   [U.P.T.U. II Sem., 2003-04]

Solution:

Data given,

$$V_1 = 0.5 \text{ m}^3$$
$$p_1 = 1.5 \text{ bar} = 150 \text{ kPa}$$

$$T_1 = 100° + 273 = 373 \text{ K}$$
$$V_2 = 0.125 \text{ m}^3$$
$$p_2 = 9 \text{ bar} = 900 \text{ kPa}.$$

(i) *Mass of air*

Assume $R = 0.287$ kJ/kg-K for air

$$p_1 V_1 = m\, RT_1$$

$\therefore$
$$m = \frac{p_1 V_1}{RT_1} = \frac{150 \times 0.5}{0.287 \times 373} = \textbf{0.7 kg.} \quad \textbf{Ans.}$$

(ii) *Value of index*

$$p_1 V_1^n = p_2 V_2^n$$

$$\left(\frac{V_1}{V_2}\right)^n = \frac{p_2}{p_1}$$

$$n \ln\left(\frac{V_1}{V_2}\right) = \ln \frac{p_2}{p_1}$$

$\therefore$
$$n = \frac{\ln\left(\dfrac{p_2}{p_1}\right)}{\ln\left(\dfrac{V_1}{V_2}\right)} = \frac{\ln\left(\dfrac{900}{150}\right)}{\ln\left(\dfrac{0.5}{0.125}\right)} = \frac{1.7917595}{1.3862944}$$

$$= \textbf{1.29} \quad \textbf{Ans.}$$

(iii) *Work done*

$$W_{1-2} = \frac{p_1 V_1 - p_2 V_2}{n-1} = \frac{150 \times 0.5 - 900 \times 0.125}{1.29 - 1} = \frac{75 - 112.5}{0.29}$$

$$= -129.3 \text{ kJ}$$

Work is done on the system during compression.

(iv) *Heat supplied*

$$Q_{1-2} = \frac{\gamma - n}{\gamma - 1} W_{1-2} = -\frac{1.4 - 1.29}{1.4 - 1} \times (129.3)$$

$$= -35.56 \text{ kJ.}$$

Heat is rejected by the system. **Ans.**

Example 2.5: Three kg of air kept at 100 kPa and 300K is compressed polytropically to 1500 kPa and 500K. Calculate:

(i) Index n

(ii) Final volume

(iii) Work done

(iv) Heat exchanged

Solution:

(i) *Index n*

$$p_1 = 100 \text{ kPa} = 100 \times 10^3 \text{ N/m}^3$$
$$T_1 = 300K$$
$$p_2 = 1500 \text{ kPa} = 1500 \times 10^3 \text{ N/m}^2$$
$$T_2 = 500K$$
$$p_1 V_1^n = p_2 V_2^n$$

and

$$\frac{p_1 V_1}{T_1} = \frac{p_2 V_2}{T_2}$$

∴

$$\frac{T_2}{T_1} = \left(\frac{p_2}{p_1}\right)^{\frac{n-1}{n}}$$

∴

$$\frac{n-1}{n} = \frac{\ln \dfrac{T_2}{T_1}}{\ln \dfrac{p_2}{p_1}} = \frac{\ln \dfrac{500}{300}}{\ln \dfrac{1500}{100}} = \frac{0.5108}{2.708} = 0.1886.$$

∴

$$n = 1.23$$

(ii) *Final volume*

$$p_1 V_1 = m R T_1$$

Take $R = 287$ J/kg-K for air

$$V_1 = \frac{m R T_1}{p_1} = \frac{3 \times 287 \times 300}{100 \times 10^3} = 2.583 \text{ m}^3$$

$$p_2 V_2 = m R T_2$$

$$V_2 = \frac{m R T_2}{p_2} = \frac{3 \times 287 \times 500}{1500 \times 10^3} = \textbf{0.287 m}^3 \quad \textbf{Ans.}$$

(iii) *Work done*

$$W_{1-2} = \frac{p_1 V_1 - p_2 V_2}{n-1} = \frac{m\,R(T_1 - T_2)}{n-1} = \frac{3 \times 287\,(300 - 500)}{1.23 - 1}$$

$$= -748696\ J = -748.7\ kJ$$

Work is done on the air during compression.

(iv) *Heat exchanged*

$$Q_{1-2} = \frac{\gamma - n}{\gamma - 1}\,W_{1-2} = \frac{1.4 - 1.23}{1.4 - 1}\,(-748.7)$$

$$= -318.2\ kJ.$$

Heat is rejected by the system.

2.6 STEADY FLOW PROCESSES

The following conditions must hold good in a steady flow process:

1. The mass flow rate through the system remains constant

$$\dot{m}_1 = \dot{m}_2 = \dot{m} = \text{constant}$$

2. The rate of heat transfer is constant

$$\dot{Q}_{1-2} = \text{constant}$$

3. The rate of work transfer remains constant

$$\dot{W}_{1-2} = \text{constant}$$

4. The state of working substance at any point in the system is the same at all points.

5. There is no change in the chemical composition of the system.

If any one condition is not satisfied, the process is called unsteady flow process.

The mechanical work in a steady flow process is compared with that in non-flow process in Table 2.2.

2.7 THROTTLING PROCESS

The expansion of gas through an obstruction in the form of a partly opened valve or orifice is called throttling.

There is reduction in pressure and increase in volume of fluid. The process is adiabatic as there is no exchange of heat but it is irreversible. No work output occurs during throttling. It is not possible to compress the fluid back to initial pressure without the aid of external work. Therefore, the process is irreversible.

Table 2.2 Mechanical work (Unit mass)

Sl. No.	Process	Non-flow work $\int pdv$	Steady flow work $-\int vdp$
1.	Isochoric ($v = c$)	0	$v(p_1 - p_2)$
2.	Isobaric ($p = c$)	$p(v_2 - v_1)$	0
3.	Isothermal ($t = c$)	$p_1 v_1 \ln \dfrac{v_2}{v_1}$	$p_1 v_1 \ln \dfrac{v_2}{v_1}$
4.	Adiabatic ($pv^\gamma = c$)	$\dfrac{p_1 v_1 - p_2 v_2}{\gamma - 1}$	$\dfrac{\gamma}{\gamma - 1}(p_1 v_1 - p_2 v_2)$
5.	Polytropic ($pv^n = c$)	$\dfrac{p_1 v_1 - p_2 v_2}{n - 1}$	$\dfrac{n}{n - 1}(p_1 v_1 - p_2 p_2)$
6.	Free expansion ($u = c$)	0	–
7.	Throttling ($h = c$)	–	0

The Joule-Thomson Porous Plug Experiment

This experiment was performed by Joule and Thomson is 1852. A gas was made to pass steadily through a porous plug placed in an insulated horizontal pipe. The inlet parameters pi and Ti are kept constant. The resistance of porous plug is varied by compressing it. The pressure drop through the plug varies. The pressure (pe) and temperature (Te) at the exit are measured.

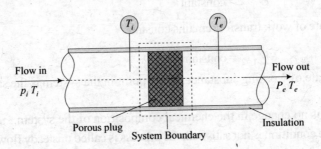

Fig. 2.9 Process Plug Apparatus

During this experiment:
1. There is no woke output
$$W_{1-2} = 0$$
2. There is no heat exchange as the surface is insulated
$$Q_{1-2} = 0$$
3. The pipe is placed horizontally
$$\therefore \qquad Z_1 = Z_2$$
$$PE_1 = PE_2$$
4. The change in velocity is negligible

$$V_1 \approx V_2$$
$$KE_1 = KE_2$$

Applying steady flow energy equation to unit mass flow

$$h_1 + \frac{v_1^2}{2} + gz_1 + q_{1-2} = h_2 + \frac{v_2^2}{2} + gz_2 + w_{1-2}$$

$$\therefore \qquad h_1 = h_2$$

Therefore, throttling process is a constant enthalpy process

$$h_1 = h_2 = h_3 = h_4 = h_5 = \ldots$$

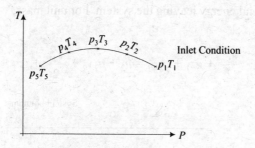

Fig. 2.10   Constant enthalpy process

If the readings of pressure and temperature of the experiment are plotted on *T-p* diagram, a constant enthalpy line is obtained.

The slope of the constant enthalpy curve is called Joule-Thomson coefficient

$$\mu = \left(\frac{dT}{dp} \right)_h$$

For perfect gas, $\mu = 0$.

Applications of Throttling

Although throttling is a energy loss process, it is used for the following.

1. To find out the dryness fraction of steam in a throttling calorimeter.
2. The speed of steam turbine is controlled in throttle governing.
3. Refrigeration effect (cooling) is obtained by throttling the refrigerant in a valve or capillary tube at inlet to evaporator.

$$\because \qquad h_1 = h_2$$
$$\therefore \qquad h_1 - h_2 = 0$$
$$Cp(T_1 - T_2) = 0$$
$$\therefore \qquad T_1 = T_2$$
$$Cv(T_1 - T_2) = 0$$
$$\therefore \qquad u_1 = u_2$$

For an ideal gas, throttling takes place at

1. Constant enthalpy
2. Constant internal energy
3. Constant temperature.

However, in a real gas or fluid (refrigerant),

$$T_1 \neq T_2.$$

2.8   STEADY FLOW ENERGY EQUATION

According to the first law of thermodynamics, the total energy entering a system must be equal to total energy leaving the system. For unit mass,

$$e_1 = e_2$$

$$u_1 + p_1 v_1 + \frac{V_1^2}{2} + g z_1 + q_{1-2} = u_2 + p_2 v_2 + \frac{V_2^2}{2} + g z_2 + w_{1-2}$$

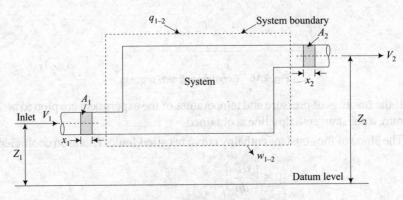

Fig. 2.11   Steady flow process

where, suffix 1 is for inlet and 2 for outlet.

u = specific internal energy

pv = Flow work

V = Fluid velocity

Z = Height

q_{1-2} = Heat exchange

w_{1-2} = Work exchange

Now, $\qquad h = u + pv$

$\therefore \qquad h_1 + \frac{V_1^2}{2} + q z_1 + q_{1-2} = h_2 + \frac{V_2^2}{2} + q z_2 + w_{1-2}$

This is called steady flow energy equation. This equation may also be written as follows:

$$q_{1-2} - w_{1-2} = (h_2 - h_1) + \frac{V_2^2 - V_1^2}{2} + q(z_2 - z_1)$$

In differential form,

$$dq - dw = dh + d(ke) + d(pe)$$

Case I. If the effect of gravity can be neglected, i.e., $z_2 \approx z_1$

$\therefore$
$$d(pe) = 0$$
$$dq - dw = dh + (dke)$$

or
$$q_{1-2} - w_{1-2} = (h_2 - h_1) + \left(\frac{V_2^2 - V_1^2}{2} \right)$$

Case II. If gravity can be neglected and the change in velocity is negligible, i.e., $V_2 \approx V_1$

$$d(pe) = 0$$
$$d(ke) = 0$$

$\therefore$
$$q_{1-2} - w_{1-2} = (h_2 - h_1).$$

Case III. Applying the steady flow energy equation to a closed system (non-flow process)

$$d(pe) = 0$$
$$d(ke) = 0$$
$$p_1 v_1 = 0 \qquad \text{(Flow energy or displacement energy at}$$
$$\qquad\qquad\qquad\qquad\qquad \text{inlet and outlet is zero)}$$
$$p_2 v_2 = 0$$

$\therefore$
$$h_1 = u_1$$
$$h_2 = u_2$$

$\therefore$
$$q_{1-2} - w_{1-2} = u_2 - u_1$$

This is called energy equation for a non-flow process.

2.8.1 Equation of Continuity

The mass flow rate ($\dot{m}$) of the working substance entering the system is same as leaving the system. The steady flow energy equation will be:

$$\dot{m} \left(h_1 + \frac{V_1^2}{2} + qZ_1 + q_{1-2} \right) = \dot{m} \left(h_2 + \frac{V_2^2}{2} + qZ_2 + W_{1-2} \right)$$

or
$$\dot{Q}_{1-2} - \dot{W}_{1-2} = \dot{m}(h_2 - h_1) + \frac{\dot{m}}{2}(V_1^2 - V_1^2) + \dot{m}g(z_2 - z_1)$$

where, $\dot{m} = \dfrac{A_1 V_1}{v_1} = \dfrac{A_2 V_2}{v_2}$ (kg/s)

This is called equation of continuity.

Example 2.6: 0.5 kg/s of a fluid flows in a steady state process. The properties of fluid at entrance are measured as p_1 = 1.4 bar, density = 2.5 kg/m³, u_1 = 920 kJ/kg while at exit the properties are p_2 = 5.6 bar, density = 5 kg/m³ and u_2 = 720 kJ/kg. The velocity at entrance is 200 m/s while at exit is 180 m/s. It rejects 60 kW of heat and rises through 60 m during the flow. Find the change of enthalpy and the rate of work done.   [U.P.T.U. II Sem., 2002-03]

Solution: The data given,

$$\dot{m} = 0.5 \text{ kg/s}$$

$$p_1 = 1.4 \text{ bar} = 140 \text{ kPa}$$

$$\rho_1 = 2.5 \text{ kg/m}^3 \quad \therefore v_1 = \frac{1}{\rho_1} = 0.4 \text{ m}^3/\text{kg}$$

$$u_1 = 920 \text{ kJ/kg}$$

$$p_2 = 5.2 \text{ bar} = 560 \text{ kPa}$$

$$\rho_2 = 5 \text{ kg/m}^3 \quad \therefore v_2 = \frac{1}{\rho_2} = \frac{1}{5} = 0.25 \text{ m}^3/\text{kg}$$

$$u_2 = 720 \text{ kJ/kg}$$

$$V_1 = 200 \text{ m/s}$$

$$V_2 = 180 \text{ m/s}$$

$$Q_{1-2} = 60 \text{ kW} = 60 \text{ kJ/s}$$

$$Z_2 - Z_1 = 60 \text{ m}$$

(i) *Change of enthalpy*

$$\Delta H = \dot{m}\,(h_2 - h_1) = \dot{m}[(u_2 + p_2 v_2) - (u_1 + p_1 v_1)]$$
$$= 0.5\,[(720 + 560 \times 0.25) - (920 + 140 \times 0.4)]$$
$$= 0.5\,[860 - 976] = -\,58 \text{ kJ}$$

There is a decrease in the enthalpy.

(ii) *Rate of work done*

Apply steady flow energy equation to the system

$$\dot{Q}_{1-2} - \dot{W}_{1-2} = \dot{m}\left[(h_2 - h_1) + \frac{V_2^2 - V_1^2}{2 \times 10^3} + \frac{g(Z_2 - Z_1)}{10^3}\right]$$

$$-60 - \dot{W}_{1-2} = 0.5\left[(860 - 976) + \frac{180^2 - 200^2}{2 \times 10^3} + \frac{9.81 \times 60}{10^3}\right]$$

$$= 0.5\,[-116 - 3.8 + 0.5886] = -\,59.6$$

$$\therefore \qquad \dot{W}_{1-2} = -60 + 59.6 = \textbf{0.4 kW} \quad \textbf{Ans.}$$

Work is done on the system.

2.9 APPLICATIONS OF STEADY FLOW ENERGY EQUATION

The steady flow energy equation can be applied to various energy systems and devices such as boilers, condensers, evaporators, nozzles, turbines, compressors etc.

2.9.1 Heat Exchanger

A heat exchanger is a device in which heat is transferred from one fluid to another. There are two steady flow streams, one of heating fluid and other fluid to be heated. The flow through a heat exchanger is characterized by,

1. No work exchange,

$$W_{1-2} = 0$$

2. No change in potential energy

$$Z_2 = Z_1$$

3. No change in kinetic energy

$$V_2 \approx V_1$$

4. Normally no external heat interaction, if heat exchanger is insulated.

Heat gained by cold fluid = Heat lost by hot fluid

$$\dot{m}_c(h_1 - h_2) = \dot{m}_h(h_4 - h_3)$$

where, $\dot{m}_c$ and $\dot{m}_h$ are flow rate of cold fluid and hot fluid respectively.

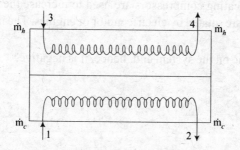

Fig. 2.12 Heat exchanger

The boiler, condenser, evaporator, etc. are designed in a similar manner by using steady flow energy equation. The heat supplied or removed is equal to change of enthalpy,

$$q_{1-2} = (h_2 - h_1)$$

Example 2.7: In a heat exchanger, 50 kg of water is heated per minute from 50°C to 110°C by hot gases which enter the heat exchanger at 250°C. If the flow rate of gases is 100 kg/min, estimate the net change of enthalpy. For water $C_p = 4.186$ kJ/kg-K and for air, $C_p = 1$ kJ/kg-K.

Solution:

$$\dot{m}_w = 50 \text{ kg/min}$$
$$t_{w1} = 50°C$$
$$t_{w2} = 110°C$$
$$\dot{m}_g = 100 \text{ kg/min}$$
$$t_{g1} = 250°C$$
$$Cp_w = 4.186 \text{ kJ/kg-K}$$
$$Cp_g = 1 \text{ kJ/kg-K}$$

Using energy balance,

Heat gained by water = Heat lost by hot gas

$$\dot{m}_w \, Cp_w \, (t_{w2} - t_{w1}) = \dot{m}_g \, Cp_g (t_{g1} - t_{g2})$$
$$50 \times 4.186 \, (110 - 50) = 100 \times 1 \, (250 - t_{g2})$$

∴
$$t_{g2} = 125°C$$

Net change of enthalpy of gas = $100 \times 1 \times (250 - 125)$
$$= 12500 \text{ kJ/min}$$

2.9.2 Compressor

Rotary and reciprocating compressors are used to increase the pressure of air and other gases. These are rotated by electric motor or engines. The main characteristics of the system are:

1. Work is done on the system and, hence, it is negative.

$\dot{W}_{1-2}$ is –ve.

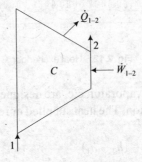

Fig. 2.13 Compressor

2. Potential energy and kinetic energy can normally be neglected.

$$Z_2 \approx Z_1$$
$$V_2 \approx V_1$$

3. Heat is lost from the compressor either by radiation or through a coolant (air or water).

The heat exchange is –ve.

$\dot{Q}_{1-2}$ is –ve or zero.

$$-\dot{Q}_{1-2} - (-\dot{W}_{1-2}) = \dot{m}(h_2 - h_1)$$

$$\therefore \qquad \dot{W}_{1-2} = \dot{Q}_{1-2} + \dot{m}(h_2 - h_1)$$

Therefore, work is done on the system to increase the enthalpy of the fluid.

Example 2.8: 0.8 kg of air flows through a compressor under steady state conditions. The properties of air at entry are: pressure 1 bar, velocity 10 m/s, specific volume 0.95 m³/kg and internal energy 30 kJ/kg. The corresponding values at exit are: 8 bar, 6 m/s, 0.2 m³/kg and 124 kJ/kg. Neglecting the change in potential energy, determine the power input and pipe diameter at entry and exit.

[U.P.T.U. CO. 2005-06]

Solution: The data given,

$$\dot{m} = 0.8 \text{ kg/s}$$
$$p_1 = 1 \text{ bar} = 100 \text{ kPa}$$
$$V_1 = 10 \text{ m/s}$$
$$v_1 = 0.95 \text{ m}^3/\text{kg}$$
$$u_1 = 0.9 \text{ kJ/kg}$$
$$p_2 = 8 \text{ bar} = 800 \text{ kPa}$$
$$V_2 = 6 \text{ m/s}$$
$$v_2 = 0.2 \text{ m}^3/\text{kg}$$
$$u_2 = 124 \text{ kJ/kg}$$
$$h_1 = u_1 + p_1 v_1 = 30 + 100 \times 0.95 = 125 \text{ kJ/kg}$$
$$h_2 = u_2 + p_2 v_2 = 124 + 800 \times 0.2 = 284 \text{ kJ/kg}$$

(i) Apply steady flow energy equation to the compressor,

$$Q_{1-2} - W_{1-2} = \dot{m}\left[(h_2 - h_1) + \frac{V_2^2 - V_1^2}{2 \times 10^3} + \frac{g}{10^3}(Z_2 - Z_1)\right]$$

$$Q_{1-2} = 0$$

W_{1-2} is –ve.

$$Z_2 = Z_1$$

$$\therefore \quad W_{1-2} = 0.8\left[(284-125) + \frac{6^2 - 10^2}{2 \times 10^3}\right] = \textbf{127 kW} \quad \textbf{Ans.}$$

(ii) Apply equation of continuity,

$$\dot{m} = \frac{A_1 V_1}{v_1} = \frac{A_2 V_2}{v_2}$$

$$\therefore \quad A_1 = \frac{\pi}{4} D_1^2 = \frac{\dot{m} v_1}{V_1}$$

$$\therefore \quad D_1 = \sqrt{\frac{4\,\dot{m}\,v_1}{\pi V_1}} = \sqrt{\frac{4 \times 0.8 \times 0.95}{\pi \times 10}} = 0.311 \text{ m}$$

$$= \textbf{311 mm} \quad \textbf{Ans.}$$

$$D_2 = \sqrt{\frac{4\,\dot{m}\,v_2}{\pi V_2}} = \sqrt{\frac{4 \times 0.8 \times 0.2}{\pi \times 6}} = 0.184 \text{ m}$$

$$= \textbf{184 mm} \quad \textbf{Ans.}$$

Example 2.9: An air compressor compresses atmospheric air at 0.1 MPa and 27°C by ten times of inlet pressure. During compression, the heat loss to the surrounding is estimated to be 5% of compression work. Air enters the compressor with a velocity of 40 m/s and leaves with 100 m/s. Inlet and exit cross-sectional areas are 100 cm^2 and 20 cm^2 respectively. Estimate the temperature of air at exit from compressor and power input to the compressor.

[U.P.T.U. II Sem., 2001-02]

Solution:

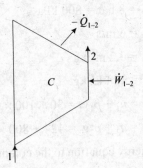

1. *Inlet conditions*

$$p_1 = 0.1 \text{ MPa} = 0.1 \times 10^3 \text{ kN/m}^2$$

$$T_1 = 27°C + 273 = 300 \text{ K}$$

$$V_1 = 40 \text{ m/s}$$
$$A_1 = 100 \text{ cm}^2 = 100 \times 10^{-4} \text{ m}^2$$

2. *Outlet conditions*

$$p_2 = 10 \times 0.1 \times 10^3 \text{ kN/m}^2 = 10^3 \text{ kN/m}^2$$
$$V_2 = 100 \text{ m/s}$$
$$A_2 = 20 \text{ cm}^2 = 20 \times 10^{-4} \text{ m}^2$$

3. *Equation of continuity*

The equation of state,

$$p_1 V_1 = mRT_1$$

or $\qquad p_1 v_1 = RT_1$

$\therefore \qquad v_1 = \dfrac{RT_1}{p_1} = \dfrac{0.287 \times 300}{100}$

$$= 0.861 \text{ m}^3/\text{kg} \qquad\qquad [R = 0.287 \text{ kJ/kg-K for air}]$$

$$\dot{m} = \frac{A_1 V_1}{v_1} = \frac{(100 \times 10^{-4}) \times 40}{0.861} = 0.4646 \text{ kg/s}$$

$$\dot{m} = \frac{A_2 V_2}{v_2}$$

$\therefore \qquad v_2 = \dfrac{A_2 V_2}{\dot{m}} = \dfrac{(20 \times 10^{-4}) \times 100}{0.4646} = 0.43 \text{ m}^3/\text{kg}$

$$p_2 v_2 = RT_2$$

$$T_2 = \frac{p_2 v_2}{R} = \frac{10^3 \times 0.43}{0.287} = 1498 \text{ K}$$

4. *Steady flow energy equation*

$$Q_{1-2} - W_{1-2} = \dot{m}\left[(h_2 - h_1) + \frac{V_2^2 - V_1^2}{2 \times 10^3} + \frac{g}{10^3}(Z_2 - Z_1)\right]$$

$$-0.05\,W_{1-2} - W_{1-2} = 0.43\left[1.005(1498 - 300) + \frac{(100)^2 - (40)^2}{2 \times 10^3} + 0\right]$$

$$W_{1-2} = \textbf{546.86 kW} \quad \textbf{Ans.}$$

2.9.3 Gas Turbine

A gas turbine converts the heat energy of hot gases into mechanical work. A compressor driven by the gas turbine compresses air or gas to a higher pressure. The high pressure air or gas is heated by combustion of fuel. The high pressure and high temperature air or gas is admitted to a gas turbine. Power is produced in the turbine at the expense of enthalpy drop of the gas. The main characteristics of the system are:

1. The heat loss by radiation to the surrounding is $-$ve.
2. Normally, *ke* and *pe* are neglected.

$$V_2 \approx V_1$$

$$Z_2 = Z_1$$

$$\dot{W}_{1-2} = Q_{1-2} + \dot{m}(h_1 - h_2)$$

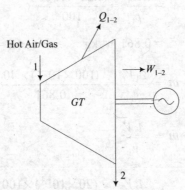

Fig. 2.14   Gas turbine

Example 2.10:   Air passes through a gas turbine system at the rate of 4.5 kg/s. It enters the turbine system with a velocity of 90 m/s and a specific volume of 0.85 m^3/kg. It leaves the turbine system with a specific volume of 1.45 m^3/kg. The exit area of the turbine system is 0.38 m^2. In its passage through the turbine, the specific enthalpy of air is reduced by 200 kJ/kg and there is a heat loss of 40 kJ/kg. Determine:

(i) The inlet area of turbine

(ii) The exit velocity of air in m/s.

(iii) Power developed by the turbine system in kW.

Solution:

1. *Inlet conditions*

$$\dot{m} = 4.5 \text{ kg/s}$$

$$V_1 = 90 \text{ m/s}$$

$$v_1 = 0.85 \text{ m}^3/\text{kg}$$

2. *Outlet conditions*

$$v_2 = 1.45 \text{ m}^3/\text{kg}$$

$$A_2 = 0.38 \text{ m}^2$$

$$Q_{1-2} = -40 \text{ kJ/kg}$$

$$\Delta h = 200 \text{ kJ/kg}$$

3. *Equation of continuity*

$$\dot{m} = \frac{A_1 V_1}{v_1} = \frac{A_2 V_2}{v_2}$$

$$\therefore \qquad A_1 = \frac{\dot{m} V_1}{v_1} = \frac{4.5 \times 0.85}{90} = \mathbf{0.0425 \ m^2} \quad \textbf{Ans.}$$

$$V_2 = \frac{\dot{m} V_2}{A_2} = \frac{4.5 \times 1.45}{0.38} = \mathbf{17 \ m/s} \quad \textbf{Ans.}$$

4. *Steady flow energy equation*

$$Q_{1-2} - W_{1-2} = \dot{m}\left[(h_2 - h_1) + \frac{V_2^2 - V_1^2}{2 \times 10^3} + \frac{g}{10^3}(Z_2 - Z_1)\right]$$

$$-40 \times 4.5 - W_{1-2} = 4.5\left[200 + \frac{(17)^2 - (90)^2}{2 \times 10^3} + 0\right]$$

$$W_{1-2} = \mathbf{702.43 \ kW} \quad \textbf{Ans.}$$

2.9.4 Steam Turbine

High pressure and high temperature steam from a steam generator or a boiler is admitted through a steam turbine. During expansion of steam through the turbine, there is enthalpy drop. The turbine gives out positive mechanical work. Although it is insulated, there can be some heat loss due to radiation. The main characteristics of the system are:

1. There is enthalpy drop through the turbine.
2. Change of potential energy is usually neglected

$$Z_2 = Z_1$$

3. If $V_2 = V_1$, change of kinetic energy is neglected.
4. The heat loss due to radiation is –ve.

$$W_{1-2} = \dot{m}(h_1 - h_2) + Q_{1-2}$$

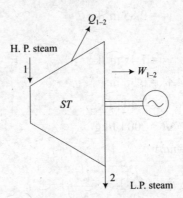

Fig. 2.15 Steam turbine

Example 2.11: The following details refer to a steam turbine:

Steam flow rate = 1 kg/sec

Inlet velocity of steam = 100 m/sec

Exit velocity of steam = 150 m/sec

Enthalpy at inlet = 2900 kJ/kg

Enthalpy at outlet = 1600 kJ/kg

Write the steady flow energy equation. Assuming that the change in potential energy is negligible, determine the power available from the turbine.

[U.P.T.U. II Sem., (CO) 2003-04]

Solution:

1. *Inlet Data*

$$\dot{m} = 1 \text{ kg/sec}$$
$$V_1 = 100 \text{ m/s}$$
$$h_1 = 2900 \text{ kJ/sec}$$

2. *Outlet Data,*

$$V_2 = 150 \text{ m/sec}$$
$$h_2 = 1600 \text{ kJ/sec}$$

3. *Steady flow energy equation*

$$Q_{1-2} - W_{1-2} = \dot{m}\left[(h_2 - h_1) + \frac{V_2^2 - V_1^2}{2 \times 10^3} + \frac{g}{10^3}(Z_2 - Z_1)\right]$$

Here
$$Q_{1-2} = 0$$
$$Z_2 - Z_1 = 0$$

$$\therefore \qquad -W_{1-2} = \dot{m}\left[(h_2 - h_1) + \frac{V_2^2 - V_1^2}{2 \times 10^3}\right]$$

$$= 1\left[(1600 - 2900) + \frac{(150)^2 - (100)^2}{2 \times 10^3}\right]$$

$$= -1300 + 6.25$$

$$\therefore \qquad W_{1-2} = 1293.75 \text{ kW}$$

Example 2.12: A steam turbine operating under steady flow conditions receives 3600 kg of steam per hour. The steam enters the turbine at a velocity of 80 m/s an elevation of 10 m and specific enthalpy of 3276 kJ/kg. It leaves the turbine at a velocity of 150 m/sec, an elevation of 3 m and a specific enthalpy of 2465 kJ/kg. Heat losses from the turbine to the surrounding amount to 36 MJ/hr. Estimate the power output of turbine.   [U.P.T.U. I/II Sem., 2001-02]

Solution:

1. *Inlet Data,*

$$\dot{m} = 3600 \text{ kg/hour} = 1 \text{ kg/sec}$$

$$V_1 = 80 \text{ m/s}$$

$$Z_1 = 10 \text{ m}$$

$$h_1 = 3276 \text{ kJ/kg}$$

$$Q_{1-2} = -36 \text{ MJ/hour} = -\frac{36 \times 10^3}{3600} \text{ kJ/sec}$$

$$= -10 \text{ kJ/sec}$$

2. *Outlet conditions,*

$$V_2 = 150 \text{ m/sec}$$

$$Z_2 = 3 \text{ m}$$

$$h_2 = 2465 \text{ kJ/kg}$$

3. *Steady flow energy equation,*

$$Q_{1-2} - W_{1-2} = \dot{m}\left[(h_2 - h_1) + \frac{V_2^2 - V_1^2}{2 \times 10^3} + \frac{g}{10^3}(Z_2 - Z_1)\right]$$

$$-10 - W_{1-2} = 1\left[(2465 - 3276) + \frac{(150)^2 - (80)^2}{2 \times 10^3} + \frac{9.81}{10^3}(3 - 10)\right]$$

$$-10 - W_{1-2} = -811 + 8.05 - 0.06867$$

$$W_{1-2} = \textbf{793 kW} \quad \textbf{Ans.}$$

Example 2.13: The steam supply to an engine comprises two streams which mix before entering the engine. One stream is supplied at the rate of 0.01 kg/s with an enthalpy of 2950 kJ/kg and a velocity of 20 m/s. The other stream is supplied at the rate of 0.1 kg/s with an enthalpy of 2665 kJ/kg and a velocity of 120 m/s. At the exit from the engine the fluid leaves as two streams, one of water at the rate of 0.001 kg/s with an enthalpy of 421 kJ/kg and the other of steam. The fluid velocities at the exit are negligible. The engine develops a shaft power of 25 kW. The heat transfer is negligible. Evaluate the enthalpy of the second exit stream.

[U.P.T.U. II Sem., 2004-05]

Solution:

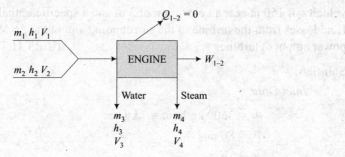

1. *Inlet data,*

$$m_1 = 0.01 \text{ kg/s}$$

$$h_1 = 2950 \text{ kJ/kg}$$

$$V_1 = 20 \text{ m/s}$$

$$m_2 = 0.1 \text{ kg/s}$$

$$h_2 = 2665 \text{ kJ/kg}$$

$$V_2 = 120 \text{ m/s}$$

Total inlet energy

$$E_i = m_1 h_1 + \frac{m_1 V_1^2}{2 \times 10^3} + \frac{g}{10^3} Z_1 + Q_{1-2} + m_2 h_2 + \frac{m_1 V_2^2}{2 \times 10^3} + \frac{g}{10^3} Z_2$$

$$= 0.01 \times 2950 + \frac{0.01 \times (20)^2}{2 \times 10^3} + 0 + 0 + 0.1 \times 2665 + \frac{0.1 \times 120^2}{2 \times 10^3} + 0$$

$$E_i = 29.5 + 0.002 + 266.5 + 0.72 = 296.722 \text{ kJ}$$

$$m_i = 1\ 0.01 + 0.1 = 0.11 \text{ kg/s}$$

2. *Outlet Data*

$$m_3 = 0.001 \text{ kg/s}$$

$$h_3 = 421 \text{ kJ/kg}$$

$$V_3 = 0$$

$$Z_3 = 0$$

$$m_4 = m_i - m_3 = 0.11 - 0.001 = 0.109 \text{ kg/sec}$$

Total energy,

$$E_e = m_3 h_3 + m_4 h_4 + W_{1-2}$$

$$= 0.001 \times 421 + 0.109 h_4 + 25$$

$$= 0.109 h_4 + 25.421$$

For steady flow conditions,

$$E_i = E_e$$

∴ $$296.722 = 0.109 h_4 + 25.421$$

∴ $$h_4 = \frac{296.722 - 25.421}{0.109} = \textbf{2489 kJ/kg} \quad \textbf{Ans.}$$

2.9.5 Nozzle

The nozzle converts the pressure energy of a stream into its kinetic energy and as a result the velocity of the stream increases. The enthalpy drop of the fluid is used to accelerate the flow. Nozzles are used in steam turbines, gas turbines, pumps, etc. The operating characteristics of a nozzle are:

1. There is no work output,

$$W_{1-2} = 0$$

2. The heat loss is normally absent,

$$Q_{1-2} = 0$$

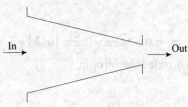

Fig. 2.16 Nozzle

3. The change of potential energy is negligible,

$$Z_2 = Z_1$$

4. The steady flow energy equation,

$$\frac{V_2^2 - V_1^2}{2} = (h_1 - h_2)$$

$$V_2 = \sqrt{2(h_1 - h_2) + V_1^2}$$

If $V_2 >> V_1$ and V_1 can be neglected,

$$V_2 = \sqrt{2(h_1 - h_2)}$$

The mass flow rate,

$$\dot{m} = \frac{A_1 V_1}{v_1} = \frac{A_2 V_2}{v_2}$$

The area of nozzle at the inlet and outlet can be estimated.

Example 2.14:   In an isentropic flow through a nozzle, air flows at the rate of 600 kg/hr. At inlet to nozzle, pressure is 2 MPa and temperature is 127°C. The exit pressure is 0.5 MPa. If the initial air velocity is 300 m/s, determine:

   (i) Exit velocity of air, and

   (ii) Inlet and exit area of nozzle.   (U.P.T.U. I Sem., 2000-01)

Solution: •

1. *Inlet conditions,*

$$p_1 = 2 \text{ MPa} = 2 \times 10^3 \text{ kPa}$$
$$T_1 = 127°C + 273 = 400 \text{ K}$$

$$\dot{m} = 600 \text{ kg/hr} = \frac{600}{3600} = \frac{1}{6} \text{ kg/s}$$

$$V_1 = 300 \text{ m/s}$$
$$Q_{1-2} = 0$$

2. *Outlet conditions,*

$$p_2 = 0.5 \text{ MPa} = 0.5 \times 10^3 \text{ kPa}$$

The flow through nozzle is isentropic,

$\therefore$

$$\frac{T_2}{T_1} = \left(\frac{p_2}{p_1}\right)^{\frac{\gamma-1}{\gamma}}$$

For air $\gamma = 1.4$

$$\therefore \qquad T_2 = 400 \left(\frac{0.5}{2} \right)^{\frac{1.4-1}{1.4}} = 269.18 \text{ K}$$

$$W_{1-2} = 0$$

3. *Steady flow energy equation,*

$$Q_{1-2} - W_{1-2} = \dot{m} \left[(h_2 - h_1) + \frac{V_2^2 - V_1^2}{2 \times 10^3} + \frac{g}{10^3} (Z_2 - Z_1) \right]$$

$$h_2 - h_1 = C_p (T_2 - T_1)$$

$$\therefore \qquad 0 - 0 = \frac{1}{6} \left[1.005 (400 - 269.18) + \frac{V_2^2 - (300)^2}{2 \times 10^3} + 0 \right]$$

$$\therefore \qquad 0 = 131.474 + \frac{V_2^2 - (300)^2}{2 \times 10^3}$$

$$\therefore \qquad V_2 = \sqrt{131.474 \times 2 \times 10^3 + 300^2}$$

$$= \textbf{594 m/s} \quad \textbf{Ans.}$$

4. *Equation of continuity*

$$\dot{m} = \frac{A_1 V_1}{v_1} = \frac{A_2 V_2}{v_2}$$

From equation of state,

$$p_1 v_1 = R T_1$$

$$v_1 = \frac{R T_1}{p_1} = \frac{0.287 \times 400}{2 \times 10^3} = 0.0574 \text{ m}^3/\text{kg}$$

For air, $\qquad R = 0.287$ kJ/kg-K

$$\therefore \qquad A_1 = \frac{\dot{m} V_1}{v_1} = \frac{\frac{1}{6} \times 0.0574}{300} = 3.1888 \times 10^{-5} \text{ m}^2$$

$$= 31.88 \text{ mm}^2$$

$$v_2 = \frac{R T_2}{p_2} = \frac{0.287 \times 269.18}{0.5 \times 10^3} = 0.1545 \text{ m}^3/\text{kg}$$

$$\therefore \qquad A_2 = \frac{\dot{m}\,v_2}{V_2} = \frac{\dfrac{1}{6} \times 0.1545}{594} = 4.335 \times 10^{-5}\ \text{m}^2$$

$$= \textbf{43.35 mm}^2 \quad \textbf{Ans.}$$

The exit area of the nozzle is more than inlet area. Therefore, the nozzle is a convergent-divergent nozzle.

2.9.6 Diffuser

A diffuser has varying cross-section and reduces the velocity of the flowing fluid. There are two types of diffusers:

1. *Subsonic Diffuser*

   The velocity of the fluid is less then sonic speed and the area of cross-section of diffuser increases from inlet to exit.

2. *Supersonic Diffuser*

   The velocity of fluid is more than sonic velocity and the area of diffuser decreases.

The operating characteristics of a diffuser are similar to that for a nozzle.

Example 2.15: Water vapour at 90 kPa and 150°C enters a subsonic diffuser with a velocity of 150 m/s and leaves the diffuser at 190 kPa with a velocity of 55 m/s and during the process 1.5 kJ/kg of heat is lost to the surrounding. Determine:

(i) The final temperature

(ii) The mass flow rate

(iii) The exit diameter, assuming the inlet diameter as 10 cm and steady flow.

[U.P.T.U. II Sem., 2001-02]

Solution:

1. *Inlet condition,*

$$p_1 = 90\ \text{kN/m}^2$$
$$T_1 = 150°\text{C} + 273 = 423\ \text{K}$$
$$V_1 = 150\ \text{m/s}$$

2. *Outlet condition,*

$$p_2 = 190\ \text{kN/m}^2$$
$$V_2 = 55\ \text{m/s}$$

For water vapours take Cp = 2.1 kJ/kg-K

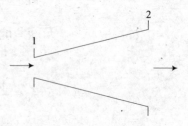

3. *Steady flow energy equation,*

$$Q_{1-2} - W_{1-2} = (h_2 - h_1) + \frac{V_2^1 - V_1^2}{2 \times 10^3} + \frac{g}{10^3}(Z_2 - Z_1)$$

For the given diffuser,

$$W_{1-2} = 0$$
$$Z_2 = Z_1$$

$$\therefore \qquad q_{1-2} = (h_2 - h_1) + \frac{V_2^1 - V_1^2}{2 \times 10^3}$$

$$-1.5 = Cp(T_2 - T_1) + \frac{(55)^2 - (150)^2}{2 \times 10^3}$$

$$Cp(T_2 - T_1) = -1.5 + 9.7375 = 8.2375$$

$$T_2 = \frac{8.2375}{2.1} + 423 = 427 \text{ K}$$

or $\qquad t_2 = 427 - 273 = \textbf{154°C} \quad \textbf{Ans.}$

4. *Mass flow rate,*

From equation of state,

$$p_1 v_1 = RT_1$$

Take, $\qquad R = \dfrac{\overline{R}}{M} = \dfrac{8.314}{18} = 0.4619 \text{ kJ/kg-K}$

$$\therefore \qquad v_1 = \frac{0.4619 \times 423}{90} = 2.17 \text{ m}^3/\text{kg}$$

$$v_2 = \frac{0.4619 \times 427}{190} = 1.038 \text{ m}^3/\text{kg}$$

$$\dot{m} = \frac{A_1 V_1}{v_1} = \frac{\dfrac{\pi}{4} \times (0.10)^2 \times 150}{2.17} = \textbf{0.543 kg/s} \quad \textbf{Ans.}$$

5. *Exit diameter,*

$$A_2 = \frac{\dot{m} v_2}{V_2} = \frac{0.543 \times 1.038}{55} = 0.010248 \text{ m}^2$$

$$\therefore \qquad D_2 = \sqrt{\frac{A_2 \times 4}{\pi}} = \sqrt{\frac{0.010248 \times 4}{\pi}} = 0.1142 \text{ m}$$

$$= \textbf{11.42 cm} \quad \textbf{Ans.}$$

2.10 PERFORMANCE OF HEAT ENGINE

A heat engine is a thermodynamic device for conversion of heat energy into mechanical work. It requires a source for continuous supply of heat. It also requires a sink to reject that part of heat which cannot be converted into work.

The heat engine can also be operated in reverse direction. The refrigerator and heat pump are reversed heat engines. These reversed engines transport heat energy from low temperature to higher temperature. For operation of these devices, power is required to operate them.

2.10.1 Thermal Reservoirs

A thermal reservoir exchanges heat with the system. It has following characteristics:

1. It is a part of the environment.
2. It has sufficiently large capacity for supply or receipt of heat from the system.

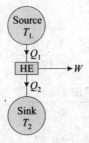

Fig. 2.17 Thermal reservoirs

3. The temperature of heat reservoir remains constant.
4. The temperature of heat reservoir is not affected by the quantity of heat transferred to or from it.
5. The changes that take place in the thermal reservoir are very slow as heat enters or leaves it.
6. The processes within the thermal reservoirs are considered quasi-static.
7. There are two types of heat reservoirs; a source and a sink.

8. *Heat source* is the reservoir at high temperature and supplies heat to the system.

   Examples: boiler furnace, combustion chamber, nuclear reactor, sun, etc.

9. *Heat sink* is the reservoir at a low temperature and receives heat from the system.

   Examples: river, lake, ocean, atmospheric air, etc.

2.10.2   Heat Engine

It is a thermodynamic device which can produce mechanical work from heat continuously. It works on a cyclic process. Examples are steam engine, steam turbine, gas turbine, petrol engine, diesel engine, gas engine.

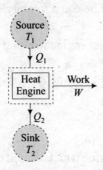

Fig. 2.18   Heat engine

The schematic diagram of a heat engine is shown in Fig. 2.18. The heat and work interactions take place across the boundary of the engine. There is a working substance which continuously flows through the engine to ensure continuous/cyclic operation. It operates as follows:

1. It receives heat Q_1 from a high temperature thermal reservoir (source) at temperature T_1.
2. It converts part of the heat into mechanical work, W.
3. It rejects remaining heat Q_2 to a low temperature heat reservoir (sink) at temperature T_2.

   From 1st law of thermodynamics,
   $$W = Q_1 - Q_2.$$

Thermal efficiency

The performance of a heat engine is measured by its thermal efficiency which is the degree of useful conversion of heat received into work.

The thermal efficiency,

$$\eta_{th} = \frac{\text{Network output}}{\text{Total heat supplied}} = \frac{\dot{W}}{\dot{Q}_1} = \frac{\dot{Q}_1 - \dot{Q}_2}{\dot{Q}_1}$$

$$\eta_{th} = 1 - \frac{\dot{Q}_2}{\dot{Q}_1}$$

For a reversible heat engine,

$$\frac{\dot{Q}_1}{\dot{Q}_2} = \frac{T_1}{T_2}$$

$$\therefore \qquad \eta_{th} = 1 - \frac{T_2}{T_1}$$

Observations

1. Thermal efficiency could be equal to unity, if $Q_1 \to \infty$ or $Q_2 \to 0$, which is practically not possible.
2. Thermal efficiency of a heat engine working between two thermal reservoirs is always less than unity.
3. If $\dot{Q}_2$ can be reduced by keeping $\dot{Q}_1$ constant, thermal efficiency can be increased.

2.10.3 Heat Pump

Heat pump is a reversed heat engine which is used to remove heat from a body at a low temperature and to transfer heat to a body at higher temperature. It is used to keep the room warm in winter.

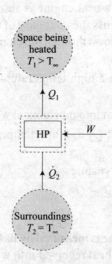

Fig. 2.19 Heat pump

If T_∞ is the atmospheric temperature and T_1 is the temperature of space to be heated, then

$$T_\infty = T_2$$
$$T_\infty < T_1$$

From 1st law of thermodynamics,

$$\dot{W} = \dot{Q}_1 - \dot{Q}_2$$

The schematic diagram of a heat pump is shown in Fig. 2.18. It operates as follows:

1. Heat pump receives heat $\dot{Q}_2$ for atmosphere which is at a temperature T_2 equal to atmospheric temperature, T_∞.
2. It receives external work, $\dot{W}$ and transfers heat from low temperature to a higher temperature.
3. It supplies heat $\dot{Q}_1$ to the space to be heated which is maintained at a temperature T_1 which is higher than atmospheric temperature.

Coefficient of performance (COP)

The performance of a heat pump is measured by its coefficient of performance. *COP* is the ratio of amount of heat supplied to the space and mechanical work received.

$$(COP)_{HP} = \frac{\dot{Q}_1}{\dot{W}} = \frac{\dot{Q}_1}{\dot{Q}_1 - \dot{Q}_2} = 1 + \frac{\dot{Q}_2}{\dot{Q}_1 - \dot{Q}_2}$$

For a heat pump working as reversible reversed heat engine.

$$COP = \frac{T_1}{T_1 - T_2}$$

2.10.4 Refrigerator

A refrigerator is a reversed heat engine which removes heat from a body at low temperature and transfers heat to a body at higher temperature. The objective of the system is to produce refrigeration effect and is used to preserve food and drugs by storing at low temperatures.

The schematic diagram of a refrigerator is shown in Fig. 2.20.

$$T_1 = T_\infty$$
$$T_\infty > T_2$$
$$W = Q_1 - Q_2 \text{ as per 1st law of thermodynamics.}$$

The refrigerator works as follows:

1. It removes heat $\dot{Q}_2$ from space being cooled at low temperature T_2 which is lower than atmospheric temperature, T_∞.

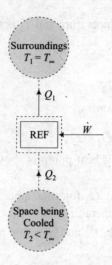

Fig. 2.20 Refrigerator

2. It rejects heat $\dot{Q}_1$ to the surrounding at temperature T_1 which is equal to atmospheric temperature, T_∞.

3. It receives power in the form of work, $\dot{W}$ to transfer heat from a lower temperature to higher temperature.

Coefficient of performance (COP)

The performance of a refrigerator is measured by its coefficient of performance. COP is the ratio of amount of heat extracted called *refrigerated effect* and work input

$$(CPO)_{REF} = \frac{\dot{Q}_2}{\dot{W}} = \frac{\dot{Q}_2}{\dot{Q}_1 - \dot{Q}_2}$$

Observation,

$$(COP)_{HP} = \frac{\dot{Q}_1}{\dot{Q}_1 - \dot{Q}_2} = 1 + \frac{\dot{Q}_2}{\dot{Q}_1 - \dot{Q}_2}$$

$$(COP)_{REF} = \frac{\dot{Q}_2}{\dot{Q}_1 - \dot{Q}_2}$$

$$\therefore \qquad (COP)_{HP} = 1 + (COP)_{REF}.$$

If refrigerator works as reversible reversed heat engine, $COP = \dfrac{T_2}{T_1 - T_2}$.

Example 2.16: A reversible heat engine operates between reservoirs at 420 K and 280 K. If the output from the engine is 2.5 kJ, determine the efficiency of the engine and its heat interactions with the two heat reservoirs.

Subsequently, the engine is reversed and made to operate as heat pump between the same reservoirs. Calculate the coefficient of the heat pump and the power input required when the heat transfer rate from the 280 K reservoir is 5 kW.

[U.P.T.U. II Sem., (CO) 2003-04]

Solution:

1. Draw the schematic diagram of heat engine,

$T_1 = 420$ K

$T_2 = 280$ K

$W = 2.5$ kJ

$$\eta = \frac{T_1 - T_2}{T_1} = \frac{420 - 280}{420}$$

$$= 0.3333 = 33.33\%$$

$$\eta = \frac{W}{Q_1}$$

∴ $$Q_1 = \frac{W}{\eta} = \frac{2.5}{0.3333} = 7.5 \text{ kJ}$$

$$W = Q_1 - Q_2$$

∴ $$Q_2 = Q_1 - W = 7.5 - 2.5 = 5 \text{ kJ}$$

2. Draw the schematic diagram of heat pump

$$\text{COP} = \frac{T_1 - T_2}{T_1} = \frac{420 - 280}{420}$$

$$= 3$$

$$Q_2 = 5 \text{ kW} = 5 \text{ kJ/s}$$

$$\frac{Q_1}{T_1} = \frac{Q_2}{T_2}$$

∴ $$Q_1 = \frac{T_1}{T_2} Q_2 = \frac{420}{280} \times 5 = 7.5 \text{ kJ/s}$$

$$W = Q_1 - Q_2 = 7.5 - 5$$

$$= 2.5 \text{ kJ/s} = \textbf{2.5 kW}$$

Heat Engine

Heat Pump

Example 2.17: A reversed Carnot engine is used for heating a building. It supplies 210×10^3 kJ/hr of heat to the building at 20°C. The outside air is at –5°C. Find the heat taken from the outside air per hour and power of the driving motor.

Solution:

1. Draw the schematic diagram.
2. Given data,

$$T_1 = 20°C + 273 = 293 \text{ K}$$

$$T_2 = -5°C + 273 = 268 \text{ K}$$

$$Q_1 = 210 \times 10^3 \text{ kJ/hr} = \frac{210 \times 10^3}{3600} = 58.33 \text{ kJ/sec}$$

3.

$$(COP)_{HP} = \frac{T_1}{T_1 - T_2} = \frac{Q_1}{Q_1 - Q_2} = \frac{Q_1}{W}$$

∴

$$W = Q_1 \left(\frac{T_1 - T_2}{T_1} \right)$$

$$= 58.33 \left(\frac{293 - 268}{293} \right)$$

$$= 4.98 \text{ kW}$$

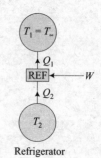

Heat Pump

4. Heat taken from outside air,

$$Q_2 = Q_1 - W = 58.33 - 4.98 = 53.33 \text{ kJ/sec}$$

$$= 53.35 \times 3600 = \textbf{192060 kJ/hr} \quad \textbf{Ans.}$$

Example 2.18: A domestic food refrigerator maintains a temperature of –15°C. The air temperature is 30°C. If heat leaks into the freezer is 1.75 kJ/sec. continuously, then what is the least power necessary to pump the heat out continuously.

Solution:

1. Draw the schematic diagram of refrigerator.
2. The refrigerator must work as reversible device so that power required is minimum.
3. Given data,

$$T_1 = 30°C + 273 = 303 \text{ K}$$

$$T_2 = -15 + 273 = 258 \text{ K}$$

$$Q_2 = 1.75 \text{ kJ/sec}$$

Refrigerator

4. $$(COP)_{REF} = \frac{T_2}{T_1 - T_2} = \frac{Q_2}{Q_1 - Q_2} = \frac{Q_2}{W}$$

∴ $$W = Q_2 \left(\frac{T_1 - T_2}{T_2} \right) = 1.75 \left(\frac{303 - 258}{303} \right)$$

$$= 0.30 \text{ kW} \quad \textbf{Ans.}$$

Example 2.19: Obtain the COP of the composite refrigerator system in which two reversible refrigerators A and B are arranged in series in terms of the COP of refrigerator A and COP of refrigerator B only. [U.P.T.U. (CO) 2005-06]

Solution:

1. Draw the schematic diagram of composite refrigerator system.

    Assume COP of refrigerator $1 = R_1$
    COP of refrigerator 2 $\quad = R_2$
    COP of composite refrigerator $= R_{12}$

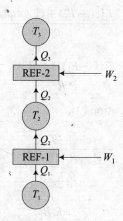

2. The refrigerators are reversible,

∴ $$R_1 = \frac{T_1}{T_2 - T_1} \qquad \qquad \text{...(1)}$$

$$R_2 = \frac{T_2}{T_3 - T_1} \qquad \qquad \text{...(2)}$$

$$R_{12} = \frac{T_1}{T_3 - T_1} \qquad \qquad \text{...(3)}$$

From equation (1),

$$T_2R_1 - T_1R_1 = T_1$$

$$T_2R_1 = T_1 + T_1R_1 = T_1(1 + R_1)$$

$$\therefore \qquad T_2 = \frac{1 + R_1}{R_1} \cdot T_1$$

From equation (2)

$$R_2 = \frac{\left(\dfrac{1 + R_1}{R_1}\right)T_1}{T_3 - \left(\dfrac{1 + R_1}{R_1}\right)T_1} = \frac{(1 + R_1)T_1}{T_3R_1 - (1 + R_1)T_1}$$

$$\therefore \qquad T_3R_1R_2 - (1 + R_1)\,T_1R_2 = (1 + R_1)T_1$$

$$T_3\,R_1R_2 = (1 + R_1)T_1 + (1 + R_1)R_2T_1$$

$$\therefore \qquad T_3 = \frac{T_1[(1 + R_1) + (1 + R_1)R_2]}{R_1R_2}$$

$$= \frac{T_1(1 + R_1)(1 + R_2)}{R_1R_2}$$

Substituting T_3 into equation (3),

$$R_{12} = \frac{T_1}{\dfrac{T_1(1 + R_1)(1 + R_2)}{R_1R_2} - T_1}$$

$$R_{12} = \frac{R_1R_2}{(1 + R_1)(1 + R_2) - R_1R_2}$$

Example 2.20:   A Carnot engine E_1 operates between temperatures T_1 and T_2 and engine E_2 operates between temperatures T_2 and T_3 receiving heat rejected from engine E_1. What is the relationship of η_1 and η_2 with η_3 of engine E_3 working between T_1 and T_3.

Solution:

1. Draw schematic diagram. The engines are reversible.

$$\therefore \qquad \eta_1 = \frac{T_1 - T_2}{T_1} = 1 - \frac{T_2}{T_1}$$

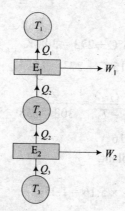

Composite Engine

$$1 - \eta_1 = \frac{T_2}{T_1} \qquad \qquad ...(1)$$

Similarly, $\qquad 1 - \eta_2 = \dfrac{T_3}{T_2} \qquad \qquad ...(2)$

Multiply equations (1) and (2),

$$\therefore \quad (1 - \eta_1)(1 - \eta_2) = \frac{T_2}{T_1} \cdot \frac{T_3}{T_2} = \frac{T_3}{T_1} \qquad ...(3)$$

But, $\qquad \eta_3 = \dfrac{T_1 - T_3}{T_1} = 1 - \dfrac{T_3}{T_1}$

$$\therefore \quad 1 - \eta_3 = \frac{T_3}{T_1} \qquad \qquad ...(4)$$

From equations (3) and (4),

$$1 - \eta_3 = (1 - \eta_1)(1 - \eta_2) = \frac{T_3}{T_1}$$

$$= 1 - \eta_2 - \eta_1 + \eta_1 \eta_2$$

$$\therefore \qquad \eta_3 = \eta_1 + \eta_2 - \eta_1 \eta_2.$$

Example 2.21: A cold storage of 100 tonnes of refrigeration capacity runs at 1/4th of its Carnot COP. Inside temperature is –15°C and atmosphesic temperature is 35°C. Determine the power required to run the plant. Take one tonne of refrigeration as 3.52 kW.
 [U.P.T.U. I Sem., (CO), 2003]

Solution:

$$T_1 = 35°C + 273 = 308 \text{ K}$$

$$T_2 = -15°C + 273 = 258 \text{ K}$$

$$(COP)_{Carnot} = \frac{T_2}{T_1 - T_2} = \frac{258}{308 - 258}$$

$$= 5.16$$

COP of actual refrigeration plant,

$$(COP)_{REF} = \frac{1}{4} \times 5.16 = 1.25$$

Refrigeration load,

$$Q_2 = 3.52 \times 100 = 352 \text{ kW} = 352 \text{ kJ/s}$$

$$(COP)_{REF} = \frac{Q_2}{Q_1 - Q_2} = \frac{Q_2}{W}$$

$$\therefore \qquad W = \frac{Q_2}{(COP)_{REF}} = \frac{352}{1.25} = \textbf{281.6 kW} \quad \textbf{Ans.}$$

Example 2.22: Show that the minimum work needed by the heat pump for the heat transfer between two bodies at temperature T_1 initially, such that one body cools down to temperature T_2 shall be:

$$W = Cp\left[\frac{T_1^2}{T_2} + T_2 - 2T_1\right]$$

Consider the specific heat of the two bodies as Cp.

[U.P.T.U. I Sem., 2002-03]

Solution: Draw the schematic diagram of heat pump.

To satisfy the condition of minimum work requirement, the heat pump must be reversible.

$$\therefore \qquad COP = \frac{T_2}{T_1 - T_2}$$

For a unit mass, the heat to be extracted will be

$$Q_2 = Cp\,(T_1 - T_2)$$

Now, $\qquad COP = \dfrac{Q_2}{W}$

Heat Pump

$$\therefore \quad W = \frac{Q_2}{\text{COP}} = \frac{Cp\,(T_1 - T_2)}{\left(\dfrac{T_2}{T_1 - T_2}\right)}$$

$$= Cp\left(\frac{T_1 - T_2}{T_2}\right)^2 = Cp\left[\frac{T_1^2 + T_2^2 - 2T_1T_2}{T_2}\right]$$

$$\therefore \quad W = Cp\left[\frac{T_1^2}{T_2} + T_2 - 2T_1\right]$$

Example 2.23: Two identical bodies of constant heat capacity are at the same initial temperature T_i. A refrigerator operates between these two bodies until one body is cooled to temperature T_2. If the bodies remain at constant pressure and undergo no change of phase, find the minimum amount of work needed to do this in terms of T_i, T_2 and heat capacity. [U.P.T.U. II Sem., 2004-05]

Solution: Draw the schematic diagram of refrigerator.

The refrigerator is reversible,

$$\therefore \quad \text{COP} = \frac{T_2}{T_1 - T_2}$$

Heat to be extracted to cool the body from T_i to T_2

$$Q_2 = Cp(T_i - T_2)$$

$$\text{COP} = \frac{Q_2}{W}$$

Refrigerator

$$\therefore \quad W = \frac{Q_2}{\text{COP}} = \frac{Cp\,(T_i - T_2)}{\left(\dfrac{T_2}{T_i - T_2}\right)}$$

$$= Cp\,\frac{(T_i - T_2)^2}{T_2}$$

$$= Cp\left[\frac{T_i^2}{T_2} + T_2 - 2T_i\right]$$

Example 2.24: Using an engine of 30% thermal efficiency to drive a refrigerator having a COP of 5, what is the heat input into the engine for each MJ removed from the cold body by the refrigerant. If the system is used as heat pump, how many MJ of heat would be available for heating for each MJ of heat input to the engine.

Solution:

(i) *Engine-Refrigerator system,*

The schematic diagram is shown.

The thermal efficiency of engine,

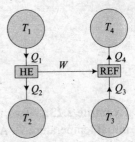

$$\eta = \frac{W}{Q_1} = 0.3$$

$$(COP)_{REF} = \frac{Q_3}{W} = 5$$

Engine-Refrigerator System

∴ $$W = 0.3Q_1$$

Also, $$W = \frac{Q_3}{5}$$

∴ $$0.3Q_1 = \frac{Q_3}{5}$$

$$Q_1 = \frac{Q_3}{5 \times 0.3}$$

For $$Q_3 = 1 \text{ MJ}$$

$$Q_1 = \frac{1}{5 \times 0.3} = \textbf{0.67 MJ} \quad \textbf{Ans.}$$

(ii) *Engine-Heat Pump system,*

$$(COP)_{HP} = (COP)_{REF} + 1 = 5 + 1 = 6$$

$$= \frac{Q_4}{W}$$

But from engine efficiency,

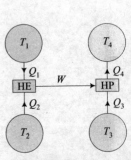

$$W = 0.3 \, Q_1$$

If $$Q_1 = 1 \text{ MJ}$$

$$W = 0.3 \text{ MJ}$$

∴ $$Q_4 = W(COP)_{HP} = 0.3 \times 6$$

$$= \textbf{1.8 MJ} \quad \textbf{Ans.}$$

Engine-Heat Pump system

Example 2.25: Heat pump is used for heating the premises in winter and cooling the same in summer such that temperature inside remains 25°C. Heat transfer across the walls and roof is found 2 MJ per hour per degree temperature difference between interior and exterior. Determine the minimum power required for operating the pump in winter when outside temperature is 1°C and also give the maximum temperature in summer for which the device shall be capable of maintaining the premises at desired temperature for same power input.

[U.P.T.U. II Sem., 2001-02]

Solution:

(i) In winter, the device works as heat pump.

$$T_1 = 25°C + 273 = 298 \text{ K}$$

$$T_2 = 1°C + 273 = 274 \text{ K}$$

For a reversible heat pump,

$$(COP)_{HP} = \frac{T_1}{T_1 - T_2} = \frac{298}{298 - 274} = 12.4$$

Also, $\quad COP = \dfrac{Q_1}{W}$

$$Q_1 = \frac{2 \times 10^6}{3600}(298 - 274) = 13333.33 \text{ J} = 13.33 \text{ kJ}$$

$$W = \frac{13.33}{12.42} = \textbf{1.074 kW} \quad \textbf{Ans.}$$

(ii) In summer, the device works as air-conditioner,

$$(COP)_{REF} = \frac{T_2}{T_1 - T_2} = \frac{Q_2}{W}$$

$$\therefore \quad \frac{298}{(298 - T_2)} = \frac{\dfrac{2 \times 10^6}{3600}(298 - T_2)}{1.074}$$

$$(298 - T_2)^2 = \frac{298 \times 1.074 \times 3600}{2 \times 10^6}$$

$$= 576$$

$$298 - T_2 = \sqrt{576} = 24$$

$$\therefore \quad T_2 = 298 + 24 = 322 \text{ K}$$

$$= \textbf{49°C} \quad \textbf{Ans.}$$

Heat Pump

Refrigerator (Air-conditioner)

2.11 EQUIVALENCE OF SECOND LAW STATEMENTS

The Kelvin-Planck statement and Clausius statements are two statements of second law of thermodynamics (see para 2.1). These statements look different but these are two parallel statements and are equivalent in all respects. The equivalence of these statements will be proved by the logic that violation of one statement leads to violation of second statement and vice-versa.

2.11.1 Violation of Clausius Statement

A cyclic heat engine (HE) operates between two reservoirs drawing heat Q_1 and producing work, W as shown in Fig. 2.20(a).

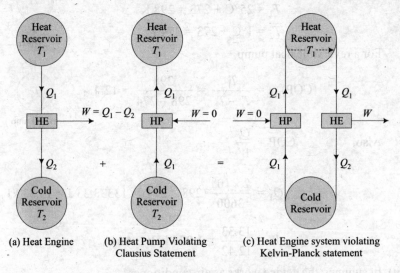

(a) Heat Engine   (b) Heat Pump Violating Clausius Statement   (c) Heat Engine system violating Kelvin-Planck statement

Fig. 2.21   Violation of Clausius statement

A cyclic heat pump (HP) is shown in Fig. 2.21(b) which is transferring heat from cold reservoir (T_2) to a hot reservoir (T_1) with no work input $(W = 0)$. This violates Clausius statement as heat cannot flow from a cold temperature to a higher temperature without any external work input.

Figure 2.21(c) shows a heat engine (HE) and heat pump (HP) combined to form a heat engine system. The heat pump is supplying heat Q_1 to hot reservoir which is the requirement of heat engine. Therefore, hot reservoir can be by-passed. The heat pump and heat engine T form a system operating in cycles and producing work W continuously while exchanging heat with one reservoir (cold) only. Thus, the Kelvin-Planck statement is violated. Hence, when Clausius statement is violated, simultaneously, Kelvin-Plank statement is also violated.

2.11.2 Violation of Kelvin-Planck Statement

A heat pump (HP) is extracting heat Q_1 from low temperature reservoir (T_2) and discharging heat to high temperature reservoir (T_1) getting work from outside.

A heat engine (HE) is producing work W by exchanging heat with one reservoir at temperature T_1 only. The Kelvin-Planck statement is violated which requires two heat reservoirs for an engine to produce work continuously.

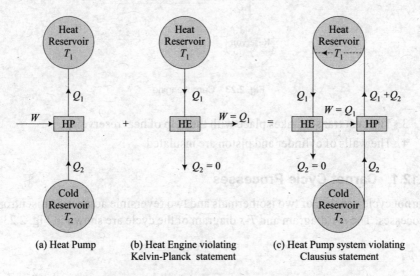

| (a) Heat Pump | (b) Heat Engine violating Kelvin-Planck statement | (c) Heat Pump system violating Clausius statement |

Fig. 2.22 Violation of Kelvin-Planck statement

The heat pump and heat engine are combined in Fig. 2.22(c) to create a heat pump system. The heat pump is getting work W from heat engine. Therefore, heat pump system is transferring heat from a lower temperature (T_2) to a higher temperature (T_1) without any external work. Therefore, the Clausius statement is also violated.

From the above two cases, it is clear that when one of the statements of second law is violated, the other statement is also violated. Therefore, the two statements are equivalent.

2.12 CARNOT CYCLE

Sadi Carnot proposed the concept of Carnot engine in 1824 which works on reversible cycle called Carnot cycle.

The *Carnot engine* has air enclosed in a cylinder. The cylinder receives heat Q_1 from hot reservoir and rejects heat Q_2 to cold reservoir. The cylinder head can be covered with insulating cap at will. The following assumptions are made:

1. The working substance is air.
2. Piston-cylinder arrangement is weightless and frictionless.

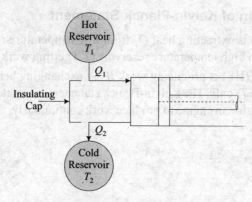

Fig. 2.23 Carnot engine

3. The heat transfer takes place with the help of heat reservoirs.
4. The walls of cylinder and piston are insulated.

2.12.1 Carnot Cycle Processes

Carnot cycle consists of two isothermals and two reversible adiabatic or isentropic processes. The *p-v* diagram and *T-s* diagram of the cycle are shown in Fig. 2.23.

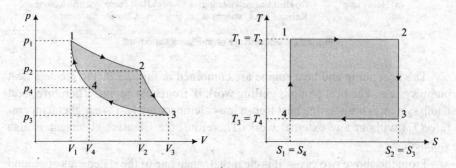

Fig. 2.24 Carnot cycle

1. *Process 1–2: Isothermal expansion*

The cylinder is brought in contact with hot reservoir. The air expands at constant temperature ($T_1 = T_2$) from volume V_1 to V_2. The heat supplied to the air is utilized in doing external work.

Applying first law of thermodynamics,

$$Q_{1-2} = (U_2 - U_1) + W_{1-2}$$

$$\therefore \qquad T_1 = T_2$$

$$\therefore \qquad U_2 - U_1 = 0$$

$$Q_{1-2} = W_{1-2} = p_1 V_1 \ln\left(\frac{V_2}{V_1}\right)$$

$\because \qquad p_1 V_1 = mRT_1$

$\therefore \qquad Q_{1-2} = mRT_1 \ln\left(\frac{V_2}{V_1}\right)$

Let $\qquad \dfrac{V_2}{V_1} = r = $ expansion ratio

$\therefore \qquad Q_{1-2} = mRT_1 \ln r$

Also $\qquad Q_{1-2} = T_1(S_2 - S_1)$

2. *Process 2–3: Reversible Adiabatic Expansion or Isentropic Expansion*

The cylinder is disconnected from hot reservoir and insulation cap is put on. The air expands from V_2 to V_3 and T_3.

$$Q_{2-3} = 0$$

Applying first law of thermodynamics,

$$Q_{2-3} = (U_3 - U_2) + W_{2-3} = 0$$

$\therefore \qquad (U_3 - U_2) = \dfrac{p_2 V_2 - p_3 V_3}{\gamma - 1}$

$$pV = mRT$$

$\therefore \qquad (U_3 - U_2) = \dfrac{mRT_2 - mRT_3}{\gamma - 1}$

$$= \frac{mR(T_1 - T_3)}{\gamma - 1} \qquad\qquad [\because T_1 = T_2]$$

Also, $\qquad S_2 = S_3.$

3. *Process 3–4: Isothermal Expansion*

Insulating cap is removed and the cylinder head is brought in contact with cold reservoir. The air is compressed at constant temperature ($T_3 = T_4$) from volume V_3 to V_4.

Applying 1st law of thermodynamics,

$$Q_{3-4} = (U_4 - U_3) + W_{3-4}$$

$\therefore \qquad T_3 = T_4$

$\therefore \qquad U_4 - U_3 = 0$

$$\therefore \qquad Q_{3-4} = W_{3-4} = p_3 V_3 \ln \frac{V_3}{V_4}$$

$$= mRT_3 \ln \frac{V_3}{V_4} \qquad\qquad [\because pV = mRT]$$

$$= mRT_3 \ln r$$

where, r = compression ratio = $\dfrac{V_3}{V_4} = \dfrac{V_2}{V_1}$

Also, $\qquad Q_{3-4} = T_3(S_3 - S_4).$

4. *Process 4–1: Reversible Adiabatic or Isentropic Compression*

The cylinder head is disconnected from cold reservoir and covered with insulating cap. The air is allowed to be compressed isentropically from V_4 to V_1.

$$Q_{4-1} = 0$$

Applying 1st law of thermodynamics,

$$Q_{4-1} = (U_4 - U_1) + W_{4-1}$$

$$\therefore \qquad (U_4 - U_1) = \frac{p_1 V_1 - p_2 V_2}{\gamma - 1}$$

$$= \frac{mR(T_1 - T_2)}{\gamma - 1}$$

But, $\qquad T_3 = T_4$

$$\therefore \qquad U_4 - U_1 = \frac{mR(T_1 - T_3)}{\gamma - 1}$$

$$\therefore \qquad U_3 - U_2 = U_4 - U_1$$

Also, $\qquad S_4 = S_1.$

2.12.2 Efficiency of Carnot Cycle

Work done,

$$W = \text{Heat supplied} - \text{Heat rejected}$$

$$= mRT_1 \ln r - mRT_3 \ln r$$

$$= mR \ln r (T_1 - T_3)$$

Efficiency,

$$\eta = \frac{\text{Work done}}{\text{Heat supplied}} = \frac{mR \ln r (T_1 - T_3)}{mR \ln r T_1} = \frac{T_1 - T_3}{T_1}$$

From *T-s* diagram,

$$W = T_1(S_2 - S_1) - T_3(S_3 - S_4)$$

$$\eta = \frac{T_1(S_2 - S_1) - T_3(S_3 - S_4)}{T_1(S_2 - S_1)}$$

But, $S_3 = S_2$ and $S_1 = S_4$

$\therefore$ $$\eta = \frac{T_1 - T_3}{T_1}$$

For process 2–3, $$\frac{T_2}{T_3} = \left(\frac{V_3}{V_2}\right)^{\gamma-1}$$

But, $T_2 = T_1$

$\therefore$ $$\frac{T_1}{T_3} = \left(\frac{V_3}{V_2}\right)^{\gamma-1}$$

For process 4–1, $$\frac{T_1}{T_4} = \left(\frac{V_4}{V_1}\right)^{\gamma-1}$$

But, $T_4 = T_3$

$\therefore$ $$\frac{T_1}{T_3} = \left(\frac{V_4}{V_1}\right)^{\gamma-1}$$

$\therefore$ $$\frac{V_3}{V_2} = \frac{V_4}{V_1}$$

or $$\frac{V_2}{V_1} = \frac{V_3}{V_4} = r$$

$\therefore$ $$\frac{T_1}{T_3} = (r)^{\gamma-1}$$

$\therefore$ $$\eta = 1 - \frac{T_3}{T_1} = 1 - \left(\frac{1}{r}\right)^{\gamma-1}$$

$$= 1 - \frac{1}{r^{\gamma-1}}$$

2.12.3 Observations

1. Carnot efficiency, $\eta = 1 - \dfrac{T_2}{T_1}$.

   Thermal efficiency, $\eta = \dfrac{\text{Work done}}{\text{Heat supplied}} = \dfrac{W}{Q_1} = \dfrac{Q_1 - Q_2}{Q_1} = 1 - \dfrac{Q_2}{Q_1}$.

   $\therefore \qquad \dfrac{Q_2}{Q_1} = \dfrac{T_2}{T_1}$

   or $\qquad \dfrac{Q_1}{T_1} = \dfrac{Q_2}{T_2} = \text{constant.}$

   The heat transfer from a heat reservoir is proportional to its temperature. This is called *Carnot principle*.
2. Carnot efficiency is independent of working fluid.
3. It is a function of absolute temperature of heat reservoirs.
4. Carnot efficiency increases with the decrease of sink temperature T_2.

   If $T_2 = 0$, η is 100%. It violates Kelvin-Planck statement as no cold reservoir is required.
5. The sink temperature T_2 cannot be lower than atmospheric temperature or temperature of sea or river or lake which can act as heat sink.
6. Carnot efficieny increases with the increase of source temperature T_1. This requires high temperature resisting materials which are very costly and not easily available. Therefore, T_1 is restricted by metallurgical considerations and is called *metallurgical limitation*.
7. Carnot efficiecy depends upon temperature difference $(T_1 - T_2)$. Higher the temperature difference between source and sink, higher the Carnot efficiency.
8. For same degree increase of source temperature or decrease in sink temperature, Carnot efficiency is more sensitive to change in sink temperature.

2.12.4 Impracticability of Carnot Engine (Carnot Cycle)

Although Carnot cycle is most efficient, it is not practical and Carnot engine is only a hypothetical device. The main reasons for impracticability of Carnot cycle are listed below:

1. All the processes are taken as reversible in a Carnot cycle. Practically processes are not reversible due to the presence of internal fluid friction of working substance and mechanical friction between piston and cylinder.
2. There is also irreversibility due to heat absorption and rejection taking place with finite temperature difference between reservoir and working substance.

3. The cycle demands that the engine speed should be very slow to ensure iso-thermal process and very fast for adiabatic process. Such a large variation of piston speed is not practical.

4. The slope difference of isothermal and adiabatic processes is very small on *p-v* diagram. The area of *p-v* diagram and, hence, work done is very small per cycle. Therefore, piston-cylinder arrangement has to be very big and bulky to get sufficient work output as the engine has to work with large range of pressures and volumes.

Though, it is not practicable to build an engine which can work on Carnot cycle, however, it serves as a standard for comparison and improvement of practical cycles. The Carnot cycle has maximum efficiency for given temperatures of source and sink.

2.12.5 Carnot Refrigerator and Heat Pump

The coefficient of performance of a refrigerator working as reversed engine can be expressed as:

$$(\text{COP})_{\text{REF}} = \frac{Q_2}{Q_1 - Q_2}$$

From Carnot principle,

$$\frac{Q_1}{T_1} = \frac{Q_2}{T_2}$$

The COP of a Cannot refrigerator will be

$$\text{COP} = \frac{T_2}{T_1 - T_2}$$

The coefficient of performance of a heat pump is expressed as

$$(\text{COP})_{\text{HP}} = \frac{Q_2}{Q_1 - Q_2}$$

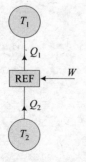

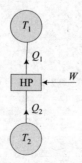

Fig. 2.25 Carnot refrigerator   **Fig. 2.26** Carnot heat pump

From Carnot principle,

$$\frac{Q_1}{T_1} = \frac{Q_2}{T_2}$$

The COP of a Carnot heat pump will be:

$$COP = \frac{T_1}{T_1 - T_2}$$

2.13 CARNOT THEOREM

The Carnot theorem states:

1. "No heat engine operating in a cycle between two given thermal reservoirs, with fixed temperatures, can be more efficient than a reversible engine operating between the same thermal reservoirs".
2. "The efficiency of all reversible heat engines operating between the same temperature levels is the same".
3. "The efficiency of a reversible engine is independent of the nature and amount of the working substance undergoing the cycle".

E_A is any engine and E_B is a reversible engine working between the same thermal reservoirs at temperatures T_1 and T_2.

We have to prove that $\eta_B > \eta_A$. If it is not true, then $\eta_A > \eta_B$.

1. *Assume,*

$$Q_{1A} = Q_{1B} = Q_1$$

and

$$\eta_A > \eta_B$$

∴

$$\frac{W_A}{Q_{1A}} > \frac{W_B}{Q_{1B}}$$

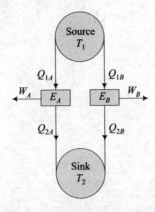

Fig. 2.27 Two engines working between same reservoir

$\therefore$ $$\frac{W_A}{Q_1} > \frac{W_B}{Q_1}$$

or $$W_A > W_B$$

2. *Reverse Engine E_B*

The engine E_B is reversible. Reverse the engine B so that all input and output energy values are same but directions are reversed. Since W_A is more than W_B, therefore, part of W_A (equal to W_B) may be used to drive the reversed heat engine $\exists_B$.

The heat Q_{1B} discharged by $\exists_B$ may be supplied to E_A. The source can be eliminated.

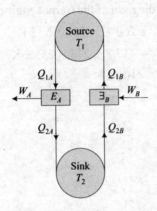

Fig. 2.28 Engine E_B is reversed

3. *Combined Engines $(E_A + \exists_B)$*

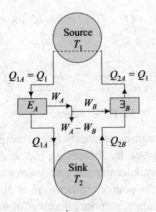

Fig. 2.29 Combined operation of $E_A + \exists_B$

The combined engines $E_A + \exists_B$ constitute a heat engine system operating on a single reservoir (T_2) and still produces work equal to $W_A - W_B$. This system violates the Kelvin-Planck statement of second law of thermodynamics. Therefore, our assumption that $\eta_A > \eta_B$ is wrong.

$$\therefore \qquad \eta_B \geq \eta_A$$

The efficiency of an irreversible engine cannot be greater than that of the reversible engine if both operate between same thermal reservoirs with fixed temperatures.

Example 2.26: A Carnot engine operates between source and sink temperature of $260°C$ and $-17.8°C$. If the system receives 100 kJ from the source, find the net work transfer, heat rejected to sink and efficiency of the system.

Solution: The schematic diagram of the Carnot engine is shown.

$$T_1 = 260°C + 273 = 533 \text{ K}$$
$$T_2 = -17.8 + 273 = 255.2 \text{ K}$$
$$Q_1 = 100 \text{ kJ}$$

From Carnot principle,

$$\frac{Q_1}{T_1} = \frac{Q_2}{T_2}$$

$$\therefore \qquad Q_2 = \frac{T_2}{T_1} Q_1$$

Carnot Engine

$$= \frac{255.2}{533} \times 100 = \textbf{47.88 kJ} \quad \textbf{Ans.}$$

The net work transfer,

$$W = Q_1 - Q_2 = 100 - 47.88 = 52.12 \text{ kJ}$$

The Carnot efficiency,

$$\eta = 1 - \frac{T_2}{T_1} = 1 - \frac{255.2}{533} = 0.5212$$

$$= \textbf{52.12\%} \quad \textbf{Ans.}$$

Exampe 2.27: A carnot engine operates between temperatures 500 K and 100 K. Its efficiency may be increased by increasing the source temperature to 600 K. Determine the reduction in sink temperature necessary to affect the same improvement in efficiency holding the source temperature at 500 K. On the basis of these calculations, which method appears more effective for increasing the effficiency of Carnot cycle.

Solution: The schematic diagram of Carnot engine is shown.

$$T_1 = 500 \text{ K}$$
$$T_2 = 100 \text{ K}$$

$$\eta_c = 1 - \frac{T_2}{T_1} = 1 - \frac{100}{500} = 80\%$$

Case I.
$$T_2 = 100 \text{ K}$$
$$T_1 = 600 \text{ K}$$

$\therefore$
$$\eta_c = 1 - \frac{T_2}{T_1} = 1 - \frac{100}{600} = 83.33\%$$

Carnot Engine

Case II.
$$T_1 = 500 \text{ K}$$
$$\eta_c = 0.8333$$

$$\eta_c = 1 - \frac{T_2}{T_1}$$

$\therefore$
$$T_2 = (1 - \eta_c)T_1 = (1 - 0.8333)500 = 83.35 \text{ K}$$

In order to achieve an increase in efficiency to 83.33%, either increase the source temperature by 100 K or decrease the sink temperature by 16.65 K, i.e., (100 – 83.35) K. Therefore, decreasing the sink temperature is more effective to increase the Carnot efficiency as we have to decrease it only by 16.65 K against 100 K increase in source temperature.

2.14 CLAUSIUS INEQUALITY

The clausius theorem states: "Whenever a closed system undergoes a cyclic process, the cyclic integral $\oint \frac{dQ}{T}$ is less than zero (i.e., negative) for an irreversible cyclic process and equal to zero for a reversible cyclic process".

Mathematically

$$\oint \frac{dQ}{T} < 0 \text{ for an irreversible process}$$

$$\oint \frac{dQ}{T} = 0 \text{ for reversible process}$$

Combining the equations for reversible and irreversible processes

$$\oint \frac{dQ}{T} \geq 0$$

The equation for irreversible cyclic process may be witten as,

$$\oint \frac{dQ}{T} + I = 0$$

where, I = Amount of irreversibility of a cyclic process

$I = 0$ for a reversible cyclic process.

1. *Reversible Heat Engine* (RHE)

The efficiency of a reversible heat engine operating between temperature limits T_1 and T_2 is given by,

$$\eta = \frac{Q_1 - Q_2}{Q_1} = \frac{T_1 - T_2}{T_1} = 1 - \frac{T_2}{T_1}$$

$$\therefore \qquad \frac{T_2}{T_1} = \frac{Q_2}{Q_1}$$

or

$$\frac{Q_1}{T_1} = \frac{Q_2}{T_2}$$

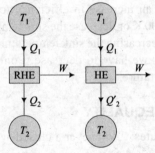

Fig. 2.30 Clausius inequality

The heat rejected by the system is negative

$$\therefore \qquad \left(\frac{Q_1}{T_1}\right) - \left(-\frac{Q_2}{T_2}\right) = 0$$

or

$$\frac{Q_1}{T_1} + \frac{Q_2}{T_2} = 0$$

Extending the equation to any number of heat reservoirs,

$$\oint \frac{dQ}{T} = 0 \text{ for a reversible cycle.}$$

2. *Irreversible Heat Engine* (HE)

The efficiency of an irreversible heat engine operating between the same temperature limits T_1 and T_2.

$$\eta < \eta_c$$

$$\therefore \qquad \frac{Q_1 - Q_2'}{Q_1} < \frac{T_1 - T_2}{T_1}$$

$$\left(1 - \frac{Q_2'}{Q_1}\right) < \left(1 - \frac{T_2}{T_1}\right)$$

$$\frac{Q_2'}{Q_1} > \frac{T_2}{T_1}$$

or

$$\frac{Q_2'}{T_2} > \frac{Q_1}{T_1}$$

or

$$\frac{Q_1}{T_1} < \frac{Q_2'}{T_2}$$

If Q_2' is $-$ve, being heat rejected by the system,

$$\frac{Q_1}{T_1} - \left(\frac{-Q_2'}{T_2}\right) < 0$$

$$\frac{Q_1}{T_1} + \frac{Q_2'}{T_2} < 0$$

Extending the case to arbitrary number of heat reservoirs

$$\oint \frac{dQ}{T} < 0 \text{ for an irreversible cycle.}$$

Statement for Clausius Inequality:

"When a system operates on a cycle (reversible or irreversible) having heat interaction with a number of heat reservoirs, then the algebraic sum of the quotients of heat transfers from heat reservoirs to their respective absolute temperature is always equal to or less than zero (respectively)".

Example 2.28: A heat engine is supplied with 2512 kJ/min of heat at 650°C. Heat rejection takes place at 100°C. Specify the following results of heat rejection:

  (i) 867 kJ/min
 (ii) 1015 kJ/min
(iii) 1494 kJ/min

Solution: The schematic diagram of the engine is shown,

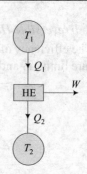

$$T_1 = 650°C + 273 = 923 \text{ K}$$
$$T_2 = 100°C + 273 = 373 \text{ K}$$
$$Q_1 = 2512 \text{ kJ/min}$$

(i) $Q_2 = 867$ kJ/min

Applying Clausius inequality,

$$\frac{Q_1}{T_1} - \frac{Q_2}{T_2} = \frac{2512}{923} - \frac{867}{373} = 0.34770 > 0$$

The cycle is not possible.

(ii) $Q_2 = 1015$ kJ/min

$$\frac{2512}{923} - \frac{1015}{373} = 0$$

The cycle is reversible.

(iii) $Q_2 = 1494$ kJ/min

$$\frac{2512}{923} - \frac{1494}{373} = -1.284 < 0$$

The cycle is irreversible and possible.

Example 2.29:   A reversible heat engine receives heat from two thermal reservoirs at 750 K and 500 K. The engine develops 100 kW and rejects 3600 kJ/min of heat to the sink at 250 K. Determine thermal efficiency of the engine and heat supplied by each thermal reservoir.

Solution: The schematic diagram of the engine is shown,

$$T_1 = 750 \text{ K}$$
$$T_2 = 500 \text{ K}$$
$$T_3 = 250 \text{ K}$$

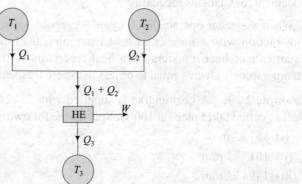

$$W = 100 \text{ kW} = 100 \times 60 = 6000 \text{ kJ/min}$$

$$Q_3 = 3600 \text{ kJ/min}$$

As per 1st law of thermodynamics,

$$Q_1 + Q_2 = W + Q_3 = 6000 + 3600$$

$$= 9600 \text{ kJ/mm}$$

$$Q_2 = 9600 - Q_1$$

As per Clausius theorem,

$$\frac{Q_1}{T_1} + \frac{Q_2}{T_2} - \frac{Q_3}{T_3} = 0 \qquad \text{(for a reversible heat engine)}$$

$$\frac{Q_1}{750} + \frac{(9600 - Q_1)}{500} - \frac{3600}{250} = 0$$

$$Q_1 + 14400 - 1.5Q_1 - 10800 = 0$$

$$Q_1 = 7200 \text{ kJ/min}$$

$$Q_2 = 9600 - 7200 = 2400 \text{ kJ/min}$$

Example 2.30: A heat engine working on Carnot cycle absorbs heat from three reservoirs at 1000 K, 800 K and 600 K. The engine does 10 kW of net work and rejects 400 kJ/min of heat to a heat sink at 300 K. If heat supplied by the reservoir at 1000 K is 60% of the heat supplied by reservoir at 600 K. Calculate the quantity of heat absorbed from each reservoir.

Solution: The schematic diagram of engine is shown,

$$T_1 = 600 \text{ K}$$

$$T_2 = 800 \text{ K}$$

$$T_3 = 1000 \text{ K}$$

$$T_4 = 300 \text{ K}$$

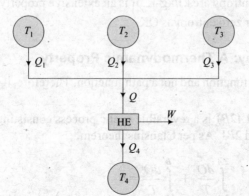

$$Q_4 = 400 \text{ kJ/min}$$

$$W = 10 \text{ kW} = 10 \times 60 = 600 \text{ kJ/min}$$

$$Q = Q_1 + Q_2 + Q_3 = W + Q_4$$

$$= 600 + 400$$

$$= 1000 \text{ kJ/min}$$

$$Q_3 = 0.6 Q_1$$

$$Q_2 = Q - (Q_1 + Q_3) = 1000 - Q_1 - 0.6 Q_1$$

$$= 1000 - 1.6 Q_1$$

Apply Clausius theorem,

$$\frac{Q_1}{T_1} + \frac{Q_2}{T_2} + \frac{Q_3}{T_3} - \frac{Q_4}{T_4} = 0$$

$$\frac{Q_1}{600} + \frac{(1000 - 1.6 Q_1)}{800} + \frac{0.6 Q_1}{1000} - \frac{400}{300} = 0$$

$$Q_1 = \mathbf{312.5 \text{ kJ/min}}$$

$$Q_2 = 1000 - 1.6(312.5) = \mathbf{500 \text{ kJ/min}}$$

$$Q_3 = 0.6 (312.5) = \mathbf{187.5 \text{ kJ/min}} \quad \textbf{Ans.}$$

2.15 CONCEPT OF ENTROPY

Entropy is a thermodynamic property of a working substance which increases with the addition of heat and decreases with its removal. Heat absorbed by the working substance can be expressed as:

$$dQ = TdS$$

where, T = Absolute temperature

dS = Change in entropy.

The units of entropy are kJ/kg-K. It is an extensive property.

The datum for zero entropy is OK.

2.15.1 Entropy: A Thermodynamic Property

Entropy is a point function and not a path function. Therefore, it is a property of the system.

In Fig. 2.31, 1*A*2*B*1 is a reversible cyclic process consisting of two reversible processes 1*A*2 and 2*B*1. As per Clausius theorem,

$$\int\limits_{1A}^{2A} \frac{dQ}{T} + \int\limits_{2B}^{1B} \frac{dQ}{T} = 0 \qquad \qquad \text{...(1)}$$

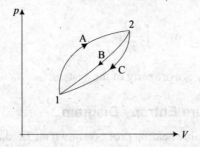

Fig. 2.31 Entropy

For another reversible cyclic process $1A\,2\,C1$ consisting of processes $1A2$ and $2C1$, Clausius theorem gives,

$$\int_{1A}^{2B} \frac{dQ}{T} + \int_{2C}^{1C} \frac{dQ}{T} = 0 \qquad\qquad ...(2)$$

Subtracting equation (2) from equation (1),

$$\int_{2B}^{1B} \frac{dQ}{T} - \int_{2C}^{1C} \frac{dQ}{T} = 0$$

Changing the limits of integral

$$\int_{1B}^{2B} \frac{dQ}{T} - \int_{1C}^{2C} \frac{dQ}{T} = 0$$

or

$$\int_{1B}^{2B} \frac{dQ}{T} = \int_{1C}^{2C} \frac{dQ}{T}$$

Therefore, $\int \dfrac{dQ}{T}$ is independent of the path of the process and is a state function. The change of entropy,

$$dS = \frac{dQ}{T}$$

For a process 1–2,

$$\int_{1}^{2} dS = \int_{1}^{2} \frac{dQ}{T}$$

$$\therefore \qquad \tilde{S}_2 - S_1 = \int_1^2 \frac{dQ}{T}$$

Therefore, entropy is a property of the system.

2.15.2 Temperature Entropy Diagram

Process 1–2 is a reversible heating process shown on *T-s* diagram.

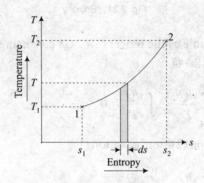

Fig. 2.32 *T-s* diagram

When a small quantity of heat dQ is supplied at temperature T, then

$$dQ = TdS$$

$$dS = \frac{dQ}{T}$$

$$\int_1^2 dS = \int_1^2 \frac{dQ}{T}$$

The area under any thermodynamic process on *T-s* diagram represents the heat added or rejected.

1. *Constant volume process*

A constant volume process 1–2 is shown on *T-s* diagram. For a constant volume process,

$$dQ = mC_v dT$$

$$\therefore \qquad \frac{dQ}{T} = mCv\frac{dT}{T}$$

$$dS = mCv\frac{dT}{T}$$

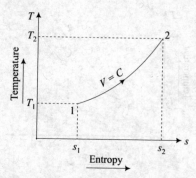

Fig. 2.33 Constant volume process

Integrating between limits 1 and 2,

$$\int_{S_1}^{S_2} ds = mCv \int_{T_1}^{T_2} \frac{dT}{T}$$

$\therefore$
$$S_2 - S_1 = mCv \ln \frac{T_2}{T_1}$$

For $V = C$,
$$\frac{T_2}{T_1} = \frac{p_2}{p_1}$$

$\therefore$
$$S_2 - S_1 = mCv \ln \left(\frac{p_2}{p_1} \right)$$

Again,
$$dS = mCv \frac{dT}{T}$$

$\therefore$
$$\frac{dT}{dS} = \frac{T}{mCv}$$

This equation gives the shape (or slope) of process 1–2 on *T-s* diagram. Equations are valid for both reversible and irreversible processes.

2. *Constant pressure process*

A constant pressure process is plotted on *T-s* diagram. For a constant pressure process,

$$dQ = mCpdT$$

$\therefore$
$$\frac{dQ}{T} = mCp \frac{dT}{T}$$

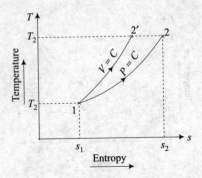

Fig. 2.34   Constant pressure process

$$dS = m\,Cp\left(\frac{dT}{T}\right)$$

$$\int_{S_1}^{S_2} dS = m\,Cp \int_{T_1}^{T_2} \frac{dT}{T}$$

$$S_2 - S_1 = m\,Cp \ln\frac{T_2}{T_1}$$

For $p = C$, $\dfrac{T_2}{T_1} = \dfrac{V_2}{V_1}$

$\therefore$ $$S_2 - S_1 = m\,Cp \ln\left(\frac{V_2}{V_1}\right)$$

This equation is valid for both reversible and irreversible processes.

Again, $$dS = m\,Cp\,\frac{dT}{T}$$

$\therefore$ $$\frac{dT}{dS} = \frac{T}{Cp} = \text{slope of process line } 1\text{--}2$$

For constant volume process,

$$\frac{dT}{dS} = \frac{T}{Cv}$$

$\because$ $$Cv < Cp$$

$\therefore$ $$\frac{1}{Cv} > \frac{1}{Cp}$$

$$\therefore \qquad \frac{T}{Cv} > \frac{T}{Cp}$$

The slope of constant volume process is higher than that for constant pressure process.

3. *Isothermal process*

An isothermal process 1–2 is shown on *T-s* diagram. For isothermal process,

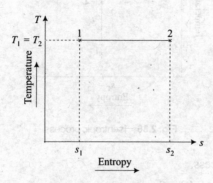

Fig. 2.35   Isothermal process

$$Q_{1-2} = W_{1-2} = mRT \ln\left(\frac{V_2}{V_1}\right)$$

$$\therefore \qquad S_2 - S_1 = \frac{mRT}{T} \ln\left(\frac{V_2}{V_1}\right)$$

$$= mR \ln\left(\frac{V_2}{V_1}\right)$$

$$= m(Cp - Cv) \ln\left(\frac{V_2}{V_1}\right)$$

For $T = C$, $\qquad \dfrac{V_2}{V_1} = \dfrac{P_1}{P_2}$

$$\therefore \qquad S_2 - S_1 = mR \ln\left(\frac{P_1}{P_2}\right)$$

or $\qquad S_2 - S_1 = m(Cp - Cv) \ln\left(\frac{P_1}{P_2}\right)$

The equation is valid for reversible and irreversible processes.

4. *Adiabatic process*

A reversible adiabatic process 1–2 is shown on *T-s* diagram whereas irreversible adiabatic process 1–2 as shown a dotted line.

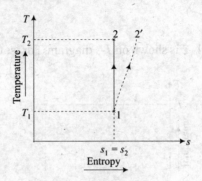

Fig. 2.36   Isentropic process

For adiabatic process,

$$dQ = 0$$

$\therefore$ $$dS = 0 \text{ for reversible process.}$$

For irreversible adiabatic process, there is increase in entropy due to internal heating.

5. *Polytropic process*

A polytropic process 1–2 is shown on *T-s* diagram. For a polytropic process,

$$dQ = \frac{\gamma - n}{\gamma - 1} dW$$

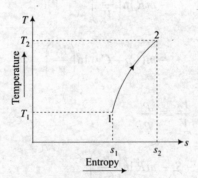

Fig. 2.37   Polytropic process

Now, $\qquad dW = pdV$

∴ $\qquad\qquad dQ = \dfrac{\gamma - n}{\gamma - 1} pdV$

$\qquad\qquad \dfrac{dQ}{T} = \dfrac{\gamma - n}{\gamma - 1}\dfrac{pdV}{T}$

∴ $\qquad\qquad dS = \dfrac{\gamma - n}{\gamma - 1} mR \dfrac{dV}{V}$

∴ $\qquad\qquad pdV = mRT$

∴ $\qquad\qquad \dfrac{p}{T} = \dfrac{mR}{V}$

Integrating,

$$S_2 - S_1 = \dfrac{\gamma - n}{\gamma - 1} mR \ln\left(\dfrac{V_2}{V_1}\right)$$

6. *Carnot cycle*

Carnot cycle is represented on *T-s* diagram.

    Process 1–2: Isothermal expansion

    Process 2–3: Isentropic expansion

    Process 3–4: Isothermal compression

    Process 4–1: Isentropic compression

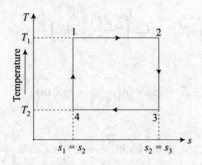

Fig. 2.38 Carnot cycle

Heat added, $\qquad Q_1 = T_1(s_2 - s_1)$

Heat rejected, $\qquad Q_2 = T_2(s_4 - s_3)$

$\qquad\qquad\qquad = -T_2(s_3 - s_4)$

$\qquad s_2 - s_1 = s_3 - s_4$

Work done during cycle,

$$W = Q_1 - Q_2 = (T_1 - T_2)(s_2 - s_1)$$

Thermal efficiency,

$$\eta = \frac{W}{Q_1} = \frac{(T_1 - T_2)(s_2 - s_1)}{T_1(s_2 - s_1)} = \frac{T_1 - T_2}{T_1}$$

2.15.3 Entropy Change for Ideal Gases

Changes of entropy of a certain quantity, m of a perfect gas during heating by any thermodynamic process from state 1 $(p_1 V_2 T_1)$ to state 2 $(p_2 V_2 T_2)$ can be found out as follows:

1. *In terms of volume and temperature*

General energy equation of 1st law of thermodynamics:

$$dQ = dU + dW = m\,Cv\,dT + pdV$$

$\therefore$
$$\frac{dQ}{T} = m\,Cv\,\frac{dT}{T} + \frac{pdV}{T}$$

From equation of state, $pV = mRT$

$$\frac{p}{T} = \frac{mR}{V}$$

$\therefore$
$$\frac{dQ}{T} = dS = m\,Cv\,\frac{dT}{T} + mR\,\frac{dV}{V}$$

$$\int_{s_1}^{s_2} dS = m\,Cv \int_{T_1}^{T_2} \frac{dT}{T} + mR \int_{V_1}^{V_2} \frac{dV}{V}$$

$\therefore$
$$S_2 - S_1 = m\,Cv\,\ln\left(\frac{T_2}{T_1}\right) + mR\,\ln\left(\frac{V_2}{V_1}\right)$$

$$= m\left[Cv\,\ln\frac{T_2}{T_1} + (Cp - Cv)\ln\left(\frac{V_2}{V_1}\right) \right]$$

2. *In terms of pressure and temperature*

From equation of state,

$$\frac{p_1 V_1}{T_1} = \frac{p_2 V_2}{T_2}$$

$$\therefore \qquad \frac{V_2}{V_1} = \frac{p_1 T_2}{p_2 T_1}$$

$$S_2 - S_1 = m Cv \ln\left(\frac{T_2}{T_1}\right) + mR\left(\frac{p_1}{p_2} \times \frac{T_2}{T_1}\right)$$

$$= m Cv \ln\left(\frac{T_2}{T_1}\right) + mR \ln\left(\frac{p_1}{p_2}\right) + mR \ln\left(\frac{T_2}{T_1}\right)$$

$$= m(Cv + R) \ln\left(\frac{T_2}{T_1}\right) + mR \ln\left(\frac{p_1}{p_2}\right)$$

Now, $\qquad\qquad R = Cp - Cv$

$$Cv + R = Cp$$

$$\therefore \qquad S_2 - S_1 = m Cp \ln\left(\frac{T_2}{T_1}\right) + m(Cp - Cv) \ln\left(\frac{p_1}{p_2}\right)$$

$$= m\left[Cp \ln\left(\frac{T_2}{T_1}\right) + (Cp - Cv) \ln\left(\frac{p_1}{p_2}\right)\right]$$

3. *In terms of pressure and volume*

$$\frac{p_1 V_1}{T_1} = \frac{p_2 V_2}{T_2}$$

$$\therefore \qquad \frac{T_2}{T_1} = \frac{p_2}{p_1} \times \frac{V_2}{V_1}$$

$$S_2 - S_1 = m Cv \ln\left(\frac{T_2}{T_1}\right) + mR \ln\left(\frac{V_2}{V_1}\right)$$

$$= m Cv \ln\left(\frac{p_2}{p_1} \times \frac{V_2}{V_1}\right) + m(Cp - Cv) \ln\left(\frac{V_2}{V_1}\right)$$

$$= m Cv \ln\left(\frac{p_2}{p_1}\right) + m Cv \ln\left(\frac{V_2}{V_1}\right) + m(Cp - Cv) \ln\left(\frac{V_2}{V_1}\right)$$

$$= m Cv \ln\left(\frac{p_2}{p_1}\right) + m Cv \ln\left(\frac{V_2}{V_1}\right) + m Cp \ln\left(\frac{V_2}{V_1}\right) - m Cv \ln\left(\frac{V_2}{V_1}\right)$$

$$S_2 - S_1 = m \left[Cv \ln \frac{p_2}{p_1} + Cp \ln \frac{V_2}{V_1} \right]$$

2.15.4 Entropy Change during Steam Generation

(i) 1 kg of water at temperature T_1 is heated to boiling temperature T_2. Heat supplied,

$$Q_{1-2} = h_f - h_1 = C_w(T_2 - T_1)$$

where, h_f = enthalpy of saturated water,

$$C_w = \text{specific heat of water}$$
$$= 4.186 \text{ kJ/kg-K}$$

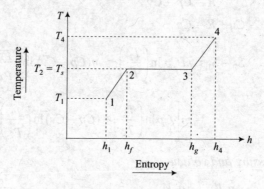

Fig. 2.39 *T-h* diagram

Change of entropy,

$$S_2 - S_1 = \int_{T_1}^{T_2} \frac{dQ}{T} = \int_{T_1}^{T_2} \frac{C_w \, dT}{T} = C_w \ln\left(\frac{T_2}{T_1}\right)$$

(ii) During process 2–3 there is evaporation of water. The temperature is constant at saturation temperature T_s.

$$S_3 - S_2 = \int_{2}^{3} \frac{dQ}{T} = \frac{Q_{2-3}}{T_S} = \frac{h_{fg}}{T_S}$$

(iii) During process 3–4, the steam is superheated. The heat added

$$dQ = mC_S \, dT$$

where, C_S = specific heat of steam
$$= 2.1 \text{ kJ/kg-K}$$

The entropy change,

$$S_4 - S_3 = \int_{T_S}^{T_4} \frac{dQ}{T} = m \int_{T_S}^{T_4} \frac{C_S \, dT}{T} = mC_S \ln\left(\frac{T_4}{T_S}\right)$$

2.15.5 Entropy Change of Thermal Reservoirs

A thermal reservoir (source or sink) has infinite capacity of heat energy and its temperature remains constant during heat exchange.

$$dS = -\frac{dQ}{T}$$

$$S_2 - S_1 = \int -\frac{dQ}{T} = -\frac{Q_{1-2}}{T}$$

2.15.6 Entropy Change of the Universe

The entropy change of the universe is the algebraic sum of the change of entropy of the system and change of entropy of the surrounding.

$$(\Delta S)_{universe} = (\Delta S)_{system} - (\Delta S)_{surrounding}$$

For a reversible process,

$$(\Delta S)_{system} = (\Delta S)_{surrounding}$$

But in practice, all processes are irreversible. The change of entropy of the system undergoing irreversible process will always be greater than change of entropy of the surrounding where the process is reversible.

$$(\Delta S)_{universe} \geq 0$$

The entropy of the universe goes on increasing. This is called *principle of increase of entropy.*

2.15.7 Available and Unavailable Heat Energy

The heat supplied Q_1 to a heat engine can be divided into two parts, the work done W and heat rejected Q_2. The maximum work done by an engine is called available heat energy (AHE) and the minimum heat rejected is called unavailable heat energy (UHE).

$$dQ = \text{AHE} + \text{UHE}$$

$$= \text{Work done} + \text{Heat rejected}$$

$$\eta = \frac{dW}{dQ} = \left(1 - \frac{T_2}{T_1}\right)$$

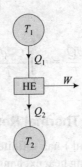

Fig. 2.40 Available energy

$$\therefore \qquad dW_{max} = dQ\left(1 - \frac{T_2}{T_1}\right)$$

But $T_2 = T_\infty$ = Temperature of surrounding for minimum heat rejection

$$\therefore \qquad dW_{max} = dQ\left(1 - \frac{T_\infty}{T_1}\right) = dQ - dQ \times \frac{T_\infty}{T_1}$$

$\therefore$ Unavailable heat energy = Heat rejected

$$= dQ\frac{T_\infty}{T_1} = T_\infty\frac{dQ}{T_1} = T_\infty dS$$

Change in entropy is a measure of unavailable heat energy or irreversibility of a process.

2.15.8   Some Important Points on Entropy

1. Entropy is a thermodynamic property of the system.
2. It is a point (state) function and not a path function.
3. It increases with the addition of heat and decreases with its removal.
4. The area under *T-s* diagram of any thermodynamic process represents the heat added or heat rejected.
5. Entropy remains constant in a reversible adiabatic process, but increases in an irreversible adiabatic process.
6. The change in entropy represents the maximum amount of work obtainable per degree drop in temperature.
7. The change in entropy is a measure of rate of availability (or unavailability) of heat for transformation into work.
8. Entropy is an extensive property of the system.
9. The specific entropy is expressed as kJ/kg-K.
10. Increase in entropy is obtained from a given quantity of heat at low temperature.

2.15.9 Important Mathematical Relations of Entropy

1. $dS = \dfrac{dQ}{T}$

2. $dQ = TdS$

3. $TdS = dU + pdV$

4. Unavailable heat energy, UHE $= T_\infty dS$.
   where, T_∞ = Ambient temperature

5. For reversible process, $dS = \dfrac{dQ}{T}$

6. For irreversible process, $dS > \dfrac{dQ}{T}$

7. For universe (isolated system), $ds \geq 0$
8. For reversible cyclic process, $ds = 0$
9. For irreversible cyclic process, $ds > 0$
10. For a constant pressure process,

$$S_2 - S_1 = mCp \ln \frac{V_2}{V_1} = mCp \ln \frac{T_2}{T_1}$$

Example 2.31: 10 m³ of air at 175°C and 5 bar is expanded to a pressure of 1 bar while temperature is 36°C. Calculate the entropy change for the process.

[U.P.T.U. II Sem., 2003-04]

Solution: *Initial conditions,*

$$V_1 = 10 \text{ m}^3$$
$$T_1 = 175°C + 273 = 448 \text{ K}$$
$$p_1 = 5 \text{ bar} = 500 \text{ kPa}$$

Assume R for air = 0.287 kJ/kg-K

$$p_1 V_1 = mRT_1$$

$$\therefore \quad m = \frac{p_1 V_1}{RT_1} = \frac{500 \times 10}{0.287 \times 448} = 38.8875 \text{ kg}$$

Final conditions,

$$p_2 = 1 \text{ bar} = 100 \text{ kPa}$$
$$T_2 = 36°C + 273 = 309 \text{ K}$$

$$S_2 - S_1 = m\left[C_p \ln\frac{T_2}{T_1} + R\ln\left(\frac{p_1}{p_2}\right)\right]$$

$$= 38.8875\left[1.005\ln\frac{309}{448} + 0.287\ln\left(\frac{500}{100}\right)\right]$$

$$= 38.8875(-0.3733 + 0.4619)$$

$$= \textbf{3.445 kJ} \quad \textbf{Ans.}$$

Example 2.32: Calculate the change in entropy during complete evaporation of water at 1 bar to dry saturated steam at the same pressure and temperature.

[U.P.T.U. (CO), 2005-06]

Solution:

$$S_2 - S_1 = \int\frac{dQ}{T} = m\int\frac{h_{fg}}{T_S}$$

From steam tables at 1 bar,

$$h_{fg} = 2258 \text{ kJ/kg}$$
$$T_S = 99.63°C + 273 = 372.63 \text{ K}$$

∴ $$s_2 - s_1 = \frac{2258}{372.63} = 6.0596 \text{ kJ/kg}$$

Example 2.33: 5 kg of ice at –10°C is kept in atmosphere which is at 30°C. Calculate the change in entropy of universe when it melts and comes into thermal equilibrium with atmosphere. Take latent heat of fusion as 335 kJ/kg and sp. heat of ice is half that of water.

[U.P.T.U. I Sem., 2005-06]

Solution:

$$m = 5 \text{ kg}$$
$$T_1 = -10°C + 273 = 263 \text{ K}$$
$$T_S = 0°C + 273 = 273 \text{ K}$$
$$T_2 = 30°C + 273 = 303 \text{ K}$$
$$h_{\text{fusion}} = 335 \text{ kJ/kg}$$

$$C_i = \frac{C_w}{2} = \frac{4.186}{2} = 2.093 \text{ kJ/kg-K}$$

The heat transfer takes place in two stages:

1. Sensible heating of ice.

2. Latent heat of fusion.

Total heat transfer, $\quad Q_{1-2} = mC_i(T_S - T_1) + mh_{fusion}$

$$= 5 \times 2.093(273 - 263) + 5 \times 335$$

$$= 104.65 + 1675$$

$$= 1779.65 \text{ kJ}$$

Change of entropy of the surrounding at constant temperature 303 K,

$$(\Delta S)_{surrounding} = -\frac{Q_{1-2}}{303} = \frac{-1779.65}{303} = -5.8734 \text{ kJ}$$

Change of entropy of the system,

$$= mC_i \ln\left(\frac{T_S}{T_1}\right) + \frac{mh_{fusion}}{T_S}$$

$$= 5 \times 2.093 \ln\left(\frac{273}{263}\right) + \frac{5 \times 335}{273}$$

$$= 0.39 + 6.1355$$

$$= 6.5255 \text{ kJ}$$

$$(\Delta S)_{universe} = (\Delta S)_{surrounding} + (\Delta S)_{system}$$

$$= -5.8734 + 6.5255 = \mathbf{0.652 \text{ kJ}} \quad \textbf{Ans.}$$

Example 2.34: 0.25 kg/s of water is heated from 30°C to 60°C by hot gases that enter at 180° and leave at 80°C. Calculate the mass flow rate of gases when its $Cp = 1.08$ kJ/kg-K. Find the entropy change of water and of hot gases. Take the specific heat of water as 4.186 kJ/kg-K. Also find the increase of unavailable energy if the ambient temperature is 27°C. [U.P.T.U. II Sem., 2002-03]

Solution:

Data on water,

$$\dot{m}_w = 0.25 \text{ kg/s}$$

$$T_{w1} = 30°C + 273 = 303 \text{ K}$$

$$T_{w2} = 60°C + 273 = 333 \text{ K}$$

$$C_w = 4.186 \text{ kJ/kg-K}$$

Data on gas

$$T_{g1} = 180°C + 273 = 453 \text{ K}$$

$$T_{g2} = 80°C + 273 = 353 \text{ K}$$

$$C_g = 1.08 \text{ kJ/kg-K}$$

$$T_\infty = 27°C + 273 = 300 \text{ K}$$

1. *Mass flow rate of gases*

   Heat gained by water = Heat lost by gases

   $$\dot{m}_w C_w (T_{w2} - T_{w1}) = \dot{m}_g C_g (T_{g1} - T_{g2})$$

   $$0.25 \times 4.186 (333 - 303) = \dot{m}_g \times 1.08 (453 - 353)$$

   $$\dot{m}_g = \frac{0.25 \times 4.186 (333 - 303)}{1.08 (453 - 353)} = 0.29 \text{ kg/s}$$

2. *Increase of entropy of water,*

   Water is heated at constant pressure

   $$S_2 - S_1 = m_w C_w \ln \frac{T_{w2}}{T_{w1}} = 0.25 \times 4.186 \ln \left(\frac{333}{303} \right)$$

   $$= 0.0988 \text{ kJ}$$

3. *Decrease in entropy of gas,*

   $$S_2 - S_1 = m_g C_g \ln \frac{T_{g2}}{T_{g1}}$$

   $$= 0.29 \times 1.08 \ln \left(\frac{353}{453} \right)$$

   $$= -0.078 \text{ kJ}$$

4. *Loss of available energy of gas,*

   $$\text{UHE} = T_\infty \, dS = 300 \times 0.078 = \mathbf{23.436 \text{ kJ}} \quad \mathbf{Ans.}$$

QUESTION BANK NO. 2

1. What are different laws of thermodynamics? What is their significance and scope?
2. State Zeroth law of thermodynamics and explain its importance in measurement of temperature.
3. Explain Clausius inequality and prove it.
4. State the 1st law of thermodynamics for a closed system undergoing a change of state. Also show that the total energy is a property of the system.
5. Show that entropy is a point function.
6. Describe Carnot cycle for an ideal gas and derive the expression of efficiency of the cycle.
7. Derive general energy equation for steady flow.
8. Derive the formula used for change of entropy of an ideal gas when its state changes from 1 to 2.

9. Classify different types of thermodynamic processes. Compare the work done in non-flow and flow processes.

10. Prove that work done for a steady flow adiabatic process is given by:

$$W_{1-2} = \frac{\gamma}{\gamma - 1} (p_1 V_1 - p_2 V_2)$$

11. Estimate the work done for an isothermal process.

12. Define engine efficiency and COP of a heat pump and a refrigerator. Prove that COP of heat pump is more than COP of refrigerator.

13. State and explain the second law of thermodynamics.

14. Write short note on "Equivalence of Kelvin-Planck and Clausius statement of second law of thermodynamics".

15. Explain Carnot cycle on p-V and T-s diagrams.

16. State and prove Carnot theorem.

17. Explain available and unavailable energy.

18. Show that the entropy change in a process when a perfect gas changes form state 1 to state 2 is given by:

$$s_2 - s_1 = Cp \ln \frac{T_2}{T_1} + R \ln \frac{p_2}{p_1}.$$

19. Are the following statements true or false?

  (i) The process that violates the second law also violates the first law of thermodynamics.

  (ii) No process in a closed system is possible in which it exchanges heat with a single reservoir and produces an equal amount of work.

  (iii) The change in entropy of a closed system must be greater than or equal to zero.

  (iv) The entropy of an isolated system always increases.

  (v) Entropy increases in the process of mixing hot water with cold water.

20. Show that the work, W in non-flow processe for polytropic process, i.e., pV^n = constant, is given by:

$$W = (p_1 V_1 - p_2 V_2)/n - 1.$$

Properties of Steam

3.1 INTRODUCTION

When water is heated in a boiler or steam generator at constant pressure, its heat content called enthalpy increases and its physical state changes. With the addition of heat, the temperature of water rises and vaporization takes place. The vapour so formed is called steam. Steam does not obey laws of perfect gases. However, in a highly superheated state, its behaviour approaches that of a perfect gas.

The different stages of steam generation are:

1. **Compressed and subcooled water:** Water (feedwater) is pumped to boiler pressure in the state of below saturation temperature. Feedwater is called compressed and subcooled water.

2. **Saturated water:** The subcooled water is heated upto saturation temperature in the economizer, a part of steam generator Saturation temperature is the temperature at which it starts to evaporate. It is a function of water pressure.

3. **Saturated steam:** The saturated water is evaporated into saturated steam in the boiler part of steam generator. Saturated steam is the steam in a state where it starts condensing on cooling.

4. **Wet steam:** It is a mixture of dry steam and water particles. Steam passes through a state of wet steam between saturated water and saturated steam.

5. **Superheated steam:** The dry and saturated steam can be heated to higher temperatures in the superheater. Steam above saturation temperature is called superheated steam.

3.2 APPLICATIONS OF STEAM

Steam is perhaps the most important working fluid used in the industry. The various applications of steam are:

1. Process heating
2. Space heating

3. Thermo-compression and vacuum generation
4. Distillation and stripping
5. Cleaning and fire-fighting
6. Displacement of liquids and gases
7. Refrigeration
8. Steam jet liquid pumping
9. Evaporation
10. Steam reforming
11. Steam locomotion and marine propulsion
12. Power generation in thermal, nuclear and geothermal power plants.

Steam finds applications in many types of industries. Some of these industries are:

1. Central power plants (thermal and nuclear)
2. Industrial power plants (combined-cycle and co-generation plants)
3. Chemical and petrochemical industries
4. Oil industries (refineries)
5. Sugar industries
6. Dairy plants
7. Distilleries
8. Fertilizer plants
9. Textile industries
10. Pharmaceutical plants
11. Food processing industries

3.3 STEAM GENERATION

Steam is generated by heating water at constant pressure in a boiler. Water passes through different states and same is shown in various property diagrams.

Process $A - B$: Feed water at pressure p_A temperature t_A is heated. The temperature will go on increasing till boiling temperature t_B is reached. This process is called sensible heating of water. The boiling point is called saturation temperature, t_s. The saturation temperature varies with pressure of water. The specific heat of water is 4.186 kJ/kg-K. At $p_s = 1.01325$ bar, $t_s = 100°C$.

Enthalpy change,

$$(dh)_{AB} = C \, dt = 4.186 \, (t_B - t_A)$$

Process $B - C$: Water is evaporated and a two-phase mixture of liquid water and vapour is formed. There is no increase in temperature and heat is utilized to break up molecular bonds of water.

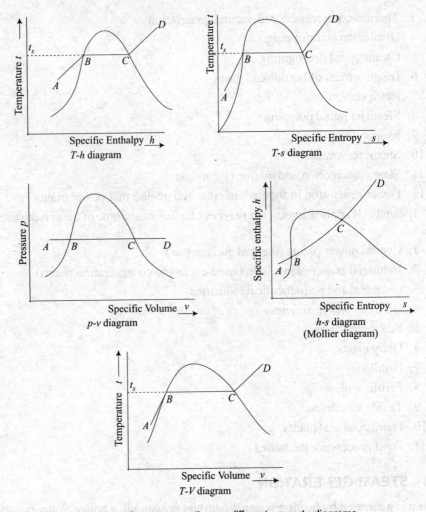

Fig. 3.1 Steam generation on different property diagrams

$$(dh)_{BC} = \text{latent heat of vaporization}$$
$$= h_C - h_B = h_g - h_f = h_{fg}$$

where, h_f = specific enthalpy of saturated water (kJ/kg)

h_g = specific enthalpy of saturated steam (kJ/kg)

h_{fg} = latent heat or specific enthalpy of vaporization (kJ/kg)

The mixture of water and steam is called wet steam.

Process C – D: The steam first becomes completely dry and at point *C*, it is called dry and saturated steam. The temperature of steam increases on further heating and the process is called superheating.

$$(dh)_{CD} = C_s \, dt = 2.1 \, (t_D - t_C)$$
$$C_s = 2.1 \text{ kJ/kg-K} = \text{specific heat of steam.}$$

3.4 IMPORTANT CONSTANTS

1. Triple point: Ice, water and water vapour coexist at the triple point of water. It is the point of intersection of sublimation line, melting line and vaporization line on *p-t* diagram. On *p – v* diagram, it is represented by triple phase line as shown in Fig. 3.3.

The triple point for water:

Pressure, $\qquad p = 0.611$ kPa

Temperature, $\qquad T = 273.16$ K or 0.01°C

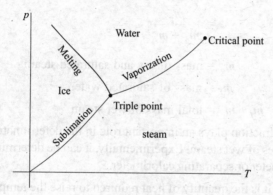

Fig. 3.2 Triple point

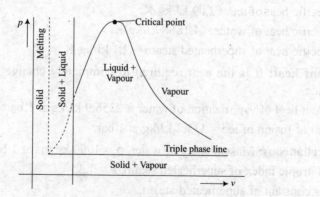

Fig. 3.3 Triple phase line

2. Critical point: At high pressure (critical pressure), saturated liquid state and saturated vapour state become identical and take the form of a point. This is called critical point. At this point, the latent heat of vaporization becomes zero and water directly flashes to vapours.

The critical point for water:

Pressure, $p_{cr} = 22.09$ MPa

Temperature, $t_{cr} = 274.14°C$

Specific volume, $v_{cr} = 0.003155$ m³/kg

The steam at pressure and temperature above critical point is called supercritical steam.

3. Dryness fraction: The wet steam is a mixture of dry steam and saturated water. The vapour fraction in the wet steam is called dryness fraction (X).

$$X = \frac{\text{Mass of actual dry steam in the mixture}}{\text{Total mass of wet steam}}$$

$$= \frac{m_g}{m_g + m_f}$$

where m_g = mass of dry and saturated steam

m_f = mass of saturated water

$m_f + m_g$ = total mass of wet steam

The dryness fraction plays an important role in the determination of thermodynamic properties of wet steam. Experimentally, it can be determined by using a throttling calorimeter or separating calorimeter.

4. Specific heat: It is the quantity of heat required to raise the temperature of one kilogram of a substance through one degree

Specific heat of ice = 2.09 kJ/kg-K.

Specific heat of water = 4.1868 kJ/kg-K.

Specific heat of superheated steam = 2.1 kJ/kg-K

5. Latent heat: It is the heat required to completely change the phase of a substance.

Latent heat of vaporization of water = 2256.9 kJ/kg at 1 bar.

Heat of fusion of ice = 335.7 kJ/kg at 1 bar.

6. Miscellaneous: Mass density of water, $\rho = 1000$ kg/m³ at 1 bar.

Polytropic index of superheated steam, $n = 1.3$

Gas constant of superheated steam,

$$R = \frac{\bar{R}}{M} = \frac{8.3143}{18}$$

$$= 0.462 \text{ kJ/kg-K}.$$

3.5 THERMODYNAMIC PROPERTIES OF STEAM

Important thermodynamic properties of steam along with their symbols and units are given in Table 3.1.

Table 3.1 Thermodynamic properties of steam

S. No.	Property	Symbol	Unit
1.	Absolute pressure	p	bar
2.	Temperature	t	°C
3.	Saturation temperature	t_s	°C
4.	Specific volume of saturated water	v_f	m³/kg
5.	Specific volume of saturated steam	v_g	m³/kg
6.	Specific enthalpy of saturated water	h_f	kJ/kg
7.	Specific enthalpy of saturated steam	h_g	kJ/kg
8.	Specific enthalpy of vaporization ($h_g - h_f$)	h_{fg}	kJ/kg
9.	Specific entropy of saturated water	s_f	kJ/kg K
10.	Specific entropy of saturated steam	s_g	kJ/kg K
11.	Specific entropy of vaporization ($s_g - s_f$)	s_{fg}	kJ/kg K
12.	Specific internal energy	u	kJ/kg

The properties of steam are required for the design of plant and equipment and study and analysis of various processes based on steam. These properties based on extensive study by different scientists are available in steam tables and Mollier diagram. Steam tables have been prepared for different states of water and steam.

1. Compressed and subcooled water: The properties of water at high pressure with temperature below saturation point are available in separate tables. The values v, u, h, s are recorded as a function of pressure and temperature.

2. Saturated water and saturated steam: The properties of saturated water v_f, h_f, and s_f are available in tables with varying temperatures and varying pressures. Similarly, in the same tables properties of saturated steam v_g, h_g, h_{fg}, s_g and s_{fg} are available.

The values of internal energy can be calculated as follows:

Internal energy of saturated water,

$$u_f = h_f - pv_f \times 10^2 \ [\text{kJ/kg}]$$

where, p is in bar.

Internal energy of saturated steam,

$$u_g = h_g - pv_g \times 10^2 \ (\text{kJ/kg})$$

3. Properties of wet steam: The properties of wet steam with dryness fraction, X can be calculated by using steam tables for saturated water and saturated steam.

(i) Specific volume

$$v = x\,v_g + (1 - x)\,v_f$$
$$= v_f + x\,(v_g - v_f)$$

The values of v_g and v_f are read from steam tables.

(ii) Specific enthalpy,

$$h = x\,h_g + (1 - x)\,h_f$$
$$= h_f + x\,(h_g - h_f)$$
$$= h_f + x\,h_{fg}.$$

The values of h_g and h_{fg} are read from steam tables.

(iii) Specific entropy,

$$s = xs_g + (1 - x)s_f$$
$$= s_f + x\,s_{fg}$$

The values of s_g and s_{fg} are read from steam tables.

(iv) Specific internal energy,

$$v = h - pv \times 10^2$$

where, h = specific enthalpy of wet steam

$$= h_f + xh_{fg}$$
$$v = \text{specific valume of wet steam}$$
$$= v_f + x\,(v_g - v_f)$$
$$p = \text{pressure of steam in bar.}$$

4. Superheated steam: All power plants and industries use superheated steam for the following advantages:

   i. More work is obtained per kg of steam without increasing pressure.

   ii. The plant or process efficiency increases due to higher temperature of steam.

   iii. There is less corrosion and erosion of equipment due to absence of moisture in the steam.

   iv. Superheating is mostly done from waste heat of boiler without additional cost of fuel.

The properties of steam can be found out:

(1) From tables for superheated steam

(2) Mollier diagram

(3) From tables for saturated steam.

    (i) Specific volume of superheated steam at pressure p and temperature T

$$v = v_g \times \frac{T}{T_s}\ (\text{m}^3/\text{kg})$$

$T_s = t_s + 273$ = saturated temperature of steam (K)

v_g can be read from steam tables.

(ii) Specific enthalpy of superheated steam at pressure p and temperature t.

$$h = h_g + C_s (T - T_s)$$

where C_s = specific heat of superheated steam

= 2.1 kJ/kg-K or as given

h_g can be read from steam tables for saturated steam.

(iii) Entropy of superheated steam at pressure p and temperature t,

$$s = s_g + C_s \ln \left(\frac{T}{T_s} \right)$$

(iv) Internal energy of superheated steam at pressure p and temperature t.

$$v = h - pv \times 10^2 \text{ (kJ/kg)}$$

where, h = specific enthalpy of superheated steam

p = pressure in bar

v = specific volume of superheated steam

5. Supercritical steam: The thermodynamic properties of supercritical steam at parameters above critical point (221.2 bar and 374.15°C) are read from steam tables for supercritical steam for varying pressure and temperature.

6. Linear interpolation of properties: The steam tables are sufficiently elaborate, the values of properties are given for each degree of change in temperature and 1 bar increase in pressure in saturated steam tables. But in case of superheated steam tables, values for a given pressure and temperature may not be given. The principle of linear interpolation can be used.

Example 3.1:   Find the specific enthalpy of superheated steam for a pressure of 150 bar and a temperature 520°C.

Solution: Find the specific enthalpy of steam at 150 bar and 520°C by using steam tables for superheated steam. Referring to steam table, specific enthalpy is not available at 520°C.

Read the following values:

Pressure	Temperature	Specific enthalpy
150 bar	500°C	3308.6 kJ/kg.
150 bar	550°C	3448.6 kJ/kg.
Subtract	50°C	140.0 kJ/kg
For temperature difference	20°C	$\frac{140.0}{50} \times 20 = 56$ kJ/kg

Add to the value for 500°C   500 + 20°C = 520°C   3308.6 + 56.0

$$= 3364 \text{ kJ/kg}$$

The specific enthalpy for steam parameters of 150 bar and 520°C is 3364.6 kJ/kg.

Note. Similarly, values can be found out by double interpolation for steam parameter of 155 bar and 520°C. Sometimes, it is more convenient to use the procedure given in pare (4) of # 3.5.

7. Summary: The important properties of wet steam and superheated steam along with the governing equations are listed in Table 3.2.

Table 3.2   Important steam properties and calculation formulae

S. No.	Properties	Wet steam	Superheated steam
1.	Specific volume (m³/kg)	$^{x}v = (1 - x)\, v_f$ $+ {}^{x}v_g \approx xv_g$	$v^{xx} = v_g \times \dfrac{T}{T_{\text{sat}}}$
2.	Specific enthalpy (kJ/kg)	$h^{x} = h_f + h_{fg}$	$h^{xx} = h_g + C_s\,(T - T_{\text{sat}})$ $[C_s = 2.1 \text{ to } 2.3 \text{ kJ/kg-K}]$
3.	Specific entropy (kJ/kg)	$s = s_f + x\, s_{fg}$	$s = s_g + C_s \ln \dfrac{T}{T_{\text{sat}}}$
4.	Specific internal energy (kJ/kg)	$u = h^{x} - pv^{x} \times 10^2$	$v = h^{xx} - pv^{xx} \times 10^2$

1. The values v_f, h_f, s_g, h_{fg}, s_{fg} and T_{sat} can be read from steam tables for saturated steam.

2. The pressure p is given in bar.

Example 3.2:   Calculate the enthalpy of 1 kg of steam at a pressure of 8 bar and dryness fraction of 0.8.

Solution:

$$p = 8 \text{ bar}$$
$$x = 0.8$$

From steam tables of saturated steam, read the following against a pressure of 8 bar:

$$h_f = 721.1 \text{ kJ/kg}$$
$$h_{fg} = 2048.04 \text{ kJ/kg}$$

Therefore, specific enthalpy of steam

$$h = h_f + x\, h_{fg}$$
$$= 721.1 + 0.8\,(2048.04)$$
$$= \mathbf{2359.5 \text{ kJ/kg.}}$$

Example 3.3:   Calculate the specific volume of steam at 250°C and dryness fraction of 0.8.

Solution: $t = 250°C$

$x = 0.8$

From steam tables of saturated steam (temperature varying), read the following values against $t = 250°C$

$$v_f = 0.001251 \text{ m}^3/\text{kg}$$
$$v_g = 0.05013 \text{ m}^3/\text{kg}$$

The specific volume of wet steam,

$$v = (1 - x) v_f + x v_g$$
$$= 0.2 \times 0.001251 + 0.8 \times 0.05013$$
$$= \textbf{0.0403542 m}^3\textbf{/kg} \quad \textbf{Ans.}$$

Example 3.4: Calculate the enthalpy of steam at 30 bar if

(a) its dryness fraction is 0.75

(b) it is superheated at 400°C

Solution:

(a) $\qquad\qquad p = 30 \text{ bar}$

$x = 0.75$

From steam tables for saturated steam (pressure varying) read the following against a pressure of 30 bar,

$$h_f = 1008.\ 41 \text{ kJ/kg}$$
$$h_{fg} = 1795.73 \text{ kJ/kg.}$$

Specific enthalpy for wet steam,

$$h = h_f + x\, h_{fg}$$
$$= 1008.41 + 0.75\ (1795.73)$$
$$= \textbf{2355.2 kJ/kg} \quad \textbf{Ans.}$$

(b) $\qquad\qquad p = 30 \text{ bar}$

$t = 400°C$

From steam tables at 30 bar,

$$t_s = 233.9°C$$
$$h_g = 2804.14 \text{ kJ/kg.}$$

The specific enthalpy for superheated steam,

$$h = h_g + C_s(t - t_{sat})$$

Assume $\qquad\qquad C_s = 2.3 \text{ kJ/kg-K.}$

$\therefore \qquad\qquad h = 2804.14 + 2.3\ (400 - 233.9)$
$$= \textbf{3186.17 kJ/kg} \quad \textbf{Ans.}$$

Example 3.5: Determine the enthalpy, volume, internal energy and entropy of superheated steam at 15 bar pressure and a temperature of 220°C. The volume of water may be neglected. Take specific heat of superheated steam equal to 2.2 kJ/kg-K.

Solution:
$$p = 15 \text{ bar}$$
$$t = 220°C.$$

From steam tables for saturated steam, read the following against a pressure of 15 bar,

$$h_f = 844.7 \text{ kJ/kg}$$
$$h_{fg} = 1945.3 \text{ kJ/kg}$$
$$t_s = 198.3°C$$
$$v_g = 0.13167 \text{ m}^3/\text{kg}$$
$$s_f = 2.314 \text{ kJ/kg-K}$$
$$s_{fg} = 4.127 \text{ kJ/kg-K}$$

(i) specific enthalpy of superheated steam,

$$h = h_f + h_{fg} + C_s (t - t_s)$$
$$= 844.7 + 1945.3 + 2.2 (220 - 198.3)$$
$$= 2837.54 \text{ kJ/kg}.$$

(ii) specific volume of superheated steam,

$$v = v_g \frac{T}{T_s} = 0.13167 \frac{(220 \times 273)}{(198.3 + 273)}$$

$$= 0.1377 \text{ m}^3/\text{kg}.$$

(iii) internal energy,

$$u = h - pv \times 10^2$$
$$= 2837.54 - 15 \times 10^2 \times 0.1377$$
$$= 2630.94 \text{ kJ/kg}$$

(iv) specific entropy of superheated steam,

$$s = s_f + s_{fg} + C_s \text{ lsn } \frac{T}{T_s}$$

$$= 2.314 + 4.127 + 2.2 \text{ lsn } \frac{(220 \times 273)}{(273 + 198.3)}$$

$$= 6.84 \text{ kJ/kg-K}.$$

Example 3.6: A sample of wet steam exists at 5 bar and possesses, dryness fraction of 0.98. With the help of steam tables, determine its temperature, enthalpy and specific volume.   [U.P.T.U. II Sem, 2003-04]

Solution: $p = 5$ bar

$x = 0.98$

From steam tables for saturated steam (pressure varying) read the following properties against a pressure of 5 bar.

$$t_s = 151.8°C$$
$$v_f = 0.001093 \text{ m}^3/\text{kJ}$$
$$v_g = 0.37466 \text{ m}^3/\text{kJ}$$
$$h_f = 640.1 \text{ kJ/kg}$$
$$h_{fg} = 2107.4 \text{ kJ/kg}.$$

(i) The steam temperature, $t_s = 151.8°C$.

(ii) The specific enthalpy of wet steam,

$$h = h_f + x h_{fg} = 640.1 + 0.98(2107.4)$$
$$= 2705.35 \text{ kJ/kg}.$$

(iii) Specific volume of wet steam

$$v = (1-x)\, v_F + x\, v_g = (1-0.98)$$
$$0.001093 + 0.98\,(0.37466)$$
$$= 0.36719 \text{ m}^3/\text{kg}$$

Example 3.7: A sample of superheated steam exists at 8 bar and a temperature of 280°C. Using steam tables, find the enthalpy, specific volume and entropy of this sample on per kg basis.   [U.P.T.U. I Sem. 2003-04]

Solution: $p = 8$ bar

$$t = 280°C$$
$$T = 280 + 273 = 553 \text{ K.}$$

Using steam tables for saturated steam, read the following properties against a pressure of 8 bar.

$$v_g = 0.24026 \text{ m}^3/\text{kg}$$
$$t_s = 170.4°C \quad \therefore\ T_s = 170.4 + 273 = 443.4 \text{ K.}$$
$$h_g = 2767.4 \text{ kJ/kg.}$$
$$s_g = 6.66 \text{ kJ/kg.}$$

Assume specific heat of superheated steam,

$$C_s = 2.2 \text{ kJ/kg K.}$$

(i) The specific enthalpy of superheated steam,

$$h = h_g + C_s\,(T - T_s)$$
$$= 2767.4 + 2.2\,(553 - 443.4)$$
$$= 3008.52 \text{ kJ/kg.}$$

(ii) The specific volume of superheated steam,

$$v = v_g \frac{T}{T_s}$$

$$= 0.24026 \times \frac{553}{443.4} = 0.2996 \text{ m}^3/\text{kg}$$

(iii) The specific entropy of superheated steam

$$s = s_g + C_s \ln \frac{T}{T_s}$$

$$= 6.66 + 2.2 \ln 553/443.3$$

$$= 7.146 \text{ kJ/kg-K.}$$

The above-mentioned values can also be directly read from steam tables for superheated steam. As values for 8 bar and 280°C are not directly available in the tables the principle of linear interpolation can be used. Read the values for 8 bar and temperature of 250°C and 300°C.

Pressure	$p = 8$ bar			
Temperature	250°C	300°C	$\Delta t = 50°C$	280°C
1. Specific volume (m³/kg)	0.2932	0.3241	0.0309	0.2932

$$\times \frac{0.0309 \times 30}{50} = 0.31174$$

2. Specific enthalpy (kJ/kg)	2950.4	3057.3	106.9	2950.4

$$\times \frac{106.9 \times 30}{50} = 3014.54$$

3. Specific entropy (kJ/kg-K)	7.040	7.235	0.195	7.040

$$\times \frac{0.195 \times 30}{50} = 7.1757$$

Example 3.8: Using steam tables, determine the volume, enthalpy and internal energy on per kg basis for steam at 12 bar and 0.95 dryness fraction. Also find its temperature. [U.P.T.U.II Sem, 2002-2003]

Solution:

Pressure, $p = 12$ bar

Dryness fraction, $x = 0.95$

Using steam tables of saturated steam (pressure varying) read the following values,

$$t_s = 188°C$$

$$v_f = 0.001139 \text{ m}^3/\text{kg}$$

$$v_g = 0.16321 \text{ m}^3/\text{kg}$$
$$h_f = 798.4 \text{ kJ/kg.}$$
$$h_{fg} = 1984.3 \text{ kJ/kg}$$

(i) The temperature of steam $= t_s = 188°C$

(ii) The specific volume of wet steam,

$$v = (1 - x) \, v_f + x.v_g$$
$$= (1 - 0.95) \, 0.001137 + 0.95 \times 0.16321$$
$$= 0.1551 \text{ m}^3/\text{kg.}$$

The specific enthalpy of wet steam,

$$h = h_f + x \, h_{fg} = 798.4 + 0.95 \, (1984.3)$$
$$= \mathbf{2683.485 \text{ kJ/kg Ans.}}$$

Example 3.9: 5 kg of steam is generated at a pressure of 10 kg bar from feedwater at a temperature of 25°C. Starting from the basic principles and taking the help of steam tables only, calculate the enthalpy and entropy of steam, if

(i) steam is dry and saturated

(ii) steam is superheated upto a temperature of 300°C. Take C_p for steam as 2.1 kJ/kg-K and C_p for water as 4.187 kJ/kg-K.

Solution:

(i) Mass of steam $\qquad\qquad m = 5$ kg

$$p = 10 \text{ bar}$$

Final condition = steam dry and saturated.

Steam tables are prepared with datum temperature at 0°C.

Enthalpy of water at $\qquad 25°C = m \, Cp_w \, (t - 0)$
$$= 5 \times 4.187 \times 25$$
$$= 523.375 \text{ kJ}$$

Entropy of water at $\qquad 25°C = m \, Cp_w \ln \dfrac{T_2}{T_1}$

$$= 5 \times 4.187 \ln \dfrac{273 + 25}{273}$$

$$= 1.83436 \text{ kJ/kg-K.}$$

The effect of pressure is neglected.

Enthalpy of saturated water $h_f = 762.8$ kJ/kg at 10 bar from steam tables.

Entropy of saturated water $s_f = 2.139$ kJ/kg-K.

For dry and saturated steam at 10 bar, $h_g = 2778.1$ kJ/kg

∴ Enthalpy of saturated steam $= m\, h_g$

$$= 5 \times 2778.1$$
$$= 1389.5 \text{ kJ}$$

Entropy of dry saturated steam at 10 bar $= m\, s_g = 5 \times 6.587$
$$= 32.935 \text{ kJ/K}.$$

(ii) Specific enthalpy of superheated steam,
$$h = h_g + C_s\,(t - t_s)$$

From steam tables at 10 bar, $h_g = 2778.1$ kJ/kg

$$t_s = 179.7°C$$

∴
$$h = 2778.1 + 2.1(300 - 179.7)$$
$$= 3030.73 \text{ kJ/kg}$$

Total enthalpy $= m h = 5 \times 3030.73 = 15153.65$ kJ.

Entropy of superheated steam

$$s = s_g + C_s \ln \frac{T}{T_s} = 6.587 + 2.1 \ln \frac{300 + 273}{179.7 + 273}$$

$$= 7.0818777 \text{ kJ/kg-K}$$

Total entropy $= ms = 35.41$ kJ/K

Example 3.10: Determine the state of steam, i.e., whether it is wet, dry or superheated in the following cases:

(i) Pressure 10 bar and specific volume 0.175 m³/kg.

(ii) Pressure 15 bar and temperature 220°C.

Solution:

(i)
$$p = 10 \text{ bar}$$
$$v = 0.175 \text{ m}^3/\text{kg}.$$

From steam tables v_g at 10 bar,
$$v_g = 0.1993 \text{ m}^3/\text{kg}$$

The steam is wet as $v_g > v$

$$v = (1 - x)\, v_g + x\, v_g$$
$$\approx x\, vg$$

∴
$$x = \frac{v}{vg} = \frac{0.175}{0.1993} = 0.9.$$

(ii)
$$p = 15 \text{ bar}$$
$$t = 220°C$$

From steam tables, saturated temperature

$$t_s = 198.3°C \text{ against a pressure of 15 bar}$$
$$t > t_s$$

$\therefore$ Steam is superheated.

Degree of superheated $= t - t_s = 220 - 198.3$

$= 21.7°C.$

Example 3.11: A mixture of wet steam has 1.25 kg of water in suspension with 40 kg of steam. Calculate the dryness fraction of steam.

Solution:

$$m_f = 1.25 \text{ kg}$$
$$m_g = 40 \text{ kg}$$

$$x = \frac{m_g}{m_f + m_g} = \frac{40}{1.25 + 40} = 0.9.$$

Example 3.12: Total heat of 2700 kJ/kg was supplied to generate steam from 0°C and 20 bar. Find the final state of steam.

Solution:

From steam tables at 20 bar,

$$h_g = 2797.2 \text{ kJ/kg}$$
$$h_f = 908.6 \text{ kJ/kg}$$
$$h_{fg} = 1888.6 \text{ kJ/kg}$$

$\because$ $h < h_g$, the steam is wet

$$h = h_f + x h_{fg}$$

$\therefore$

$$x = \frac{h - h_f}{h_{fg}} = \frac{2700 - 908.6}{1888.6} = 0.948$$

Example 3.13: A sample of steam at 5 bar has an enthalpy of 2350 kJ/kg. Find the specific volume, internal energy and entropy of this sample of steam.

Solution:

$$p = 5 \text{ bar}$$
$$h = 2350 \text{ kJ/kg.}$$

From steam tables, for a pressure of 5 bar

$$h_f = 640.1 \text{ kJ/kg}$$
$$h_{fg} = 2107.4 \text{ kJ/kg}$$
$$h_g = 2747.5 \text{ kJ/kg}$$
$$v_g = 0.0375 \text{ kJ/kg}$$
$$s_f = 1.8604 \text{ kJ/kg-K}$$
$$s_{fg} = 4.9588 \text{ kJ/kg-K}$$
$$h < h_g.$$

The steam is wet.

$$h = h_f + xh_{fg}$$

$$\therefore \quad x = \frac{h - h_t}{h_{fg}} = \frac{2350 - 640.1}{2107.4} = 0.811.$$

(i) Specific volume of wet steam,

$$v = xv_g = 0.811 \times 0.0375 = 0.3041 \text{ m}^3/\text{kg}$$

(ii) Internal energy of wet steam

$$u = h - pv \times 10^2 = 2350 - 5 \times 10^2 \times 0.3041 = 2197.95 \text{ kJ/kg}$$

(iii) Specific entropy of wet steam

$$s = s_f + xs_{fg} = 1.8604 + 0.811 \times 4.9586$$
$$= 5.882 \text{ kJ/kg-K}.$$

Example 3.14: Determine the mean specific heat for superheated steam at 1 bar between temperatures 150°C and 200°C.

Solution:

From superheated steam tables, specific enthalpy at 1 bar and 150°C,

$$h = 2776.4 \text{ kJ/kg}$$

Specific enthalpy at 1 bar and 200°C,

$$h = 2875.3 \text{ kJ/kg}$$

Now, specific enthalpy of superheated steam,

$$h = h_g + C_s (t - t_s)$$

At 1 bar and 150°C $\quad h_1 = h_{g1} + C_s (t_1 - t_s)$

At 1 bar and 200°C $\quad h_2 = h_{g2} + C_s (t_2 - t_s)$

But, $\qquad\qquad h_{g1} = h_{g2}$ at 1 bar.

$$\therefore \quad (h_2 - h_1) = C_s (t_2 - t_1)$$
$$2875.3 - 2776.4 = C_s (200 - 150)$$

$$\therefore \quad C_s = \frac{2875.3 - 2776.4}{50}$$

$$= 1.978 \text{ kJ/kg-K}$$

Example 3.15: A spherical shell of 30 cm in radius contains a mixture of saturated steam and water at 300°C. Water and steam have equal volume. Calculate the mass of each constituent.

Solution:

$$\text{Total volume of shell} = \frac{4}{3} \Pi r^3 \quad \frac{4}{3} \Pi (0.3)^3 = 0.113 \text{ m}^3$$

Volume of steam = volume of water = $\dfrac{0.113}{2}$ = 0.0565 m^3

From steam tables for a temperature of 300°C.

Mass of steam, $m_g = \dfrac{V}{v_g} = \dfrac{0.0565}{0.0216} = 2.616$ kg

Mass of water, $m_f = \dfrac{V}{v_f} = \dfrac{0.0565}{0.001404} = 40.246$ kg

Example 3.16: A vessel of 0.3 m^3 capacity contains 1.5 kg of mixture of saturated water and steam at a pressure of 5 bar. Calculate:

(a) Volume and mass of water, and

(b) Volume and mass of steam.

Solution:

From steam tables at a pressure of 5 bar,
$$v_f = 0.001093 \text{ m}^3/\text{kg}$$
$$v_g = 0.375 \text{ m}^3/\text{kg}$$

Specific volume of mixture of water and steam,
$$v = \frac{\text{Volume of vessel}}{\text{Mass of mixure}} = \frac{0.3}{1.5} = 0.2 \text{ m}^3/\text{kg}$$

Now $v < v_g$. Steam is wet steam,
$$v = (1 - x)v_f + xv_g$$

∴ $$x = \frac{v - v_f}{v_g - v_f} = \frac{0.2 - 0.001093}{0.375 - 0.001093} = 0.532$$

Also, $$x = \frac{m_g}{m_f + m_g}$$

∴
$$m_g = x(m_g + m_f) = 0.532 \times 1.5 = 0.798 \text{ kg}$$
$$m_f = (m_g + m_f) - m_g = 1.5 - 0.798 = 0.702 \text{ kg}$$
Volume of steam = $m_g\, v_g$ = 0.798 × 0.375 = 0.2992 m^3
Volume of liquid = $m_f\, v_f$ = 0.702 × 0.001093 = 0.00135 m^3.

3.6 MOLLIER DIAGRAM

The work done during a process can be found out by the area under the curve on a p-v diagram. Similarly, the heat supplied during a process can be found out by the area under the process curve on a *T-s* diagram. In 1904, Dr. Mollier developed Mollier diagram (*h-s* diagram) where the work done and heat supplied can be measured directly as a length of process line. It is also widely used to obtain steam properties for various engineering devices and processes. The engineering devices can be nozzles, turbines, steam compressors, condensers, boilers, throttling calorimeter, etc.

The important processes involved in steam engineering are:

1. Heating and cooling at constant volume in vessels.
2. Heating at constant pressure in economizers, boilers and superheaters.
3. Cooling at constant pressure in condensers.
4. Isentropic (or irreversible) expansion of steam in nozzles, turbines, etc.
5. Throttling of steam in calorimeters, valves, pipes, etc.

A skeleton diagram of Mollier chart is shown in Fig. 3.4.

1. The constant enthalpy lines are horizontal and constant entropy lines are vertical.
2. The constant temperature lines are straight in wet steam region and horizontal in superheated steam region.
3. The constant pressure lines do not change shape in wet steam region and superheated steam region.
4. The constant volume lines are shown broken lines in wet steam region and superheated steam region. The slope of constant specific volume lines is more than that of pressure lines.
5. Constant dryness fraction lines or quality lines (*x*) are also shown in wet steam region.

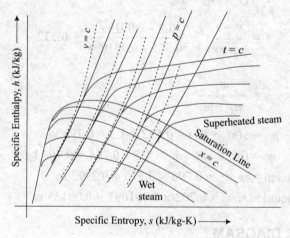

Fig. 3.4 Mollier diagram

Example 3.17: Show the following processes on a skeleton Mollier diagram:
   (i) Isentropic expansion
   (ii) Adiabatic expansion with friction
   (iii) Throttling process
   (iv) Isenthalpic process
   (v) Constant pressure heat addition.   [U.P.T.U. II Sem. 2001-02]

Solution: The Skeleton Mollier diagram is shown,

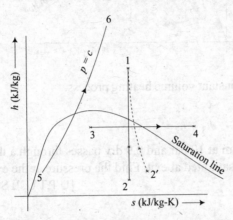

   (i) Vertical line 1–2 is an isentropic expansion line with $s_1 = s_2$
   (ii) Dotted line 1–2′ is adiabatic expansion line with friction $s_2' > s_1$
   (iii) Horizontal line 3–4 is a throttling process. $h_3 = h_4$
   (iv) Horizontal line 3–4 is also isenthalpic process $h_3 = h_4$
   (v) Line 5–6 is a constant pressure heating line $p_5 = p_6$ and $h_6 > h_5$.

Example 3.18:   Show the following process on a Skeleton Mollier diagram.
   (i) Constant pressure cooling
   (ii) Adiabatic compression with friction
   (iii) Isothermal process
   (iv) Constant volume heating process.

Solution: The process lines are shown on Skeleton Mollier diagram,
   (i) Line 1–2 is constant pressure cooling process
$$p_1 = p_2$$
$$h_2 < h_1$$
   (ii) Line 3–4 is isentropic compression process.
$$s_3 = s_4$$
   Dotted line 3–4′ is adiabatic compression with friction
$$s_4' > s_4$$
   (iii) Line 5–6 is isothermal process
$$t_5 = t_6$$

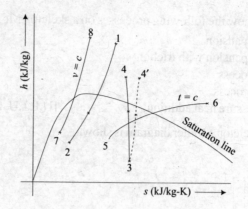

(iv)   Line 7–8 is constant volume heating process.

$$v_7 = v_8$$
$$h_8 > h_7$$

Example 3.19:   Steam at 10 bar and 0.9 dry passes through a throttle valve and becomes just dry and saturated at exit. Find the pressure at the exit of the throttle valve.   [U.P.T.U. II Sem., 2003-2004]

Solution:

On a Mollier diagram

1. Plot point 1 at intersection of p = 10 bar and x = 0.9.
2. Draw a horizontal line 1–2 till it touches the saturated line.
3. Read p_2 = 0.075 bar.

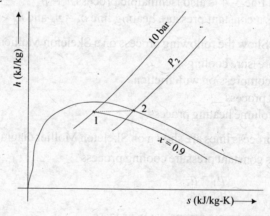

Example 3.20:   Steam expands isentropically from 4 MPa, 300°C to 50°C in a turbine. Determine the work done per kg of steam.   [U.P.T.U. I Sem., 2003]

Solution: p_1 = 4 MPa = 40 bar

t_1 = 300°C

t_2 = 50°C

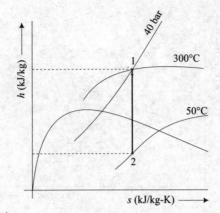

1. Plot point 1 at intersection of $p_1 = 40$ bar and $t_1 = 300°C$ on a Mollier diagram.

2. Draw a vertical line 1–2 so that point 2 lies on temperature line of 50°C.

3. Read h_1, h_2 on y-axis,

$$h_1 = 2875 \text{ kJ/kg}$$
$$h_2 = 1850 \text{ kJ/kg}$$

4. Work done
$$= 2875 - 1850$$
$$= 1025 \text{ kJ/kg}$$

Example 3.21: Two kg of steam at 12 bar pressure and 0.95 dryness fraction undergoes isothermal expansion to 1 bar pressure. Using Mollier diagram, determine the quantity of heat received, the work performed and the change in internal energy of steam.

Solution: 1. On Mollier diagram, locate point 1 at the intersection of $p_1 = 12$ bar and $x_1 = 0.95$ dry.

Read
$$t_1 = 180°C$$
$$v_1 = 0.15 \text{ m}^3/\text{kg}$$
$$h_1 = 2690 \text{ kJ/kg}$$
$$s_1 = 6.3 \text{ kJ/kg-K}$$

2. Locate point 2 at the intersection of $p_2 = 1$ bar and $t_2 (= t_1) = 180°C$.

Read
$$h_2 = 2860 \text{ kJ/kg}$$
$$v_2 = 0.75 \text{ m}^3/\text{kg}$$
$$s_2 = 7.25 \text{ kJ/kg-K}$$

For isothermal process, quantity of heat received

$$q_{1-2} = T(s_2 - s_1) = (273 + 180)(7.25 - 6.30) = 430.35 \text{ kJ}$$

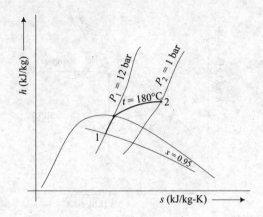

Change in internal energy,

$$u_2 - u_1 = (h_2 - p_2 v_2) - (h_1 - p_1 v_1)$$
$$= (2860 - 1 \times 10^2 \times 0.75) - (2690 - 12 \times 10^2 \times 0.15)$$
$$= 2785 - 2510$$
$$= 275 \text{ kJ/kg}$$

The work done,

$$w_{1-2} = (u_1 - u_2) + q_{1-2}$$
$$= -275 + 430.35$$
$$= 155.35 \text{ kJ/kg}$$

Quantity of heat received,

$$Q_{1-2} = mq_{1-2} = 2 \times 430.35 = 860.7 \text{ kJ}$$

The work done.

$$W_{1-2} = mw_{1-2} = 2 \times 155.35 = 310.7 \text{ kJ}$$

Total change in internal energy $U_2 - U_1 = m(u_2 - u_1)$
$$= 2 \times 275 = 550 \text{ kJ}.$$

Example 3.22: 5 kg of steam at 5 bar and 200°C is expanded adiabatically to 0.1 bar. Show the process on skeleton Mollier chart and find out the amount of work done. [U.P.T.U. I Sem., 2003-2004]

Solution:

1. Plot point 1 at the intersection of pressure p_1 = 5 bar and temperature t_1 = 200°C, on a Mollier diagram.
2. Read h_1 = 2880 kJ/kg.
3. Draw line 1–2 vertically till it intersects the pressure line p_2 = 0.1 bar.
4. Read h_2 = 2230 kJ/kg.
5. Work done = $m(h_1 - h_2)$ = 5(2880 – 2230) = 3250 kJ.

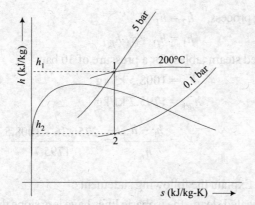

Example 3.23: For estimating the quality of steam at 3 MPa in a boiler, its sample is throttled in a throttling calorimeter where its pressure and temperature were found to be 1 bar and 140°C. What is the quality of steam.

[U.P.T.U. 2005-2006]

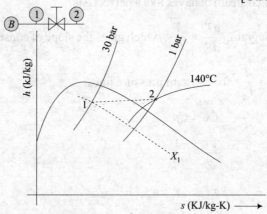

Solution:

(a) Using Mollier Diagram

1. Locate point 2 on the intersection of 1 bar and 140°C.
2. Draw line 2 – 1 horizontally so that point 1 lies on 30 bar line.
3. Read dryness fraction X_1,

$$X_1 = 0.97.$$

(b) Using Steam Tables

Steam at point 2 is superheated.

From steam table for superheated steam.

For pressure 1 bar and temperature.

t	100°C	150°C	50°C	140°C
h	2676.2	2776.4	100.2	$2676.2 + \dfrac{100.2 \times 40}{50} = 2756.36$

For throttling process, $h_2 = h_1$

$\therefore$
$$h_1 = h_{f1} + x_1 h_{fg1}$$

From saturated steam tables, for a pressure of 30 bar

$$h_{f1} = 1008.5 \text{ kJ/kg}$$
$$h_{fg1} = 1795.7 \text{ kJ/kg}$$

$\therefore$
$$X_1 = \frac{h_1 - h_{g1}}{h_{fg1}} = \frac{2756.36 - 1008.5}{1795.7} = 0.973.$$

Example 3.24: Explain the following statements:

(a) For superheated steam, the isobaric lines have less slope than isochoric lines on a Mollier diagram.

(b) For superheated steam, isothermal lines tend to be horizontal.

Solution:

(a) Superheated steam behaves like a perfect gas.

On $T\text{--}s$ diagram, $\dfrac{dT}{dS} = \dfrac{T}{Cv}$ which gives the slope of constant volume lines

and $\dfrac{dT}{dS} = \dfrac{T}{Cp}$ for constant pressure lines.

$\because$
$$Cv < Cp$$

$\therefore$
$$\frac{1}{Cv} > \frac{1}{Cp}$$

or
$$\frac{T}{Cv} > \frac{T}{Cp}$$

The slope of constant volume process is more than that for constant pressure percess.

(b) Further superheated steam behaves like a perfect gas.

For superheated steam,

$$h = h\,(T)$$

Therefore, temperature lines tend to follow enthalpy lines which are horizontal on Mollier diagram.

Example 3.25: Dry and saturated steam at 15 bar is throttled to a pressure of 1.5 bar. With the help of Mollier diagram find the final temperature and entropy change.

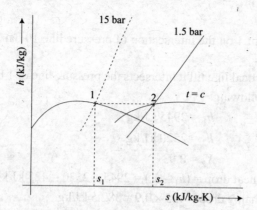

Solution:

1. On a Mollier diagram, plot point 1 on the intersection of pressure line of 15 bar with saturation line.
2. Draw a horizontal line till it intersects the pressure line of 1.5 bar at point 2.
3. Read:

$$s_1 = 6.45 \text{ kJ/kg-K}$$
$$s_2 = 7.48 \text{ kJ/kg-K}$$
$$t_2 = 162°C$$

4. The final temperature, $t_2 = 162°C$.

   Change of entropy,

$$s_2 - s_1 = 7.48 - 6.45 = 1.03 \text{ kJ/kg-K}.$$

The final state of steam is superheated with parameter 1.5 bar and 162°C.

Example 3.26: Steam at 10 bar and 250°C expands in a steam turbine to a pressure of 1 bar. Find the power developed for a steam flow rate of 500 kg/min.

(a) For isentropic expansion.

(b) For adiabatic expansion with 10% friction loss.

Use Mollier diagram.

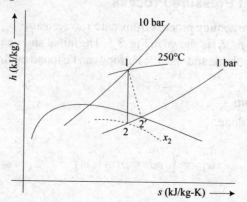

Solution:

1. Locate point 1 on the intersection of pressure line 10 bar and temperature line 250°C.
2. Draw a vertical line till it intersects the pressure line of 1 bar at point 2.
3. Read the following:

$$h_1 = 2945 \text{ kJ/kg}$$
$$h_2 = 2520 \text{ kJ/kg}$$
$$X_2 = 0.93$$

4. Isentropic heat drop = $(h_1 - h_2) = 2945 - 2520 = 425$ kJ/kg
   Actual enthalpy drop = $425 \times 0.9 = 382.5$ kJ/kg.

(a) Isentropic power developed = $\dfrac{500}{60} \times 425 = 3536.5$ kW

(b) Actual power developed = $\dfrac{500}{60} \times 382.5 = 3183$ kW.

3.7 VAPOUR PROCESSES

Water vapour or steam can be subjected to heating, cooling, expansion or compression but the results obtained are different from these processes for gases. The quality of steam continuously changes.

Thermodynamic laws are applicable to steam but gas laws and gas equations do not hold good for steam.

For closed system or non-flow process:

$$dq = du + pdv.$$

For open system and steady flow percess.

$$dq = dh + v\, dp$$

Also

$$dq = Tds$$

3.7.1 Constant Pressure Process

A constant pressure heating process from state 1, wet steam (p_1, x_1) to superheated steam state 2 $(p_2 = p_1, t_2)$ is shown in Fig. 3.5. The initial and final values of specific enthalpy, specific volume and specific entropy can be found out from steam tables or Mollier diagram.

(a) Closed system

(i) Work done,

$$w_{1-2} = \int_1^2 pdv = p(v_2 - v_1)$$

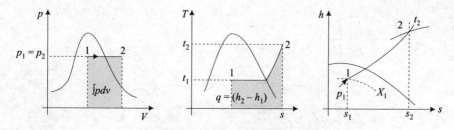

Fig. 3.5 Constant pressure process

If v is in m³/kg and p in bar,

$$w_{1-2} = p \times 10^2 (v_2 - v) \quad [\text{kJ}]$$
$$= (p_2 v_2 - p_1 v_1) \times 10^2 \qquad [\because \ p_1 = p_2]$$

(ii) Change in internal energy,

$$(u_2 - u_1) = (h_2 - p_2 v_2) - (h_1 - p_1 v_1)$$
$$= (h_2 - h_1) - (p_2 v_2 - p_1 v_1)$$

$$v_1 = x_1 v_{g1}$$

$$v_2 = v_{g2} \frac{T_2}{T_{s_2}}$$

(iii) Heat transfer,

Using 1st law of thermodynamics,

$$q_{1-2} = w_{1-2} + (u_2 - u_1)$$
$$= (p_2 v_2 - p_1 v_1) + [(h_2 - h_1) - (p_2 v_2 - p_1 v_1)]$$
$$= (h_2 - h_1)$$

(b) Open system (Steady flow process)

(i) Work done,

$$w_{1-2} = \int_1^2 - v dp$$

But $\qquad\qquad dp = 0$

$\therefore \qquad\qquad w_{1-2} = 0$

(ii) Heat transferred,

$$q_{1-2} = w_{1-2} + (h_2 - h_1) = (h_2 - h_1)$$

Example 3.27: 5 kg of steam is condensed in a condenser following reversible constant pressure process from 0.75 bar and 150°C. At the end of process, steam gets completely condensed. Determine the heat removed from steam and change in entropy. Also sketch the process on $T - s$ diagram and shade the area representing heat removed. [U.P.T.U. II Sem., 2000-2002]

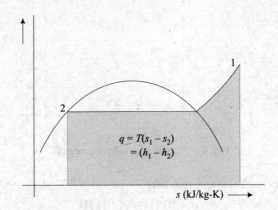

$$q = T(s_1 - s_2)$$
$$= (h_1 - h_2)$$

s (kJ/kg-K) ⟶

Solution: The process is marked on $T - s$ diagram,

$$m = 5 \text{ kg}$$
$$p_1 = p_2 = 0.75 \text{ bar}$$
$$t_1 = 150°C$$

Steam is superheated at state 1. From superheated steam tables for a temperature of 150°C and

Pressure	h	s
0.70 bar	2778.6	7.783
0.80 bar	2777.8	7.720
0.75 bar		

$\left(\dfrac{0.80 + 0.70}{2}\right)$	$\dfrac{2778.6 + 2777.8}{2}$	$\dfrac{7.783 + 7.720}{2}$
	$= 2778.2 \text{ kJ/kg}$	$= 7.7515 \text{ kJ/kg}$

∴ $h_1 = 2778.2 \text{ kJ/kg}$

$s_1 = 7.7515 \text{ kJ/kg-K}$

Point 2 is the saturated liquid at 0.75 bar,

∴ $h_2 = h_{f2}$

Pressure	h_f	s_f
0.74 bar	382.99	1.209
0.76 bar	385.9	1.217
0.75 bar	$\dfrac{382.9 + 385.9}{2}$	$\dfrac{1.209 + 1.217}{2}$
	$= 384.4 \text{ kJ/kg}$	$= 1.213 \text{ kJ/kg}$

∴ $h_2 = 384.4 \text{ kJ/kg}$

$s_2 = 1.213 \text{ kJ/kg-K}$

(i) Heat removed,

$$Q_{1-2} = m(h_1 - h_2) = 5(2778.2 - 384.4)$$
$$= \mathbf{1196.9 \ kJ \ Ans.}$$

(ii) Change in entropy,

$$S_1 - S_2 = m(s_1 - s_2) = 5(7.7515 - 1.213)$$
$$= \mathbf{32.6925 \ kJ/K.}$$

3.7.2 Constant Volume Process

Heating or cooling of steam in a closed vessel is the example of constant volume process. 1 kg of wet steam (p_1, X_1) at state 1 is heated at constant volume to super-heated state 2 (p_2, t_2). The process is shown in Fig. 3.6.

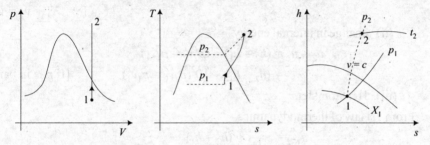

Fig. 3.6 Constant volume process

The initial volume,

$$v_1 = x_1 v_{g1}$$

The final volume,

$$v_2 = v_1$$

(i) If $v_2 < v_{g2}$: steam is wet

$\therefore$
$$v_2 = x_2 v_{g2}.$$

(ii) If $v_2 = v_{g2}$: steam is dry and saturated

(iii) If $v_2 > v_{g2}$: steam is superheated

$$v_2 = v_{g2} \frac{T_2}{T_{s_2}}$$

where, T_{s2} is saturation temperature at p_2.

(a) Closed system

(i) Work done,

$$w_{1-2} = \int_1^2 p dv = 0 \quad [\because dv = 0]$$

(ii) Change in internal energy

$$u_2 - u_1 = (h_2 - p_2 v_2) - (h_1 - p_1 v_1)$$
$$= (h_2 - h_1) - 10^2 (p_2 v_2 - p_1 v_1)$$

where, p is in bar

(iii) Heat transfer,

$$q_{1-2} = w_{1-2} + (u_2 - u_1)$$
$$= 0 + [(h_2 - h_1) - 10^2 (p_2 v_2 - p_1 v_1)]$$

(b) Open system (Steady flow process)

(i)   Work done, $w_{1-2} = \int\limits_1^2 -vdp = -v \int\limits_1^2 = -v (p_2 - p_1) = (p_1 v_1 - p_1 v_2)$

$$[\because \quad v_1 = v_2 = v]$$

(ii) change in internal energy,

$$u_2 - u_1 = (h_2 - h_1) - (p_2 v_2 - p_1 v_1)$$
$$= (h_2 - h_1) - 10^2 (p_2 v_2 - p_1 v_1) \qquad \text{[If } p \text{ is in bar]}$$

(iii) Heat transfer,

From 1st law of thermodynamics,

$$q_{1-2} = w_{1-2} + (h_2 - h_1)$$
$$= (p_1 v_1 - p_2 v_2) + [(u_2 + p_2 v_2) - (u_1 + p_1 v_1)]$$
$$= u_2 - u_1.$$

Example 3.28:   A rigid vessel of $1\ m^3$ contains steam at 20 bar and 400°C. The vessel is cooled until the steam is just dry and saturated. Calculate the mass of steam, the final pressure and heat removed during the process.

Solution:

$$p_1 = 20 \text{ bar}$$
$$t_1 = 400°C$$
$$v_1 = 1 m^3$$
$$v_2 = v_1 = 1\ m^3.$$

1. Mass of steam: From superheated steam tables at 20 bar and 400°C

$$v = 0.1513 \ m^3/kg$$

Mass of steam,

$$m = \frac{V_1}{v} = \frac{1}{0.1513} = 6.61 \text{ kg}.$$

2. Final pressure: $v_2 = v_1 = 0.1513 \ m^3/kg$

The steam is dry and saturated

$\therefore$ $v_2 = v_{g2}.$

From steam tables for $v_{g2} = 1513$ m^3/kg

$$p_2 = 13 \text{ bar.}$$

3. Heat removed: $\quad Q_{1-2} = m\,(u_2 - u_1)$

$$u_2 - u_1 = (h_2 - h_1) - 10^2\,(p_2 v_2 - p_1 v_1)$$

From superheated steam tables, for $p_1 = 20$ bar and $t_1 = 400°C$

$$h_1 = 3248.7 \text{ kJ/kg}$$
$$v_1 = 0.1513 \text{ m}^3/\text{kg}$$

From saturated steam tables for $p_2 = 13$ bar

$$h_2 = hg_2 = 2785.4 \text{ kJ/kg}$$
$$v_2 = vg_2 = 0.1513 \text{ m}^3/\text{kg}$$

$\therefore \qquad u_2 - u_1 = (h_2 - h_1) - 10^2\,(p_2 v_2 - p_1 v_1)$

$$= (2785.4 - 3248.7) - 10^2\,(13 \times 0.1513 - 20 \times 0.1513)$$
$$= -357.53 \text{ kJ/kg.}$$

Heat removed,

$$Q_{1-2} = m(u_2 - u_1) = 6.61\,(-357.53)$$
$$= -2363.27 \text{ kJ. Ans.}$$

3.7.3 Constant Temperature Process

1 kg of wet steam at state 1 (p_1, X_1) is heated at constant temperature to a state 2 (p_2). The heating of steam in wet region takes place simultaneously at constant pressure and temperature as clear from curve $1 - a$ in $T - s$ diagram. Heating of superheated steam at constant pressure is accompanied by decrease in pressure. It is not practical to heat steam at constant temperature with simultaneous expansion

(a) Heat transfer,

$$q_{1-2} = T_1(s_2 - s_1)$$

(b) Change in internal energy

$$u_2 - u_1 = (h_2 - p_2 v_2) - (h_1 - p_1 v_1)$$

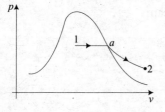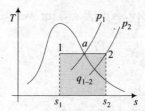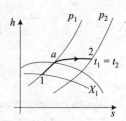

Fig. 3.7 Constant temperature process

(c) Work done,

From 1st law of thermodynamics,

$$q_{1-2} = w_{1-2} + (u_2 - u_1)$$

$$\therefore \qquad w_{1-2} = q_{1-2} - (u_2 - u_1)$$

A constant temperature process for steam is same as constant pressure process in wet region and is similar to hyperbolic process in superheated region since steam behaves like a gas in this region.

3.7.4 Hyperbolic Process (*pv = c*)

1 kg of wet steam at state 1 (p_1, X_1) is hyperbolically expanded to state 2 (P_2, t_2) as shown in Fig. 3.8.

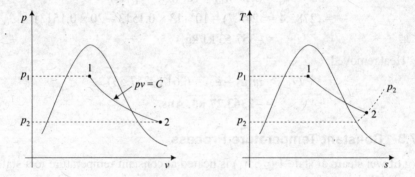

Fig. 3.8 Hyperbolic process

$$v_1 = x_1 v_{g1}$$

$$v_2 = v_{g2} \cdot \frac{T_2}{T_{s_2}} \quad \text{(if final state of steam is superheated)}$$

$$v_2 = x_2 v_{g2} \text{ (if final state of steam is wet)}$$

For hyperbolic process,

$$p_1 v_1 = p_2 v_2$$

$$p_1 v_1 = p_2 (X_2 v_{g2}) \text{ for wet steam}$$

$$X_2 = \frac{p_1 x_1 v_{g1}}{p_2 v_{g2}}$$

X_2 can be determined.

For superheated steam at final state,

$$p_1 X_1 v_{g1} = p_2 \cdot v_{g2} \frac{T_2}{T_{s_2}}$$

T_2 can be determined.

(a) Closed system

    (i) Work done,

$$w_{1-2} = p_1 v_1 \ln \frac{v_2}{v_1}$$

    (ii) Change in internal energy,

$$u_2 - u_1 = (h_2 - h_1) - (p_2 v_2 - p_1 v_1)$$

$$p_2 v_2 = p_1 v_1$$

∴

$$v_2 - v_1 = h_2 - h_1$$

    (iii) Heat removed,

From 1st law of thermodynamics,

$$q_{1-2} = w_{1-2} + u_2 - u_1$$

$$= p_1 v_1 \ln \frac{v_2}{v_1} + (h_2 - h_1)$$

(b) Open system

    (i) Work done,

$$w_{1-2} = \int - v \, dp = p_1 v_1 \ln \frac{v_2}{v_1}$$

    (ii) Heat removed,

$$q_{1-2} = w_{1-2} + (h_2 - h_1)$$

$$= p_1 v_1 \ln \frac{v_2}{v_1} + (h_2 - h_1)$$

Example 3.29: Steam at a pressure of 8 bar and 0.9 dry is expanded hyperbolically in a cylinder upto 1 bar pressure. Determine the work done and heat transfer during the process. Assume $C_s = 2.0 \, \text{kJ/kg}$.

Solution:

$$p_1 = 8 \text{ bar}$$

$$X_1 = 0.9$$

$$p_2 = 1 \text{ bar}$$

From saturated steam tables for pressure 8 bar,

$$v_{g1} = 0.24 \text{ m}^3/\text{kg}$$

∴

$$v_1 = x_1 v_{g1} = 0.9 \times 0.24 = 0.216 \text{ m}^3/\text{kg}$$

The process is hyperbolic,

$$\therefore \qquad p_1 v_1 = p_2 v_2$$

$$\therefore \qquad v_2 = \frac{p_1 v_1}{p_2} = \frac{8 \times 0.216}{1} = 1.728 \text{ m}^3/\text{kg}$$

From saturated steam tables for pressure 1 bar,

$$v_{g2} = 1.694 \text{ m}^3/\text{kg}$$

$$t_{s2} = 99.63°\text{C}.$$

$$\therefore \qquad v_2 > v_{g2}, \text{ the steam is superheated.}$$

$$v_2 = v_{g2} \cdot \frac{T_2}{T_{s2}}$$

$$\therefore \qquad T_2 = \frac{v_2}{v_{g2}} \cdot T_{s2} = \frac{1.728}{1.694} \times (273 + 99.63)$$

$$= 380.1 \text{ K}.$$

From saturated steam tables for $p_1 = 8$ bar,

$$h_{f1} = 7209 \text{ kJ/kg}$$

$$h_{fg1} = 2046.5 \text{ kJ/kg}$$

$$\therefore \qquad h_1 = h_{f1} + x_1 h_{fg1} = 720.9 + 0.9 \times 2046.5$$

$$= 2562.8 \text{ kJ/kg}.$$

$$h_2 = h_{g2} + C_s(T_2 - T_{s2})$$

$$= 2670.8 + 2(380.1 - 372.63)$$

$$= 2685.7 \text{ kJ/kg}.$$

(i) Change in internal energy,

$$(u_2 - u_1) = (h_2 - h_1) - (p_2 v_2 - p_1 v_1)$$

$$= (h_2 - h_1) - 0$$

$$= 2685.7 - 2562.8 = 122.9 \text{ kJ/kg}.$$

(ii) Work done,

$$w_{1-2} = p_1 v_1 \ln \frac{v_2}{v_{1_s}} = p_1 v_1 \ln \frac{p_1}{p_2} = 8 \times 10^5 \times 0.216 \ln \frac{8}{1}$$

$$= 3.593 \times 10^3 \text{ kJ/kg}$$

$$= 359.3 \text{ kJ/kg}$$

(iii) Heat transfer,

$$q_{1-2} = w_{1-2} + (v_2 - v_1)$$

$$= 359.3 + 122.9$$

$$= 482.2 \text{ kJ/kg}$$

3.7.5 Polytropic Process ($pv^n = c$)

1 kg of superheated steam at state 1 (p_1, t_1) is expanded polytropically to a pressure of p_2. The process is shown in Fig. 3.9.

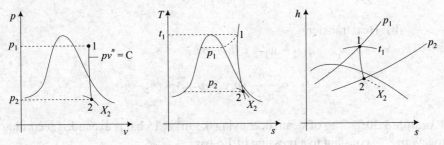

Fig. 3.9 Polytropic expansion process

v_1 can be determined from superheated steam tables against pressure p_1 and temperature t_1,

$$p_1 v_1^n = p_2 v_2^n$$

$$\therefore \qquad v_2 = \left(\frac{p_1}{p_2}\right)^{\frac{1}{n}} . v_1$$

Find v_{g2} against pressure p_2 from superheated steam tables

If $v_2 < v_{g2}$: steam is wet.

$$v_2 = x_2 \, v_{g2}$$

If $v_2 = v_{g2}$: steam is dry and saturated

If $v_2 > v_{g2}$: steam is superheated

$$v_2 = v_{g2} . \frac{T_2}{T_{s_2}} .$$

(a) Closed system

(i) Work done,

$$w_{1-2} = \frac{p_1 v_1 - p_2 v_2}{(n-1)}$$

(ii) Change is internal energy,

$$u_2 - u_1 = (h_2 - h_1) - (p_2 v_2 - p_1 v_1)$$

(iii) Heat transfer,

$$q_{1-2} = w_{1-2} + (u_2 - u_1)$$

$$= \frac{p_1 v_1 - p_2 v_2}{(n-1)} + [(h_2 - h_1) - (p_2 v_2 - p_1 v_1)]$$

(b) Open system

(a) Work done,

$$W_{1-2} = \left(\frac{n}{n-1}\right)(p_1v_1 - p_2v_2)$$

(b) Heat transfer,

$$q_{1-2} = w_{1-2} + (h_2 - h_1)$$

$$= \frac{n}{n-1}(p_1v_1 - p_2v_2) + (h_2 - h_2).$$

Example 3.30: 2 kg of steam occupying 0.3 m³ at 15 bar is expanded according to law $pv^{1.3}$ = constant to a pressure of 1.5 bar.

Calculate: (a) work done, and

(b) condition of steam after expansion.

Solution:

$$p_1 = 15 \text{ bar}$$
$$V_1 = 0.3 \text{ m}^3$$
$$m = 2 \text{ kg}$$
$$n = 1.3$$
$$p_2 = 1.5 \text{ bar}$$

$$v_1 = \frac{V_1}{m} = \frac{0.3}{2} = 0.15 \text{ m}^3/\text{kg}$$

Now,

$$p_1v_1^n = p_2v_2^n$$

∴

$$v_2 = \left(\frac{p_1}{p_2}\right)^{\frac{1}{n}} v_1$$

$$= \left(\frac{15}{1.5}\right)^{\frac{1}{1.3}} \times 0.15 = 0.8817 \text{ m}^3/\text{kg}$$

(b) From saturated steam tables for $p_2 = 1.5$ bar

$$v_{g2} = 1.1635 \text{ m}^3/\text{kg}.$$

Since,

$$v_2 < v_{g2} : \text{steam is wet.}$$

Now,

$$v_2 = x_2 v_{g2}$$

∴

$$x_2 = \frac{v_2}{v_{g2}} = \frac{0.8817}{1.1635} = 0.8595$$

(a) Work done,

$$W_{1-2} = m\left[\frac{p_1 v_1 - p_2 v_2}{n-1}\right] = 2\left[\frac{15 \times 10^5 \times 0.15 - 1.5 \times 10^5 \times 0.8817}{(1.3-1)}\right]$$

$$= 6.183 \times 10^5 \, \text{J}$$
$$= \textbf{618.3 kJ} \quad \textbf{Ans.}$$

3.7.6 Isentropic Process (Reversible Adiabatic)

Conditions of isentropic process:

1. $q_{1-2} = 0$: No heat transfer.
2. $s_2 = s_1$: No change of entropy
3. $pv^n = C$: $n = 1.135$ for wet or saturated steam initially
    : $n = 1.3$ for superheated steam initially.

One kg of steam from state 1 (p_1, t_1) expands isentropically to final pressure p_2. The process is shown in Fig. 3.10.

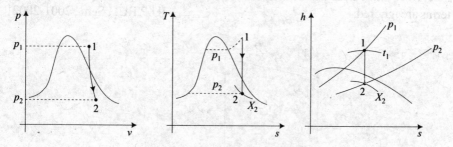

Fig. 3.10 Isentropic expansion process

Superheated steam tables can be used to find h_1 and s_1.

Now $$s_1 = s_2 = s_{f2} + x_2 s_{fg2}$$

$\therefore$ $$x_2 = \frac{s_1 - s_{f2}}{s_{fg2}}$$

s_{f2} and s_{fg2} can be found out from saturated steam tables against a pressure p_2.

$\therefore$ $$h_2 = h_{f2} + x_2 h_{fg2}$$

Again h_{f_2} and h_{fg_2} are available from saturated steam tables against pressure p_2.

(a) Closed system

   (i) Heat transfer, $q_{1-2} = 0$

   (ii) Change in internal energy,
   $$u_2 - u_1 = (h_2 - p_2 v_2) - (h_1 - p_1 v_1)$$

(iii) Work done,

$$q_{1-2} = w_{1-2} + (u_2 - u_1) = 0$$

∴
$$w_{1-2} = (u_1 - u_2) = (h_2 - p_2v_2) - (h_1 - p_1v_1)$$

(b) Open system (Steady flow process)

(i) Heat transfer,

$$q_{1-2} = 0$$

(ii) Work done,

$$q_{1-2} = w_{1-2} + (h_2 - h_1) = 0$$

∴
$$w_{1-2} = (h_1 - h_2)$$

Example 3.31: A turbine in a steam power plant operating under steady state conditions, receives superheated steam at 3 MPa and 350°C at the rate of one kg/s and with a velocity of 50 m/s at an elevation of 2 m above the ground level. The steam leaves the turbine at 10 kPa with a quality of 0.95 at an elevation of 5 m above the ground level. The exit velocity of steam is 120 m/s. The energy losses on heat from the turbine are estimated as 5 kJ/s. Estimate the power output of the turbine. How much error will be introduced if the kinetic energy and potential energy terms are ignored. [U.P.T.U. I Sem., 2001-2002]

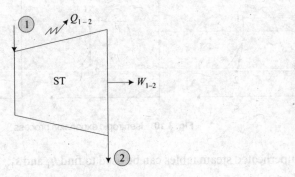

Solution: At turbine inlet,

$$p_1 = 3 \text{ Mpa} = 3 \times 10^3 \text{ kN/m}^2$$
$$t_1 = 350°\text{C}$$
$$V_1 = 50 \text{ m/s}$$
$$z_1 = 2 \text{ m}$$
$$m = 1 \text{ kg/s}$$

At turbine output,

$$p_2 = 10 \text{ kN/m}^2$$
$$X_2 = 0.95$$
$$V_2 = 120 \text{ m/s}$$

$$z_2 = 5 \text{ m}$$
$$Q_{1-2} = -5 \text{ kJ/s}$$

From superheated steam tables at 30 bar (3×10^3 kN/m^2) and 350°C
$$h_1 = 3115.25 \text{ kJ/kg}$$

From saturated steam tables at 0.1 bar (10 kN/m^2)
$$h_{f2} = 191.81 \text{ kJ/kg}$$
$$h_{fg2} = 2392.83 \text{ kJ/kg}$$

∴
$$h_2 = h_{f2} + x_2 h_{fg2} = 191.81 + 0.95(2392.82)$$
$$= 2465 \text{ kJ/kg}$$

Applying steady flow energy equation,

$$q_{1-2} - w_{1-2} = (h_2 - h_1) + \frac{V_2^2 - V_1^2}{2 \times 10^3} + \frac{g}{10^3}(z_2 - z_1)$$

$$-5 - w_{1-2} = (2465 - 3115.25) + \frac{120^2 - 50^3}{2 \times 10^3} + \frac{9.81}{10^3}(5-2) \quad (5-2)$$

$$w_{1-2} = 639 \text{ kW}$$

Neglecting the effect of *ke* and *pe*.
$$-5 - w_{1-2} = (2465 - 3115.25)$$
$$w_{1-2} = 645.25 \text{ kW}$$
$$\text{Absolute error} = 639 - 645.25 = -6.25 \text{ kW}$$

$$\% \text{ age error} = \frac{6.25}{639} \times 100 = \textbf{0.978\% Ans.}$$

Example 3.32:   Steam at 90 kPa and 150°C enters a subsonic diffuser with a velocity of 150 m/s and leaves the diffuser at 190 kPa and a velocity of 55 m/s. During the process, 1.5 kJ/kg of heat is lost to the surrounding. Determine:

(i) The final temperature.

(ii) The mass flow rate

(iii) The exit diameter assuming the inlet diameter as 10 cm and steady flow.
$$\text{[U.P.T.U., I Sem., 2001-2002]}$$

Solution:

Inlet conditions:

$$p_1 = 90 \text{ kPa} = 90 \text{ kN/m}^2$$
$$T_1 = 150 + 273 = 423 \text{ K}$$
$$V_1 = 150 \text{ m/s}$$

Outlet conditions:

$$p_2 = 190 \text{ kN/m}^2$$
$$V_2 = 55 \text{ m/s}$$

Applying steady flow energy equation for unit mass:

$$q_{1-2} - w_{1-2} = (h_2 - h_1) + \frac{V_2^2 - V_1^2}{2} + g(Z_2 - Z_1)$$

For given diffuser,

$$Z_2 = Z_1 \text{ and } w_{1-2} = 0$$

∴
$$q_{1-2} = (h_2 - h_1) + \frac{V_2^2 - V_1^2}{2}$$

∴
$$(h_2 - h_1) = -q_{1-2} - \frac{V_2^2 - V_1^2}{2}$$

∴
$$h_1 - h_2 = q_{1-2} + \frac{V_2^2 - V_1^2}{2}$$

$$= 1.5 + \frac{55^2 - 150^2}{2 \times 10^3} = 11.2375 \text{ kJ/kg}$$

From superheated steam tables against $p_1 = 90$ kPa (90×10^{-2} bar) and 150°C,

$$h_1 = 2777.1 \text{ kJ/kg}$$

∴
$$h_2 = 2777.1 - 11.2375 = 2765.86 \text{ kJ/kg}.$$

(i) Again from superheated steam tables, against $p_2 = 190$ kN/m^2 (1.9 bar) and $h_2 = 2765.86$ kJ/kg

$$t_2 = 154°C. \quad \textbf{Ans.}$$

(ii)
$$v_1 = \frac{RT_1}{p_1}$$

$$R = \frac{\bar{R}}{M} = \frac{8.314}{18} = 0.4619 \text{ kJ/kg-K}$$

$$v_1 = \frac{0.4619 \times 423}{90} = 2.17 \text{ m}^3/\text{kg}$$

$$v_2 = \frac{RT_2}{p_2} = \frac{0.4619 \times (154 + 273)}{190} = 1.038 \text{ m}^3/\text{kg}$$

∴
$$\dot{m} = \frac{A_1 V_1}{v_1} = \frac{\frac{\Pi}{4}(0.10)^2 \times 150}{2.17} = \textbf{0.543 kg/s}$$

(iii) Exit area,

$$A_2 = \frac{\dot{m} v_2}{V_2} = \frac{0.543 \times 1.038}{55} = 0.010248 \ \text{m}^2$$

$$\therefore \qquad d_2 = \sqrt{\frac{A_2 \times 4}{\Pi}} = \sqrt{\frac{0.010248 \times 4}{\Pi}} = 0.1142 \ \text{m}$$

$$= \textbf{11.42 cm} \quad \textbf{Ans.}$$

3.7.7 Throttling Process

When steam is passed through a partly opened valve or a restriction, it is throttled to a lower pressure without any work being done. Steam from state 1 (p_1, x_1) is throttled to state 2 (p_2, t_2). Applying steady flow energy equation:

$$q_{1-2} - w_{1-2} = (h_2 - h_1) + \frac{V_2^2 - V_1^2}{2} + g(Z_2 - Z_1)$$

Here, $\qquad\qquad V_1 = V_2, \quad Z_1 = Z_2, \quad q_{1-2} = 0, \quad w_{1-2} = 0$

$\therefore \qquad\qquad (h_2 - h_1) = 0$

or $\qquad\qquad h_1 = h_2 = $ constant.

In a throttling process, enthalpy of steam remains constant. The process is shown in Fig. 3.11.

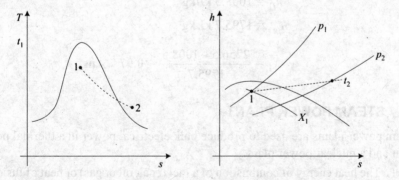

Fig. 3.11 Throttling process

The problem can be solved with the help of Mollier diagram or steam tables. Find h_2 from steam tables (usually superheated steam tables) against a pressure p_2 and temperature t_2.

$$h_2 = h_1 = h_{f1} + x_1 h_{fg1}$$

Find h_{f1} and h_{fg1} from saturated steam tables against pressure p_1.

∴ Quality of steam at inlet of throttling valve,

$$x_1 = \frac{h_2 - h_{f1}}{h_{fg1}}$$

Example 3.33: For estimating the quality of steam at 3 MPa in a boiler, its sample is throttled in a throttling calorimeter where its pressure and temperature were found to be 1 bar and 140°C. What is the quality of steam?

Solution: $p_1 = 3$ MPa $= 30$ bar

$p_2 = 1$ bar

$t_2 = 140°C$

From steam tables for superheated steam against a pressure of 1 bar and temperature of 140°C,

$$h_2 = 2756.3 \text{ kJ/kg}$$

For throttling process,

$$h_2 = h_1 = h_{g1} + x_1 h_{fg1}$$

∴ $$X_1 = \frac{h_2 - h_{f1}}{h_{fg1}}$$

From steam tables against a pressure of 30 bar,

$$h_{f1} = 1008.5 \text{ kJ/kg}$$

$$h_{fg1} = 1795.7 \text{ kJ/kg}$$

∴ $$X_1 = \frac{2756.5 - 1008.5}{1795.7} = 0.97 \quad \textbf{Ans.}$$

3.8   STEAM POWER PLANT

Steam power plants are used to produce bulk electrical power in a thermal power plant and a nuclear power plant.

1. The heat energy of combustion of a fuel (coal, oil or gas) or heat of fusion of nuclear fuels (uranium, plutonium or thorium) is converted into mechanical work which in turn is used to generate electricity.
2. The plant works on thermodynamic cycle.
3. The working fluid is water/steam.
4. 1st law of thermodynamics can be applied to the plant.

The total energy flowing into the fluid during the cycle must be equal to that flowing out of the cycle.

$$\dot{Q}_{in} + \dot{W}_{in} = \dot{Q}_{out} + \dot{W}_{out}$$

where,

$\dot{Q}$ = rate of heat transfer

$\dot{W}$ = rate of work transfer, i.e., power.

5. The boiler or nuclear reactor works as the heat source and condenser works on heat sink.

6. The performance of the plant is measured by thermal efficiency.

$$\eta_{th} = \frac{\dot{W}_{out} - \dot{W}_{in}}{\dot{Q}_{in}} = \frac{\dot{Q}_{in} - \dot{Q}_{out}}{\dot{Q}_{in}}$$

$$= 1 - \frac{\dot{Q}_{out}}{\dot{Q}_{in}}$$

The components of a steam power plant

The line diagram of a steam power plant is shown in Fig. 3.12.

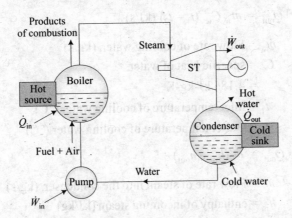

Fig. 3.12 Steam power plant

There are four main components of steam power plant:

1. Steam boiler or generator: It is the hot source reservoir. The fuel (coal, oil or gas) is burnt in the presence of air. The heat energy of combustion gases raise steam by heating feed water.

$$\dot{Q}_{in} = \dot{m}_f CV \ [kJ/s]$$

$$= \dot{m} (h_{out} - h_{in})$$

where
$$\dot{m}_f = \text{Mass rate of fuel burnt, (kg/s)}$$
$$Cv = \text{calorific value of fuel, (kJ/kg)}$$
$$\dot{m} = \text{steam generation rate, (kg/s)}$$

h_{out} and h_{in} are enthalpies of working fluid at outlet and inlet of boiler (kJ/kg).

2. Steam turbine: The steam generated in the boiler is expanded in a steam turbine adiabatically to a low pressure to produce mechanical work $\dot{W}_{out}$ which is further converted into electrical energy by the generator.

$$\dot{W}_{out} = \dot{m}\,(h_1 - h_2),\ [\text{kW}]$$

where
$$\dot{m} = \text{flow rate of steam [kg/s]}$$
$$h_1 = \text{specific enthalpy of steam at turbine inlet (kJ/kg)}$$
$$h_2 = \text{specific enthalpy of steam at turbine outlet (kJ/kg)}$$

3. Condenser: Condenser acts as cold sink into which heat is rejected. It is a tubular heat exchanger in which exhaust and low pressure steam from the turbine is cooled with the help of cooling water and condensed into water called condensate.

$$\dot{Q}_{out} = \dot{m}_w\,C_w\,(t_1 - t_2)\ (\text{kJ/s})$$

where,
$$\dot{m}_w = \text{Flow rate of cooling water, (kg/s)}$$
$$C_w = \text{specific heat of water.}$$
$$= 4.186\ \text{kJ-kg-K}$$
$$t_1 = \text{Inlet temperature of cooling water (°C)}$$
$$t_2 = \text{Outlet temperature of cooling water (°C)}$$

Also
$$s\,\dot{Q}_{out} = \dot{m}\,(h_{in} - h_{out})$$

$$\dot{m} = \text{Flow rate of steam into the condenser, (kg/s)}$$
$$h_{in} = \text{enthalpy of incoming steam (kJ/kg)}$$
$$h_{out} = \text{enthalpy of outgoing condensate (kj/kg)}$$

4. Pump: Condensate extraction pump or boiler feed pump or both return the condensate back to the boiler. This pump draws its power from the grid.

$$\dot{W}_{in} = \dot{m}\,(h_{out} - h_{in})$$

where,
$$\dot{m} = \text{Flow rate of feed water, (kg/s)}$$
$$h_{out} = \text{enthalpy of water at pump outlet (kJ/kg)}$$
$$h_{in} = \text{enthalpy of feed water at pump inlet (kJ/kg)}$$

$$\dot{W}_{in} = \text{work done by the pump}$$

$$= \dot{m}v_f(p_{out} - p_{in}) \times 10^2,\ (\text{kW})$$

where,
$$v_f = \text{specific volume of feed water which is at saturated condition}$$
$$\text{at pump inlet}$$

$$p_{out} = \text{Pump exit pressure (bar)}$$
$$= \text{Boiler pressure}$$
$$p_{in} = \text{Pump inlet pressure, (bar)}$$
$$= \text{condenser pressure.}$$

Note: Requirement of utilities

1. Fuel required in boiler

$$\dot{m}_f = \frac{\dot{Q}_{in}}{CV} = \frac{\dot{m}(h_{out} - h_{in})}{CV}$$

2. Cooling water required in condenser $\dot{m}_w = \dfrac{\dot{Q}_{out}}{Cw(t_1 - t_2)} = \dfrac{\dot{m}(h_{in} - h_{out})}{Cw(t_1 - t_2)}$

3.9   THE CARNOT VAPOUR CYCLE

It is more convenient to analyse the performance of steam power plant by means of idealised cycles which are theoretical approximations of the real cycles. The Carnot cycle is an ideal cycle, but non-practical cycle, giving the maximum possible thermal efficiency for a cycle operating on selected maximum and minimum temperature range. The line diagram of Carnot cycle along with its presentation on *T-s* diagram are shown in Fig. 3.13.

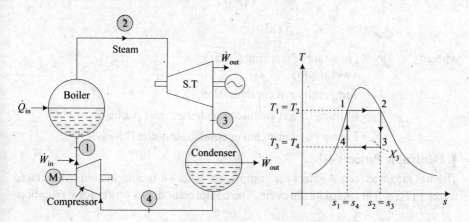

Fig. 3.13   The Carnot vapour cycle

The Carnot cycle is made up of four ideal processes:

1. Isothermal Process 1-2

Saturated water at state 1 (T_1) is fed into the boiler and evaporated into saturated steam at state 2 (T_2).

$$T_1 = T_2$$

The heat supplied in the boiler

$$\dot{Q}_{in} = \dot{m} \ T_1 \ (s_2 - s_1), \text{(kJ)}$$

where, $\dot{m}$ = Flow rate of working fluid, (kg/s)

T_1 = Temperature of evaporation, (K)

s_1 = Entropy of saturated water at boiler inlet, (kJ/kg-K)

s_2 = Entropy of saturated steam at boiler outlet, (kJ/kg-K)

2. Isentropic Process 2-3

The saturated steam from state $2(T_2, s_2)$ is expanded isentropically in the turbine upto state 3 (T_3, s_3, X_3)

$$s_2 = s_3$$

$$\dot{W}_{out} = \dot{m} \ (h_2 - h_3)$$

where, $\dot{m}$ = Flow rate of steam through turbine, (kg/s)

h_2 = Enthalpy of saturated steam at turbine inlet, (kJ/kg).

h_3 = Enthalpy of wet steam at turbine outlet, (kJ/kg).

3. Isothermal Process 3-4

The wet steam from turbine exhaust is partially condensed to state 4 (T_4, s_4, X_4) losing heat $\dot{Q}_{out}$ to cooling water

$$T_3 = T_4$$

$$\dot{Q}_{out} = \dot{m} \ T_3 \ (s_3 - s_4) \ \text{(kJ)}$$

where, $\dot{m}$ = Flow rate of working fluid (kg/s)
(wet steam)

T_3 = Temperature of condensation, (K)

s_3 = Entropy of wet steam at condenser inlet [kJ/kg-K]

s_4 = Entropy of wet steam at condenser outlet [kJ/kg-K]

4. Isentropic Process 4-1

The wet steam at state 4 enters the compressor where it is compressed to saturated water 1 $(T_1, s_1,)$ to complete the cycle. The compressor draws power for its rotation.

$$s_1 = s_4$$

$$W_{in} = \dot{m} \ (h_1 - h_4)$$

where, $\dot{m}$ = Flow rate of working fluid through the compressor (kg/s)

h_1 = Enthalpy of working fluid at compressor outlet (kJ/s)

h_4 = Enthalpy of wet steam at compressor inlet (kJ/s)

3.9.1 Cycle Analysis

1. Efficiency: The thermodynamic efficiency or thermal efficiency is the ratio of net work output of steam power plant and heat supplied.

$$\eta_{th} = \frac{\dot{W}_{out} - \dot{W}_{in}}{\dot{Q}_{in}}$$

But $\quad \dot{W}_{out} - \dot{W}_{in} = \dot{Q}_{in} - \dot{Q}_{out}$ [First law of thermodynamics]

$$\therefore \qquad \eta_{th} = \frac{\dot{Q}_{in} - \dot{Q}_{out}}{\dot{Q}_{in}} = 1 - \frac{\dot{Q}_{out}}{\dot{Q}_{in}}$$

$$\dot{Q}_{in} = \dot{m}\, T_1\, (s_2 - s_1)$$

$$\dot{Q}_{out} = \dot{m}\, T_3\, (s_3 = s_4)$$

But, $\qquad (s_2 - s_1) = (s_3 - s_4)$

$$\therefore \qquad \eta_{th} = 1 - \frac{\dot{m}\, T_3 (s_2 - s_1)}{\dot{m}\, T_1 (s_2 - s_1)}$$

$$= 1 - \frac{T_3}{T_1}$$

2. Specific steam consumption (s.s.c): Specific steam consumption or steam rate is the amount of steam required to produce 1 kWh of work.

$$s.s.c = \frac{\dot{m}}{\dot{W}}\left[\frac{kg/s}{kW}\right] = \frac{\dot{m}}{\dot{W}}\frac{kg}{kWs} = \frac{3600\,\dot{m}}{\dot{W}}\frac{kg}{kWh}$$

$$= \frac{3600}{\dot{W}/\dot{m}}\left[\frac{kg}{kWh}\right]$$

3. Work Ratio (W.R.): Work ratio is the ratio of net plant output to gross (turbine) output.

$$W.R. = \frac{\dot{W}_{out} - \dot{W}_{in}}{\dot{W}_{out}} = \frac{\dot{W}_T - \dot{W}_C}{\dot{W}_T}$$

3.9.2 Limitations of Carnot Cycle

The Carnot cycle is not practical due to the following limitations:

1. Carnot cycle is an ideal cycle consisting of four reversible processes. It is not possible to achieve reversible process in practice.

2. Carnot limitation: The Carnot efficiency depends upon boiler temperature T_1 and condenser temperature T_3.

The *condenser temperature* cannot be reduced below 300 due to the following reasons.

(i) Condesation of steam requires a continuous bulk supply of cooling water. Natural water supply below atmospheric temperature of about 15°C is unavailable.

(ii) If condenser is to be of a reasonable cost and size, the temperature difference between the condensing steam and cooling water must be at least 10°C.

The *boiler temperature* cannot more than 900 K due to nonavailability of materials of construction of boiler tubes and turbine blades to withstand higher temperatures.

The material of construction should be of high reliability when operating at high temperature and the cost should also be reasonable. This is called *metallurgical limitation.*

In fact, the steam in Carnot cycle has a maximum cycle temperature of well below the metallurgical limitation owing to the properties of steam as it is limited by *critical temperature* of 374°C (647 K) otherwise special supercritical steam boilers will be required which suffer from low plant life.

Under best circumstances, the Carnot efficiency will be:

$$\eta_{Carnot} = 1 - \frac{T_3}{T_1} = 1 - \frac{300}{900} = 66.7°C$$

Therefore, maximum cycle efficiency would be well below 100%.

3. Compression of wet steam: Compressing a very wet steam mixture would require a compressor of size and cost comparable with the turbine. Such a compressor would absorb large amount of work produced by turbine and work ratio of power plant would be very low. Such a compressor would also have a short life because of blade erosion and cavitation problems.

4. High steam rate: Due to low steam parameter used in a Carnot cycle as discussed above, steam rate will be very high to produce certain amount of power. The plant would be very large in size and very expensive.

5. Cost of power generation: Cost of power generation has two components:

(i) *Fixed charges consisting interest* and depreciation of plant cost.

(ii) *Running cost* of fuel.

The plant operating on Carnot cycle will produce power at high cost as fixed charges are high due to large size and cost of plant and running cost in high due to Carnot limitations of thermal efficiency.

6. Low plant life: A Carnot cycle works on saturated steam and the wetness of steam increases during its expansion in the turbine. The turbine blades get eroded and its efficiency decreases. It is not possible to supperheat steam at constant temperature process of Carnot cycle.

3.9.3   Uses of Carnot Cycle

The Carnot cycle is not practical due to its limitations discussed above. Still it has certain uses:

1. Design of practical cycle: The Carnot cycle is useful in helping us appreciate the desirable factors for design of practical plant, i.e.,

(i) Maximum possible temperature range of cycle.

(ii) Maximum possible heat addition into the cycle at this maximum cycle temperature.

(iii) A maximum possible temperature of heat rejection.

(iv) A maximum possible work input into the cycle.

2. Thermodynamic constraints: The Carnot cycle also helps to understand the thermodynamic constraints on the design of cycles.

(i) Carnot limitation.

The cycle efficiency will be low due to Carnot limitation.

(ii) Use of superheated steam helps not only to improve cycle efficiency but also helps to control the erosion of turbine blades and hence improve the life of plant.

3.10   THE RANKINE CYCLE

In order to overcome most of the limitations of Carnot cycle, the Rankine cycle is used as the basic cycle for steam power plants. An idealized cycle for a simple power plant is shown in Fig. 3.14. The Rankine cycle with superheated steam is 1-2-3-4-1 and that with dry and saturated steam is 1-2'-3'-4-1.

It is made up of four practical processes:

1. Isobaric heating process 1-2: Heat is added to feedwater in three steps:

(i) The water is heated upto its saturated value (process 1-A) in the economizer at constant pressure.

(ii) The saturated water is evaporated at constant temperature and pressure (process A-2) in the boiler.

(iii) The saturated steam is superheated at constant pressure (process 2'-2) in the superheater.

Applying 1st law of thermodynamics to boiler,

$$\dot{Q}_{in} = \dot{m}\,(h_2 - h_1)$$

where, $\dot{m}$ = mass flow rate of working fluid, (kg/s)

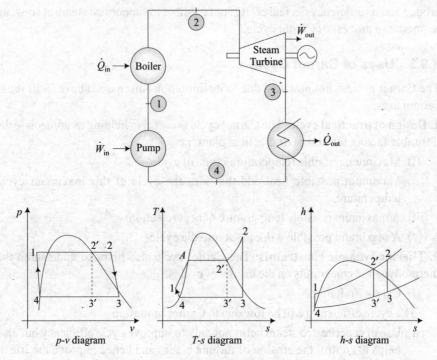

Fig. 3.14 Ideal Rankine cycle

2. Isentropic expansion process 2-3: The high pressure superheated steam is expanded to a low pressure in a steam turbine isentropically in an ideal Rankine cycle. However, in actual plant, friction of steam flow results in the expansion with increasing entropy.

Applying first law to this process

$$\dot{Q}_{out} = \dot{W}_T = \dot{m}\,(h_2 - h_3)$$

where, $\dot{W}_T$ = Turbine work [kW]

3. Isobaric cooling process 3-4: The low pressure wet steam from the steam turbine is *completely* condensed at constant condenser pressure back into saturated water. The latent heat of condensation is thereby rejected to the condenser cooling water, which, in turn, rejects this heat to the atmosphere.

Applying 1st law of thermodynamics to this process,

$$\dot{Q}_{out} = \dot{m}(h_3 - h_4)$$

4. Isentropic compression process 4-1: The low pressure saturated water is pumped back to the boiler pressure and, in doing so, it becomes sub-saturated. The water then enters the boiler and a new cycle begins.

Applying 1st law to this process,

$$\dot{W}_{in} = \dot{m}(h_1 - h_4)$$

However, $\dot{W}_{in}$ can be neglected with reasonable accuracy and it can be assumed that $h_1 = h_4$. The thermal efficiency of the cycle is given by:

$$\eta_{th} = \frac{\dot{W}_{out} - \dot{W}_{in}}{\dot{Q}_{in}} = \frac{\dot{m}(h_2 - h_3) - \dot{m}(h_1 - h_4)}{\dot{m}(h_2 - h_1)}$$

$$= \frac{(h_2 - h_3) - (h_1 - h_4)}{(h_2 - h_1)}$$

If $\dot{W}_{in}$ can be neglected,

$$\eta_{th} = \frac{\dot{W}_{out}}{\dot{Q}_{in}} = \frac{(h_2 - h_3)}{(h_2 - h_1)}$$

Specific steam consumption or steam rate is given by:

$$\text{s.s.c.} = \frac{3600}{\dot{W}/\dot{m}} \left[\frac{kg}{kwh} \right] = \frac{3600}{h_2 - h_3} \text{ kg/kWh.}$$

Work Ratio, $\qquad WR = \dfrac{\dot{W}_{out} - \dot{W}_{in}}{\dot{W}_{out}}$

$$= \frac{(h_2 - h_3) - (h_4 - h_1)}{h_2 - h_3}$$

3.11 COMPARISON OF A CARNOT CYCLE AND A RANKINE CYCLE

The advantages of a Rankine cycle over a Carnot cycle are listed in Table 3.3. The Table also explains how the limitations of Carnot cycle are overcome in a Rankine cycle. (Points 4, 5, 9, 10 of Table 3.3).

Table 3.3 Comparison of Carnot and Rankine cycle

Carnot cycle	*Rankine cycle*
1. It consists of two isothermals and two isentropic processes.	1. It consists of two isobaric and two isentropic processes.
2. It is an ideal cycle and Carnot be achieved in practice.	2. All steam power plants are desigred on Rankine cycle.
3. The steam at turbine input is saturated steam.	3. The steam at turbine input is superheated steam in modern power plants.
4. Steam is partially condensed in the condenser.	4. Steam is fully condensed in the condenser.

Table Contd.

Table Contd.

Carnot cycle	Rankine cycle
5. A compressor is used to compress two-phase wet steam. The size and power consumption are very large.	5. A pump is used to raise the pressure of water to boiler pressure. The size and power consumtion are very small.
6. The thermal efficiency of a Carnot cycle is higher than an equivalent Rankine cycle working on saturated steam.	6. The thermal efficiency of a Rankine cycle can be improved by the use of high pressure superheated steam.
7. Superheated steam cannot be used due to difficulty of superheating at constant temperature simultaneously expanding the steam.	7. Superheated steam is used in Rankine cycle as superheating can be easily carried out at constant pressure.
8. Specific steam consumption is high.	8. Specific steam consumption is low.
9. Work ratio is very low as lot of power is required in the comperssion of wet steam.	9. Work ratio is very high as pump power consumption is very low.
10. Plant life is low due to operation of steam turbine and compressor with two-phase mixture of steam and water.	10. Plant life is high as steam turbine works mainly with superheated steam, or dryness fraction is very high. Pump operation is simple and reliable.

Example 3.34: Steam is supplied to a steam turbine at 5 MPa dry saturated. Condenser pressure is 5 kPa. Showing the Rankine cycle on T-s diagram, determine the simple Rankine efficiency. [U.P.T.U. I Sem., 2004-2005]

Solution:

$$p_1 = 5 \text{ MPa} = 50 \text{ bar}$$

$$p_2 = 5 \text{ kPa} = 0.05 \text{ bar}$$

From saturated steam tables, for $p_1 = 50$ bar

$$h_1 = h_{g_1} = 2794.3 \text{ kJ/kg}$$

$$s_1 = s_{g_1} = 5.973 \text{ kJ/kg-K}$$

$$h_4 = h_{f_4} = 1154.2 \text{ kJ/kg-K}$$

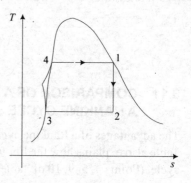

For isentropic expansion through turbine

$$s_1 = s_2 = s_{f2} + X_2 \, s_{fg2}$$

From saturated steam tables, for $p_2 = 0.05$ bar

$$s_{f2} = 0.476 \text{ kJ/kg-K}$$

$$s_{fg2} = 7.920 \text{ kJ/kg}$$

$$h_{f2} = 137.8 \text{ kJ/kg}$$

$$h_{fg2} = 2423.7 \text{ kJ/kg}$$

$$\therefore \qquad X_2 = \frac{s_1 - s_{f2}}{s_{fg2}} = \frac{5.973 - 0.476}{7.92} = 0.694$$

$$h_2 = h_{f2} + x_2\ h_{fg2} = 137.8 + 0.694\,(2423.7) = 1820 \text{ kJ/kg}.$$

$$\eta_R = \frac{h_1 - h_2}{h_1 - h_4} = \frac{h_1 - h_2}{h_1 - h_3} \qquad\qquad \text{(Neglecting pump work)}$$

$$= \frac{2794.3 - 1820}{2794.3 - 137.8} = \frac{974.3}{2656.5} = 36.67\%$$

Example 3.35: A steam power plant working on Rankine cycle has a steam supply pressure of 20 bar and condenser pressure of 0.5 bar. If the initial condition of steam is dry and saturated, calculate the Carnot and Rankine efficiencies of the cycle, neglecting pump work. [U.P.T.U. II Sem., 2005-2006]

Solution:
$$p_1 = 20 \text{ bar}$$
$$p_2 = 0.5 \text{ bar}$$

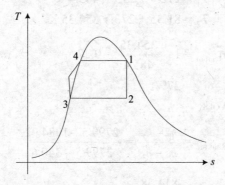

From steam tables for saturated steam, against a pressure of 20 bar,

$$h_1 = hg_1 = 2799.5 \text{ kJ/kg}$$
$$s_1 = sg_1 = 6.341 \text{ kJ/kg-K}.$$

For isentropic expansion through turbine,

$$s_1 = s_2$$

From saturated steam tables at a pressure of 0.5 bar,

$$s_{f2} = 1.091 \text{ kJ/kg-K}$$
$$s_{fg2} = 6.504 \text{ kJ/kg-K}$$
$$h_{f2} = 340.6 \text{ kJ/kg}$$
$$h_{fg2} = 2305.4 \text{ kJ/kg}.$$
$$\therefore \qquad s_1 = s_2 = s_{f2} + X_2\ s_{fg2}$$

$$\therefore \qquad X_2 = \frac{s_1 - s_{f2}}{s_{fg2}} = \frac{6.341 - 1.091}{6.504} = 0.80$$

$$\therefore \qquad h_2 = h_{f2} + X_2 h_{fg2}$$

$$= 340.6 + 0.80 \, (2305.4)$$

$$= 1984.92 \text{ kJ/kg}$$

Carnot efficiency,

$$\eta_{carnot} = 1 - \frac{T_2}{T_1}$$

From saturated steam tables:

For a pressure of 20 bar,

Saturation temperature, t_{s1}, = 212.4°C

Saturation temperature for 0.5 bar,

$$t_{s2} = 81.35°C$$

$$T_1 = 212.4 + 273 = 485.4 \text{ K}$$

$$T_2 = 81.35 + 273 = 354.35 \text{ K}$$

$$\therefore \qquad \eta_{carnot} = 1 - \frac{354.35}{485.4} = 0.27$$

$$= 27\%$$

Neglecting pump work,

$$\eta_{Rankine} = \frac{h_1 - h_2}{h_1 - h_3} = \frac{2799.5 - 1984.92}{2799.5 - 340.6}$$

$$= \frac{814.58}{2458.9} = 0.33$$

$$= 33\%$$

Example 3.36: A steam power plant working on Rankine cycle has steam parameters at turbine inlet as 100 bar and 550°C and condenser pressure of 0.05 bar. Determine the cycle efficiency, specific steam consumption and work ratio if all processes are revesible.

Solution:

(i) Steam turbine

From superheated steam tables, at p_1 = 100 bar and t_1 = 550°C,

$$h_1 = 3500.9 \text{ kJ/kg}$$

$$s_1 = 6.756 \text{ kJ/kg}$$

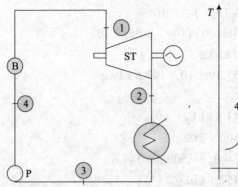 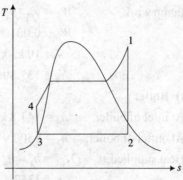

Process 1-2 is reversible and isentropic

∴ $$s_1 = s_2$$

At turbine exit, from saturated steam tables,

$$p_2 = 0.05 \text{ bar}$$
$$s_{f2} = 0.476 \text{ kJ/kg}$$
$$s_{fg2} = 7.920 \text{ kJ/kg}$$
$$h_{f2} = 137.80 \text{ kJ/kg}$$
$$h_{fg2} = 2423.7 \text{ kJ/kg}$$
$$s_1 = s_{f2} + x_2 s_{fg2}$$

∴ $$X_2 = \frac{s_1 - s_{f2}}{s_{fg2}} = \frac{6.756 - 0.476}{7.920} = 0.80.$$

∴ $$h_2 = h_{f2} + X_2\, h_{fg2} = 137.80 + 0.8\,(2423.7)$$
$$= 2076.76 \text{ kJ/kg}.$$

∴ Turbine work,

$$W_T = h_1 - h_2 = 3500.9 - 2076.76 = 1424.14 \text{ kJ/kg}$$

(ii) Condenser

At inlet, $\qquad h_2 = 2076.76 \text{ kJ/kg}$

At outlet, $\qquad h_3 = h_{f_3} = 137.8 \text{ kJ/bg}$

Heat removed $\qquad = h_2 - h_3 = 2076.76 - 137.8 = 1938.96 \text{ kJ/kg}$

(iii) Pump

At pump inlet, $\qquad h_3 = 137.80 \text{ kJ/kg}$

At pump outlet, $\qquad h_4 = h_3 + W_p$

Assuming water as incompressible,

Pump work, $\qquad W_p = v_{f3} (p_4 - p_3) \times 10^2$

$\qquad\qquad\qquad\qquad W_p = 0.001005 (100 - 0.05) \times 10^2$

$\qquad\qquad\qquad\qquad\quad = 10 \text{ kJ/kg}$

$\therefore \qquad\qquad\qquad\qquad h_4 = 137.80 + 10 = 147.8 \text{ kJ/kg}$

(iv) Boiler

At inlet of boiler, $\quad h_4 = 147.8 \text{ kJ/kg}$

At outlet of boiler, $\quad h_1 = 3500.9 \text{ kJ/kg}$

Heat supplied, $\qquad Q_{1-4} = h_1 - h_4 = 3500.9 - 147.8$

$\qquad\qquad\qquad\qquad\quad = 3357.1 \text{ kJ/kg}$

(v) Steam power plant

Cycle efficiency,

$$\eta = \frac{\text{Net work done}}{\text{Heat supplied}}$$

$$= \frac{W_T - W_P}{Q_{1-4}} = \frac{1424.14 - 10}{3353.1} = 0.43$$

$$= \mathbf{43\%} \quad \textbf{Ans.}$$

Specific steam consumptions (s.s.c),

$$\text{s.s.c.} = \frac{3600}{W_T - W_p} = \frac{3600}{1414.14} = 2.55 \text{ kg/kWh}$$

Work Ratio (*WR*),

$$WR = \frac{W_T - W_p}{W_T} = \frac{1414.14}{1424.14} = \mathbf{0.99 \text{ Ans.}}$$

QUESTION BANK NO. 3

1. With the help of $(T - v)$ and $(T - s)$ diagrams, explain the difference between 'Wet; 'Dry saturated' and 'Superheated' steams.

2. How can properties of steam be calculated with the help of Steam Tables and Mollier diagram?.

3. Draw a basic layout of a steam power plant and explain the four basic components.

4. Explain the working of a simple Rankine cycle with the help of $T - s, p - v$ and $h - s$ diagrams.

5. Discuss the generation of steam at constant pressure. Show various processes on $T - v, p - v$ and $T - s$ diagrams.

6. Give reasons why the Carnot cycle cannot be considered the practical cycle for steam power plant though its efficiency is maximum.

7. How does Rankine cycle overcome the limitations of Carnot cycle in reference to steam power plants?

8. For the superheated steam the isobaric and isothermal lines diverge on "Mollier chart." Explain the statement.

9. Show the following processes on Mollier diagram.
   - (i) Isentropic expansion
   - (ii) Adiabatic expansion with friction
   - (iii) Throttling process
   - (iv) Isenthalpic process
   - (v) Constant pressure heat addition

10. Draw the Carnot cycle on *T-s* diagram and mark the different thermodynamic processes.

11. Compare Rankine cycle and Carnot cycle for a steam power plant.

12. Explain the following:
    - (a) Triple point of water
    - (b) Critical point
    - (c) Dryness fraction

13. Write short notes on:
    - (a) Mollier diagram
    - (b) Rankine cycle

Internal Combustion Engine

4

4.1 INTRODUCTION

The internal combustion engines (briefly written as I.C. engines) are piston-cylinder arrangements. The combustion of fuel takes place inside the engine cylinder. There are gas, petrol and diesel engines. The high temperature and high pressure gas produced by combustion of fuel in air pushes the piston out and heat energy is converted into mechanical work.

The internal combustion engines are very widely used in the following fields:

1. Electrical power generation.
2. Automobile engine for driving scooters, motor cycles, cars, trucks, buses, etc.
3. Marine engine for propulsion of ships.
4. Aero-engines for aircraft.
5. Locomotive engines for railways.
6. Earth moving machinery such as bulldozer, scrapper, power shovel, cranes, etc.
7. Industrial engines for standby power and mechanical drive.

The widespread applications of I.C. engines can be due to the following features:

1. Overall high efficiency
2. Mechanical simplicity
3. Absence of elaborate auxiliary equipment such as boilers, condensers, heat exchanger, etc.
4. Low initial cost and easily available.
5. Very compact and require little space and simple foundations.
6. Quick starting.

The main limitation of this type of engine is high cost of fuel and polluting emissions. Therefore, this type of engine is not able to compete with steam turbine for bulk generation of electricity. Steam power plants can work on cheap fuels such as coals, oils and gases.

4.2 TYPES OF I.C. ENGINES

I.C. engines are made in very different designs and construction for various applications.

1. Applications

   (i) Stationary engines

   (ii) Mobile engines

  (iii) Automobile engines

  (iv) Marine engines

   (v) Aero-engines

  (vi) Industrial engines.

2. Fuels

   (i) Diesel engines

   (ii) Petrol engines

  (iii) Gas engines

  (iv) Kerosene engines

   (v) Alcohol engines

  (vi) Bio-gas engines

 (vii) Dual fuel engines.

3. Number of Operating Strokes

   (i) Two-stroke cycle petrol engine

   (ii) Four-stroke cycle petrol engine

  (iii) Two-stroke cycle diesel engine

  (iv) Four-stroke cycle diesel engine.

4. Cycle of Combustion

   (i) Otto cycle engine where petrol or gas combustion takes place at constant volume.

   (ii) Diesel cycle engine where combustion of diesel oil or gas or combination of fuels takes place at constant pressure.

  (iii) Dual cycle engines which are semi-diesel cycle engines and fuel combustion takes place partly at constant volume and partly at constant pressure.

5. Types of Construction

(i) Engines may be single cylinder engines or multi-cylinder engines with different configurations.

(ii) Air cooled engines or water cooled engines

(iii) High speed engines or low speed engines

(iv) Constant speed engines or variable speed engines

(v) Spark ignition engines (SI engines) or compression ignition engines (CI engines).

4.3   BASIC ARRANGEMENT

The main parts of I.C. engines can be grouped as:

(i) Cylinder and cylinder head.

(ii) Piston, piston rings, piston rod or connecting rod.

(iii) Crank, crankshaft and crank case.

(iv) Valves, valve operating mechanism and crankshaft.

(v) Flywheel and governor.

(vi) Engine bearings and lubrication system.

(vii) Engine cooling system, radiator, water pump.

(viii) Spark plug or fuel injectors, fuel ignition system.

(ix) Fuel pump, injector or carburettor and fuel supply system.

The cylinder is closed at one end with cylinder head and it is filled with a mixture of fuel and air. The crankshaft rotates and piston is pushed up to compress the mixture. The mixture is ignited and it burns creating a high pressure and high temperature gas on the top of the piston. The piston is pushed down the cylinder. The piston rod and crank work on the principle of wheel and axle and crankshaft is given a rotary motion. The flywheel mounted on crankshaft stores energy and keeps crankshaft turning steadily. This steady rotary motion is used to drive generator or wheels or other machinery.

There are certain terms used to describe the operation and construction of I.C. engines.

1. Dead Centres

The piston moves up and down in the cylinder during different working strokes like suction stroke, compression stroke, working stroke or exhaust stroke. The extreme positions of piston are called dead centres. The piston reverses its direction of motion at the dead centres. Piston at the extreme position at the cylinder head side is called *Top Dead Centre* (TDC) and when bottommost end as *Bottom Dead Centre* (BDC).

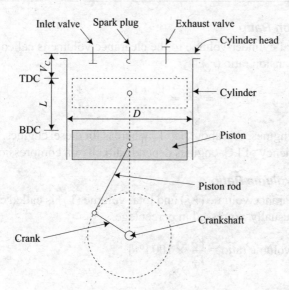

Fig. 4.1   Piston-cylinder arrangement

2. Bore (D)

The inside diameter of a cylinder is called the bore.

3. Stroke Length (L)

The distance travelled by the piston between dead centres is called stroke length.

4. Clearance Volume (V_C)

The cylinder volume between cylinder head and TDC is called clearance volume. The piston cannot be allowed to travel upto cylinder head and damage the engine. A cushion is provided. The clearance volume also forms the combustion chamber and accommodates the air-fuel mixture after compression stroke and before ignition.

5. Swept Volume (V_S)

The cylinder volume between TDC and BDC is called swept volume, displacement volume or stroke volume.

$$V_C = \frac{\Pi}{4} D^2 \times L$$

6. Total Cylinder Volume (V_T)

The volume occupied by the working fluid, when the piston is at the bottom dead centre, is called total cylinder volume.

Total cylinder volume,

$$V_T = V_S + V_C$$

7. Compression Ratio

The ratio of total cylinder volume to the clearance volume is called compression ratio (r_c) or expansion ratio (r_e).

$$r_c = \frac{V_T}{V_C} = \frac{V_S + V_C}{V_C} = 1 + \frac{V_S}{V_C}$$

For petrol engines, r_c is about 8 to 15 whereas for diesel engines, r_c is about 15 to 22. The efficiency of I.C. engines depends directly on compression ratio.

8. Clearance Volume Ratio

The ratio of clearance volume (V_C) and total volume (V_T) is called clearance volume ratio. It is usually expressed in percentage.

$$\text{Clearance volume ratio} = \frac{V_C}{V_T} \times 100 \, [\%]$$

9. Piston Speed

The piston moves with variable speed during the working cycle. The average speed of the piston is called piston speed.

Piston speed = 2 LN

where, L = length of the stroke [m]

N = speed of the engine [r.p.m]

4.4   ENGINE PERFORMANCE

The performance of the I.C. engine can be expressed in term of work output, power output, mean effective pressure, cycle efficiency, mechanical efficiency, etc.

1. Indicator Diagram

When the operations (processes) of a thermodynamic cycle are plotted on p-V coordinates, a closed figure is obtained. This is called indicator diagram. The area of the indicator diagram gives the net work done during the cycle. The indicator diagram

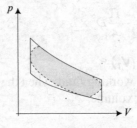

Fig. 4.2   Indicator diagram

for an engine can be plotted by an instrument called indicator. The area of the indicator diagram can be found out with the help of a *plani-meter*.

2. Mean Effective Pressure

The pressure inside the cylinder changes continuously as the engine works. It becomes difficult to find out the work done from the area of the indicator diagram. The work done can be easily found out by drawing an equivalent rectangle, i.e., a rectangle having the same length and the same area as that of indicator diagram. The height of the equivalent rectangle is called mean effective pressure.

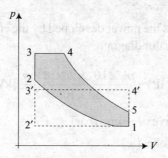

Fig. 4.3 Mean effective pressure

In Fig. 4.3, 1-2-3-4-5-1 is the indicator diagram and 1-2′-3′-4′-1 is equivalent rectangle. The mean effective pressure is given by the height 2′-3′.

Mean effective pressure is also defined as the ratio of work done per cycle to stroke volume.

$$p_m = \frac{\text{Work done per cycle}}{\text{Stroke volume}}$$

$$= \frac{\text{Heat supplied} - \text{Heat rejected}}{V_1 - V_2}$$

The mean effective pressure calculated from theoretical indicator diagram is called *theoretical* mean effective pressure. If calculated from actual indicator diagram, it is called *actual* mean effective pressure.

3. Cycle Efficiency

It is the ratio of work done to the heat supplied during a cycle.

$$\eta = \frac{\text{Work done}}{\text{Heat supplied}}$$

Since work done during a cycle is equal to heat supplied minus heat rejected, the thermal efficiency of a cycle may also be expressed as,

$$\eta = \frac{\text{Heat supplied} - \text{Heat rejected}}{\text{Heat supplied}}$$

For a petrol engine (Otto cycle), the air standard efficiency depends upon compression ratio.

$$\eta = 1 - \frac{1}{(r_c)^{\gamma-1}}$$

where, $\gamma = 1.4$ for air.

4. Indicated Power (IP)

The indicated power (IP) is the power developed by the engine cylinder. It is calculated with the help of indicator diagram.

$$\text{IP} = \frac{p_m \times 10^5 \times L \times A \times n}{60 \times 100} \ [kW]$$

where, p_m = mean effective pressure, [bar]

L = length of stroke, [m]

A = Area of piston, [m^2]

$$A = \frac{\Pi}{4} D^2$$

D = Bore [m]

n = Number of working strokes per minute

For two-stroke cycle engine,

$n = N$

For four-stroke cycle engine,

$n = N/2$

where, N = Engine speed [rpm].

5. Shaft Power or Brake Power (BP)

The shaft power is the power available at the crankshaft. It is found out from indicated power after deducting friction losses. Shaft power is measured by the help of a brake dynamometer and, therefore, it is also called brake power.

The brake power,

$$BP = \frac{2 \Pi N T}{60 \times 1000} \ [kW]$$

where, T = Torque on engine shaft, [Nm]

$T = W \times l$

W = Brake load of dynamometer [N]

l = Length of arm of dynamometer [m]

N = Engine speed [rpm]

6. Mechanical Efficiency

Some power is lost in overcoming the engine friction and is called friction power (FP),

$$FP = IP - BP$$

Mechanical efficiency is defined as the ratio of brake power to indicated power,

$$\eta_m = \frac{BP}{IP}$$

Example 4.1: A 4-stroke diesel engine has length of 20 cm and diameter of 16 cm. The engine is producing power of 25 kW when it is running at 2500 rpm. Find the mean effective pressure of the engine.   [U.P.T.U. II Sem., 2002–03]

Solution:

Stroke length, $L = 20$ cm $= 0.2$ m

$D = 16$ cm $= 0.16$ m

$P = 25$ kW

$N = 2500$ rpm.

$$P = \frac{P_m \times 10^5 \times L \times A \times n}{60 \times 1000} \text{ [kW]}$$

$$P_m = \frac{60{,}000\, P \times 2}{10^5 \times 0.2 \times \dfrac{\Pi}{4}(0.16)^2 \times 2500}$$

$$= \textbf{2.984 bar}\quad \textbf{Ans.}$$

Example 4.2: A 4-stroke petrol engine is developing mean effective pressure of 21 bar. It is developing power at the rate of 25 kW at a running speed of 1200 rpm. The ratio of stroke to bore is 1.2. Calculate the bore and stroke of the engine.

[U.P.T.U. II Sem., 2003–04]

Solution:

$P_m = 21$ bar

$P = 25$ kW

$N = 1200$ rpm

It is a 4-stroke engine,

$$\therefore \quad n = \frac{N}{2} = \frac{1200}{2} = 600 \text{ strokes per minute.}$$

$$\frac{L}{D} = 1.2.$$

$$P = \frac{p_m \times 10^5 \times L \times A \times n}{60 \times 1000}$$

$$= \frac{p_m \times 10^5 \times (1.2 \, D) \frac{\Pi}{4}(D^2) n}{60 \times 1000}$$

$$\therefore \quad D^3 = \frac{60{,}000 \, P \times 4}{p_m \times 10^5 \times 1.2 \times \Pi \times n} = \frac{60{,}000 \times 25 \times 4}{21 \times 10^5 \times 1.2 \times \Pi \times 600}$$

$$D = 0.108 \text{ m} = 108 \text{ mm}$$

$$L = 1.2 \, D = \textbf{129.7 mm} \quad \textbf{Ans.}$$

Example 4.3: A two-cylinder four-stroke cycle I.C. engine is to be designed to develop 1.5 hp at 1200 rpm. The m.e.p of the cycle is limited to 6 bar. Determine the bore diameter and stroke of the engine if stroke = 1.2 × bore diameter.

Solution:

Power to be developed per cylinder,

$$P = \frac{1.5 \times 0.735}{2} = 0.55 \text{ kW.}$$

$$P = \frac{p_m \times 10^5 \times L \times A \times n}{60 \times 1000}$$

$$= \frac{p_m \times 10^5 \times (1.2 \, D) \times \frac{\Pi}{4} D^2 \times N}{60 \times 1000 \times 2}$$

$$\therefore \quad D^3 = \frac{60{,}000 \, P \times 2 \times 4}{p_m \times 10^5 \times 1.2 \times \Pi \times N}$$

$$= \frac{60{,}000 \times 0.55 \times 2 \times 4}{6 \times 10^5 \times 1.2 \times \Pi \times 1200} = 0.1167 \text{ m}^3$$

$$D = \sqrt[3]{0.1167} = 0.0488 \text{ m}$$

$$= 48.8 \text{ mm}$$

$$L = 1.2D = 1.2 \times 48.8 = \textbf{58.6 mm} \quad \textbf{Ans.}$$

Example 4.4: An engine equipped with a cylinder having a bore of 15 cm and a stroke of 45 cm operates on air Otto cycle. If the clearance volume is 2000 cm³, compute the air standard efficiency.

Solution:

Bore, $\qquad\qquad\qquad D = 15 \text{ cm} = 0.15 \text{ m}$

Stroke, $\qquad\qquad\quad L = 45 \text{ cm} = 0.45 \text{ m}$

Stroke volume, $\qquad V_s = \dfrac{\Pi}{4}D^2 L = \dfrac{\Pi}{4}(0.15)^2 \times 0.45$

$$= 0.007952 \text{ m}^3$$

Clearance volume, $\qquad V_c = 2000 \text{ cm}^3 = 0.002 \text{ m}^3$

Compression ratio, $\qquad r_c = \dfrac{V_c + V_s}{V_c} = \dfrac{0.002 + 0.007952}{0.002} = 5$

Air standard efficiency,

$$\eta = 1 - \frac{1}{(r_c)^{\gamma - 1}} = 1 - \frac{1}{(5)^{1.4 - 1}} = 0.474$$

$$= \textbf{47.4\%} \quad \textbf{Ans.}$$

Example 4.5: In an automobile engine working on Otto cycle, energy generated per cycle is twice as much rejected through the exhaust. Calculate the thermal efficiency and compression ratio. Take $\gamma = 1.4$ for the working fluid.

Solution:

Heat supplied $\qquad = Q_1$

Heat rejected $\qquad = Q_2$

Now $\qquad\qquad Q_1 = 2Q_2$

Thermal efficiency,

$$\eta = \frac{Q_1 - Q_2}{Q_1} = 1 - \frac{Q_2}{Q_1} = 1 - \frac{1}{2} = 0.5$$

$$= 50\%$$

Also,
$$\eta = 1 - \frac{1}{(r_c)^{\gamma-1}}$$

$$r_c = \left(\frac{1}{1-\eta}\right)^{\frac{1}{\gamma-1}} = \left(\frac{1}{1-0.5}\right)^{\frac{1}{0.4}}$$

$$= 5.66 \quad \textbf{Ans.}$$

Example 4.6:   Calculate the thermal efficiency and compression ratio for an automobile working on Otto cycle, if the energy generated per cycle is thrice that of rejected during the exhaust. Consider working fluid an ideal gas with $\gamma = 1.4$.

[U.P.T.U. II Sem., 2000–01]

Solution:

Heat supplied per cycle $= Q_1$

Heat supplied per cycle $= Q_2$

Now, $Q_1 = 3Q_2$.

Thermal efficiency,

$$\eta = \frac{Q_1 - Q_2}{Q_1} = 1 - \frac{Q_2}{Q_1}$$

$$= 1 - \frac{1}{3} = 0.6667$$

$$= 66.67\%$$

Also $\eta = 1 - \frac{1}{(r_c)^{\gamma-1}}$

$$\therefore \qquad r_c = \left(\frac{1}{1-\eta}\right)^{\frac{1}{\gamma-1}} = \left(\frac{1}{1-0.6667}\right)^{\frac{1}{0.4}}$$

$$= 15.59 \quad \textbf{Ans.}$$

4.5   FOUR-STROKE PETROL ENGINE

In a petrol engine, the ignition of air and petrol inside the cylinder is initiated by a spark plug. It is also called spark ignition (SI) engine. All petrol engines work on Otto cycle. It requires four strokes of piston (travel between TDC and BDC) to complete one cycle of operation.

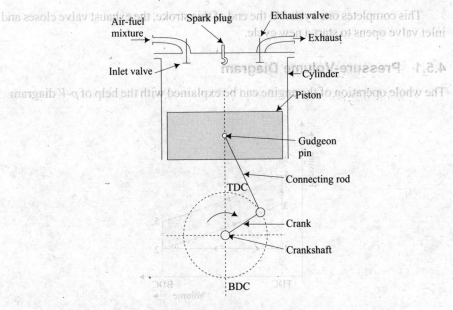

Fig. 4.4 Four-stroke petrol engine

1. Suction or Charging Stroke

The inlet valve opens and charge (fuel + air) is sucked into the cylinder. The exhaust valve remains closed. The piston moves from top dead centre (TDC) to bottom dead centre (BDC), therefore, creating low pressure at piston top. The charge is sucked in due to pressure difference between outside air and inside cylinder.

2. Compression Stroke

Both inlet and and exhaust valves are closed. The piston moves from BDC to TDC, thereby compressing the change. The pressure and temperature of the compressed charge increase. At the end of compression stroke, the charge is ignited with the help of a spark plug. There is sudden increase of pressure and temperature of burnt gases.

3. Expansion or Working Stroke

Both inlet and exhaust valves remain closed. The high pressure of the burnt gases pushes the piston downwards. During this expansion, heat energy produced by combustion of fuel is converted into mechanical work of piston. The piston moves from TDC to BDC.

4. Exhaust Stroke

Exhaust valve is opened by valve operating mechanism but inlet valve remains closed. The piston moves from BDC to TDC by inertia thereby pushing the burnt gases out of cylinder.

This completes one cycle. At the end of this stroke, the exhaust valve closes and inlet valve opens to start a new cycle.

4.5.1 Pressure-Volume Diagram

The whole operation of the engine can be explained with the help of *p-V* diagram.

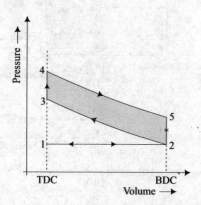

Fig. 4.5   Pressure-volume diagram

Process 1-2 : Suction or charging

Process 2-3 : Isentropic compression of the charge

Process 3-4 : Combustion at constant volume

Process 4-5 : Isentropic expansion of products of combustion

Process 5-2 : Sudden release of burnt gases through exhaust valve

Process 2-1 : Sweeping out of exhaust gases to atmosphere.

4.5.2 Valve Timing Diagram

The inlet valve, spark plug and exhaust valve open or close at specific time of cycle ensuring advance or delay operations due to time required by the working fluid and valves. The valve timing diagram is shown for a typical four-stroke petrol engine in Fig. 4.6.

TDC : Top Dead Centre

BDC : Bottom Dead Centre

IVO : Inlet valve opens (10°–20° before TDC)

IVC : Inlet valve closes (30°–40° after BDC)

IGN : Ignition (20°–30° before TDC)

EVO : Exhaust valve opens (30°–50° before BDC)

EVC : Exhaust valve closes (10°–15° after TDC).

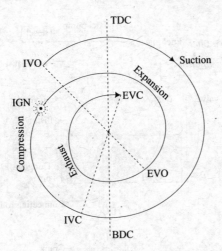

Fig. 4.6 Valve timing diagram

4.6 FOUR-STROKE DIESEL ENGINE

In a diesel engine, air is sucked into the cylinder and compressed to high pressure and temperature. The fuel is injected into the hot compressed air at the end of compression stroke. Ignition and combustion of fuel take place. There is no need of a spark plug. Therefore, diesel engine is also called compression ignition (CI) engine.

Diesel engine works on diesel cycle. It requires four strokes of the piston to complete one cycle of operation.

1. Suction or Charging Stroke

The inlet valve opens. The piston moves from TDC to BDC. The volume above the piston space increases and pressure falls. The air is sucked into the cylinder due to pressure difference between outside air and inside cylinder.

2. Compression Stroke

Both inlet and exhaust valves are closed. The piston moves from BDC to TDC. The pressure and temperature of air inside the cylinder increases. Shortly before the piston reaches TDC, fuel oil is injected into the cylinder at very high pressure. There is atomization of fuel particles and is ignited due to high temperature of air. There is sudden increase of pressure and temperature of the products of combustion above the piston.

3. Expansion or Working Stroke

Both inlet and exhaust valves remain closed. The high pressure of burnt gases pushes the piston downwards. During this expansion or working stroke, the heat energy released by combustion of fuel is converted into mechanical work of piston. The piston moves from TDC to BDC.

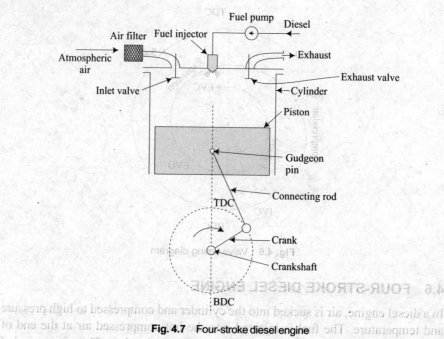

Fig. 4.7 Four-stroke diesel engine

4. Exhaust Stroke

Inlet valve remains closed and exhaust valve is opened. The piston moves from BDC to TDC due the inertia forces and pushes the burnt gases out through the exhaust valve.

This completes one cycle. At the end of this stroke, the exhaust valve closes and inlet valve opens to start a new cycle.

4.6.1 Pressure-Volume Diagram

The whole operation of the engine in one cycle can be explained with the help of pressure-volume diagram given in Fig. 4.8.

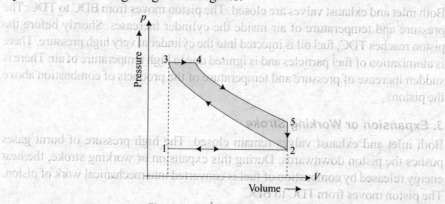

Fig. 4.8 Pressure-volume diagram

Process 1-2 : Suction of air

Process 2-3 : Isentropic compression of air

Process 3-4 : Combustion of fuel at constant pressure.

Process 4-5 : Isentropic expansion of products of combustion.

Process 5-2 : Sudden release of burnt gases through exhaust valve.

Process 2-1 : Sweeping out of exhaust gases to atmosphere.

4.6.2 Valve Timing Diagram

Valve timing diagram for a typical diesel engine is shown in Fig. 4.9.

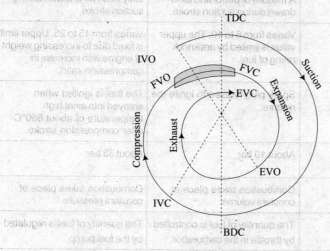

Fig. 4.9 Valve timing diagram

TDC : Top dead centre

BDC : Bottom dead centre

IVO : Inlet valve opens (10° – 20° before TDC)

IVC : Inlet valve closes (25° – 40° after BDC)

FVO : Fuel valve opens (10° – 15° before TDC)

FVC : Fuel valve closes (15° – 20° after TDC)

EVO : Exhaust valve opens (30° – 50° before BDC)

EVC : Exhaust valve closes (10° – 15° after TDC).

4.7 COMPARISON OF PETROL AND DIESEL ENGINES

The basic differences in the operation of a petrol engine and a diesel engine are given in Table 4.1.

Table 4.1 Comparison of Petrol and Diesel engines

Sl. No.	Characteristic	Petrol (SI) Engine	Diesel (CI) Engine
1.	Basic cycle of operation	Otto cycle	Diesel cycle
2.	Common fuel	Petrol (gasoline) with high self-ignition temperature	Diesel with low self-ignition temperature
3.	Fuel supply	A carburettor is used to prepare the required strength of air-fuel mixture. In MPFI system, a fuel injection is used and there is no carburettor.	Fuel is injected at high pressure by a fuel pump through fuel injector. No carburettor is required.
4.	Suction	A mixture of petrol and air is drawn during suction stroke.	Only fresh air is drawn during suction stroke.
5.	Compression ratio	Varies from 6 to 10. The upper value is limited by antiknock rating of fuel.	Varies from 15 to 25. Upper limit is fixed due to increasing weight of engine with increase in compression ratio.
6.	Ignition	Spark plug is used to ignite the mixture.	The fuel is ignited when sprayed into air at high temperature of about 600°C after compression stroke.
7.	Pressure after compression	About 10 bar	About 35 bar
8.	Combustion	Combustion takes place at constant volume.	Combustion takes place at constant pressure.
9.	Load control	The quantity of fuel is controlled by throttle in the carburettor.	The quantity of fuel is regulated by the fuel pump.
10.	Starting	Easy starting due to low compression ratio.	Difficult starting due to high compression ratio.
11.	Speed	Petrol engines are high speed engines due to light weight.	Diesel engines are low speed engines due to heavy weight.
12.	Thermal efficiency	Thermal efficiency is low around 25% due to low compression ratio.	Thermal efficiency is high around 40% due to high compression ratio.
13.	Size and weight	Small and light due to low maximum pressure of the cycle.	Bulky and heavy due to high cycle peak pressure.
14.	Space requirement	Occupies less space.	Occupies more space.
15.	Initial cost	Low	High
16.	Running cost	High due to high fuel cost.	Low due to low fuel cost.
17.	Maintenance cost	Low	High
18.	Applications	Scooters, motor cycles, cars, light duty vehicles, aeroplanes	Buses, trucks, earthmoving machinery, heavy duty vehicles and machines, power generation, ships.

4.8 TWO-STROKE PETROL ENGINE

Two-stroke petrol engine has been developed for small capacity applications where light weight is a premium. Although less efficient, the construction of the engine has been simplified to a great extent by eliminating valves and valve operating mechanism. Only two strokes of the piston are required to complete one cycle. One cycle is completed in one revolution of the crankshaft by eliminating the suction and exhaust strokes. However, these strokes are carried out simultaneously during compression and expansion strokes.

The suction and compression processes are carried out during upward movement of piston and expansion and exhaust processes are completed during downward movement of piston. There are no valves and engine has inlet port, exhaust port and transfer port.

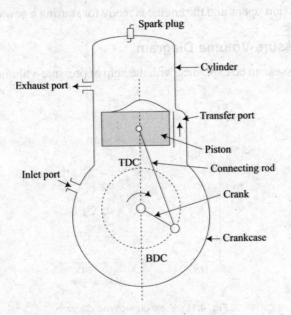

Fig. 4.10   Two-stroke petrol engine

1. Suction (Transfer) Process

The piston moves down towards BDC. The exhaust port and transfer port are uncovered. The fresh charge (fuel and air mixture) flows into the cylinder from crankcase via transfer port. The fresh charge is present in the crankcase.

2. Compression Process

The piston moves up towards TDC. The transfer port is first covered and then the exhaust port. The charge is compressed at the top of the piston.

The pressure in the crankcase falls. The inlet port opens and fresh charge enters the crankcase. This charge will be transferred to the cylinder during downward movement of the piston.

3. Expansion Process

At the end of compression stroke, the high pressure charge is ignited with the help of spark plug just before piston reaches TDC. The combustion of the charge increases its temperature and pressure. The piston is pushed down. The heat energy of combustion of fuel is converted into mechanical work in the form of movement of piston.

4. Exhaust Process

The piston moves down towards BDC during expansion process. The exhaust port is uncovered. The burnt gases are exhausted into the atmosphere.

The inlet port opens and the engine is ready for starting a new cycle.

4.8.1 Pressure-Volume Diagram

All the processes can be explained with the help of pressure-volume diagram.

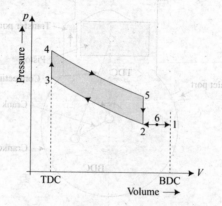

Fig. 4.11 Pressure-volume diagram

Process 1-6 : Charging of crankcase with fresh charge.

Process 6-2 : Transfer of fresh charge from crankcase to cylinder through transfer port.

Process 2-3 : Isentropic compression of charge.

Process 3-4 : Combustion at constant volume.

Process 4-5 : Isentropic expansion of products of combustion.

Process 5-2 : Release of burnt gases to atmosphere as exhaust port opens.

Process 2-1 : Sweeping out of exhaust gases of atmosphere.

4.8.2 Port Timing Diagram

The port timing diagram is shown in Fig. 4.12

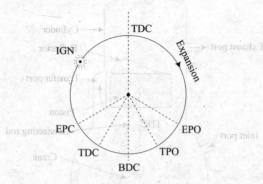

Fig. 4.12 Port timing diagram

TDC : Top dead centre

BDC : Bottom dead centre

EPO : Exhaust port opens (35° – 50° before BDC)

TPO : Transfer port opens (30° – 40° before BDC)

EPC : Exhaust port closes (35° – 40° after BDC)

IGN : Ignition (15° – 20° before TDC).

4.9 TWO-STROKE DIESEL ENGINE

It requires two strokes of piston to complete one cycle. The suction and compression processes are completed during upward movement of piston. Expansion and exhaust processes are completed during downward movement of piston. There are no valves and engine has inlet port, exhaust port, and transfer port. There is no spark plug. The fuel is injected through a fuel injection valve.

1. Suction (Transfer) Process
The piston moves down towards BDC. The exhaust port and then transfer port are uncovered. The fresh air from crankcase flows into the cylinder via transfer port.

2. Compression Process
The piston moves up towards TDC. The transfer port is first covered and then exhaust port. The air is compressed at the top of the piston.

During upward movement of piston, the inlet port opens and fresh air enters the crankcase due to pressure difference between atmosphere and crankcase.

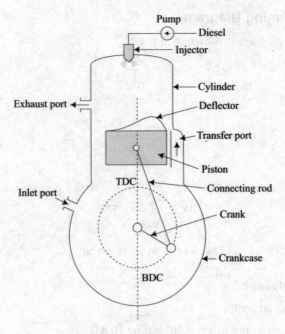

Fig. 4.13 Two-stroke diesel engine

3. Expansion Process

Shortly before the piston reaches TDC, the diesel oil is injected into the engine cylinder through the fuel injection valve. The fuel is ignited due to high temperature of air in the cylinder after compression stroke. The combustion of fuel increases the temperature and pressure of products of combustion inside the cylinder. The piston is pushed down. The heat energy released by combustion of fuel is converted into mechanical work in the form of motion of the piston.

4. Exhaust Process

The piston moves down during expansion process. The exhaust port is uncovered. The burnt gases are exhausted into the atmosphere.

The inlet port opens and the engine is ready for starting a new cycle.

4.9.1 Pressure-Volume Diagram

The complete cycle of operation can be explained with the help of pressure-volume diagram.

Process 1-6 : Charging of crankcase with fresh air.

Process 6-2 : Transfer of fresh air from crankcase to cylinder via transfer port.

Process 2-3 : Isentropic compression of air.

Process 3-4 : Injection of fuel and combustion of mixture at constant pressure.

Process 4-5 : Isentropic expansion of products of combustion.

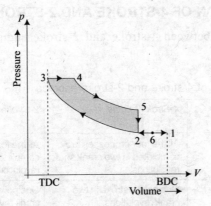

Fig. 4.14 Pressure-volume diagram

Process 5-2 : Release of burnt gases to atmosphere through exhaust port.

Process 2-1 : Sweeping out of exhaust gases to atmosphere.

4.9.2 Port Timing Diagram

The port timing diagram is shown in Fig. 4.15.

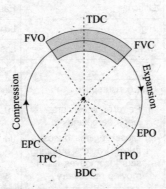

Fig. 4.15 Port timing diagram

TDC : Top dead centre

BDC : Bottom dead centre

FVO : Fuel valve opens ($10° - 15°$ before TDC)

FVC : Fuel valve closes ($15° - 20°$ after TDC)

EPO : Exhaust port opens ($35° - 50°$ before BDC)

TPO : Transfer port opens ($30° - 40°$ before BDC)

TPC : Transfer port closes ($30° - 40°$ after BDC)

EPC : Exhaust port closes ($35° - 50°$ after BDC)

4.10 COMPARISON OF 4-STROKE AND 2-STROKE ENGINES

The basic difference between 4-stroke and 2-stroke engines is compared in Table 4.2.

Table 4.2 Comparison of 4-stroke and 2-stroke engines

Sl. No.	Characteristic	4-Stroke Engine	2-Stroke Engine
1.	Processes	All the four processes are completed in two crank shaft revolutions.	All the four processes are completed in one crank shaft revolution.
2.	Power stroke	One power stroke in two crank revolutions.	One power stroke in every crank revolution.
3.	Power production	Low per cycle	70 – 80% more power per cycle.
4.	Turning moment diagram	Not uniform	Uniform
5.	Flywheel	Heavy	Light
6.	Inlet and outlet	Inlet and outlet valves	Inlet transfer and outlet ports
7.	Size and weight	Bulky and heavy	Compact and light
8.	Construction	Complicated valve mechanisms. More parts	Simple. No valves and valve mechanisms
9.	Starting	Difficult	Easy
10.	Thermal efficiency	High	Low
11.	Initial cost	More	Lass
12.	Operation	Quiet	Noisy
13.	Lubricating oil	Less consumption	More consumption
14.	Applications	(i) Petrol engines: cars, jeeps, aeroplanes (ii) Diesel engines: buses, trucks, tractors, pumping sets, locomotives, earth moving machinery	(i) Motor cycles, scooters, three wheeler and sprayers (ii) Marine propulsion due to less space requirement and lightness.

4.11 RELATIVE MERITS AND DEMERITS OF A TWO-STROKE ENGINE OVER A FOUR-STROKE ENGINE

Merits

1. The power produced by a 2-stroke engine of same size and running at same speed is about 70-80% more because there is power stroke in every revolution of crankshaft.
2. The construction of a 2-stroke engine is simpler.
3. These occupy less floor space and are less bulky.

4. No separate valve operating mechanism is required. The ports are opened and closed by piston movement itself.

5. The cost of the engine is less.

6. Maintenance cost is low due to wear of less parts.

Demerits

1. The thermal efficiency is low as complete fuel is not burnt and some fresh charge is mixed with exhaust gases. Therefore, higher specific fuel consumption.

2. The emissions are more as these carry unburned fuel also.

3. Lubricating oil consumption is more as it is mixed and carried along with fresh charge during transfer from crankcase to engine cylinder.

4. Noisy working due to sudden release of exhaust gases.

4.12 AIR STANDARD CYCLE

The operating cycle of an internal combustion engine consists of the following separate processes:

1. Intake or suction process

2. Compression process

3. Combustion process

4. Expansion process

5. Exhaust process

It is necessary to devise an engine cycle based on these processes which can be used to assess the performance of the engine. The mass of fuel in the air-fuel mixture is very small as compared to mass of air, the mixture can be assumed to obey the perfect gas laws. The engines can be assumed to be working on thermodynamic cycles. For simplification air can be taken as working substance inside the engine cylinder. The ideal engine cycle employing air as working fluid is called *air standard cycle*.

4.12.1 Assumptions

The following assumptions are usually made in the analysis of air standard cycles.

1. Air considered an ideal gas is the working medium. The composition of working medium remains constant throughout the cycle.

2. The compression and expansion processes are reversible and adiabatic. These processes are internally reversible and there is no heat loss or friction.

3. The combustion process is replaced by a reversible heat transfer process from a heat reservoir.

4. The exhaust process is replaced by a constant volume heat rejection process to a heat reservoir.

5. The specific heats are assumed constant at the ambient temperature,

$$C_p = 1.005 \text{ kJ/kg-K}$$

$$C_v = 0.718 \text{ kJ/kg-K}$$

$$\gamma = 1.4$$

6. The charging and discharging operations are omitted. Therefore, a constant mass of the medium is carried through the entire cycle.

In practice, there are following deviations from the ideal conditions assumed:

1. There are internal and external irreversibilities like fluid friction, mechanical friction, combustion, etc.

2. There are heat losses to the cooling medium and by radiation.

3. The properties of the working medium change throughout the cycle.

4. The specific heat of gases changes with temperature.

However, the air standard cycles are useful and are important tool for comparison of cycle performance.

4.12.2 Air Standard Efficiency

The efficiency of the theoretical cycle with working medium as air is called air standard efficiency or ideal efficiency. It is the ratio of ideal work done and heat supplied

$$\text{Air standard efficiency} = \frac{\text{Ideal work done}}{\text{Heat supplied}}$$

$$\text{Air standard efficiency} = \frac{\text{Heat supplied} - \text{Heat rejected}}{\text{Heat supplied}}$$

The *thermal efficiency* is the ratio of actual work done and heat supplied.

$$\text{Thermal efficiency} = \frac{\text{Actual work done}}{\text{Heat supplied}}$$

Relative efficiency is the ratio of thermal efficiency and air standard efficiency.

$$\text{Relative efficiency} = \frac{\text{Thermal efficiency}}{\text{Air standard efficiency}}$$

$$= \frac{\text{Actual work done}}{\text{Ideal work done}}$$

4.12.3   Power Cycles

Following power cycles are used depending upon type of heat addition process:

1. Otto cycle or constant volume heat addition process.
2. Diesel cycle or constant pressure heat addition process
3. Dual cycle in which heat is added partly at constant volume and partly at constant pressure.

4.13   OTTO CYCLE

The cycle was first proposed by French engineer Beau de Rochas but is named after a German Engineer Nicholas Otto who successfully built the engine working on this cycle. All petrol engines and other spark ignition engines work on Otto cycle. The air standard Otto cycle is composed of four internally reversible processes. The p-V and T-s diagrams are shown:

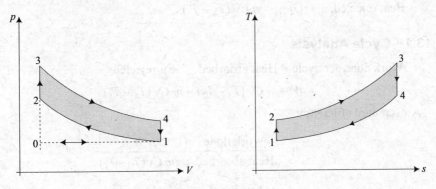

Fig. 4.16   Otto cycle

The Otto cycle consists of:

1. Process 1-2 : Isentropic compression
2. Process 2-3 : Constant volume heat addition
3. Process 3-4 : Isentropic expansion process
4. Process 4-1 : Constant volume heat rejection process

In the theoretical Otto cycle, the following processes are eliminated:

5. Process 0-1 : Suction process
6. Process 1-0 : Exhaust process.

Let engine cylinder contain m kg of air at point 1 where air parameters are p_1, V_1, and T_1.

   1. *Process 1-2: Reversible adiabatic or isentropic compression*
      No heat is added or rejected during the process,

$$Q_{1-2} = 0$$

The law of process is $pV^\gamma = C$

2. *Process 2-3: constant volume heating process*

Heat is added to air from a heat reservoir at constant volume and temperature of air changes form T_2 to T_3,

Heat added, $\quad Q_{2-3} = mCv(T_3 - T_2)$

3. *Process 3-4: Reversible adiabatic or isentropic expansion*

No heat is added or rejected during the process,

$$Q_{3-4} = 0$$

The law of process is $pV^\gamma = C$.

4. *Process 4-1: Constant volume heat rejection*

Heat is rejected to a heat reservoir at constant volume and temperature of air changes from T_4 to T_1,

Heat rejected, $\quad Q_{4-1} = m\,Cv\,(T_4 - T_1)$

4.13.1 Cycle Analysis

Work done per cycle = Heat absorbed − Heat rejected

$$W = m\,Cv\,(T_3 - T_2) - m\,Cv\,(T_4 - T_1)$$

Air standard efficiency,

$$\eta = \frac{\text{Work done}}{\text{Heat absorbed}} = \frac{W}{m\,Cv(T_3 - T_2)}$$

$$= \frac{m\,Cv(T_3 - T_2) - m\,Cv\,(T_3 - T_1)}{m\,Cv(T_3 - T_2)}$$

$$= 1 - \frac{[T_4 - T_1]}{[T_3 - T_2]}$$

$$= 1 - \frac{T_1\left[\dfrac{T_4}{T_1} - T_1\right]}{T_2\left[\dfrac{T_3}{T_2} - 1\right]}$$

Compression ratio, $\quad r_c = \dfrac{V_1}{V_2}$

Expansion ratio, $\quad r_e = \dfrac{V_4}{V_3}$

For Otto cycle, $\qquad r_c = r_e$ $\qquad\qquad [\because \quad V_1 = V_4 \quad$ and $\quad V_2 = V_3]$

For reversible adiabatic compression process 1-2,

$$\frac{T_2}{T_1} = \left(\frac{V_1}{V_2}\right)^{\gamma-1} = (r_c)^{\gamma-1} \qquad\qquad ...(1)$$

For reversible adiabatic expansion process 3-4,

$$\frac{T_3}{T_4} = \left(\frac{V_4}{V_3}\right)^{\gamma-1} = (r_e)^{\gamma-1} = (r_c)^{\gamma-1} \qquad\qquad ...(2)$$

Form equations (1) and (2),

$$\frac{T_2}{T_1} = \frac{T_3}{T_4} = (r_c)^{\gamma-1}$$

$$\therefore \qquad\qquad \frac{T_1}{T_4} = \frac{T_2}{T_3}$$

or $\qquad\qquad \dfrac{T_4}{T_1} = \dfrac{T_3}{T_2}$

$$\therefore \qquad\qquad \eta = 1 - \frac{T_1}{T_2} = 1 - \frac{1}{(r_c)^{\gamma-1}}$$

4.13.2 Observations

Efficiency of Otto cycle depends on compression ration, r_c only. As r_c increases, efficiency increases. But r_c cannot be increased beyond a value of 10 due to following reasons:

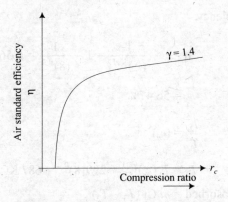

Fig. 4.17 Efficiency v/s compression ratio

(i) The cost of engine increases with increase of r_c as the engine weight and size increase. Therefore, a compromise has to be made between increase of efficiency and cost of engine.

(ii) There is a limit on r_c due to auto-ignition of gasoline. As r_c increases, there is tendency of pre-ignition and detonation of fuel. In practice, a compression ratio of 6 to 10 is used for petrol engines.

4.13.3   Mean Effective Pressure

Mean effective pressure is the ratio of work done and swept volume,

$$p_m = \frac{\text{Work done}}{\text{Swept volume}} = \frac{W}{V_s} = \frac{W}{V_1 - V_2}$$

Example 4.7:   An air standard Otto cycle has a compression ratio of 8. At the start of compression process the temperature is 26°C and pressure is 1 bar. If maximum temperature of the cycle is 1080 K. Calculate: [U.P.T.U. II Sem., 2004-05]

(i) Net output

(ii) Thermal efficiency

   Take $Cv = 0.718$ kJ/kg-K

Solution:

   Data given:

   Compression ratio,

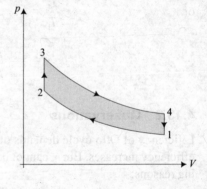

$$r_c = 8$$

$$T_1 = 26°C + 273 = 299 \text{ K.}$$

$$p_1 = 1 \text{ bar}$$

$$T_3 = 1080 \text{ K.}$$

$$Cv = 0.718 \text{ kJ/kg-K}$$

(i) Thermal efficiency,

$$\eta = 1 - \frac{1}{(r_c)^{\gamma-1}} = 1 - \frac{1}{(8)^{0.4}} = 0.5647$$

$$= 56.47\%$$

(ii) $$\frac{T_2}{T_1} = (r_c)^{\gamma-1}$$

$$T_2 = T_1 (r_c)^{\gamma-1} = 299(8)^{0.4} = 687 \text{ K.}$$

Heat absorbed $= m \, Cv(T_3 - T_2)$

$$= 1 \times 0.718 \, (1080 - 687) = 282.17 \text{ kJ.}$$

Now, $\qquad \eta = \dfrac{\text{Work done}}{\text{Heat aborded}}$

$\therefore \qquad$ Work done = Heat absorbed $\times \eta$

$$= 282.17 \times 0.5647 = \textbf{159.34 kJ.} \quad \textbf{Ans.}$$

Example 4.8: An engine of 250 mm bore and 375 mm stroke works on Otto cycle. The clearance volume is 0.00263 m^3. The initial pressure and temperature are 1 bar and 50°C. If the maximum pressure is limited to 25 bar, find:

(i) The air standard efficiency of the cycle.

(ii) The mean effective pressure for the cycle. [U.P.T.U. II Sem., 2000-01]

Solution: Bore, $\qquad D = 250$ mm $= 0.25$ m

$$L = 375 \text{ mm} = 0.375 \text{ m}$$

$$V_s = \frac{\Pi}{4} D^2 L = \frac{\Pi}{4}(0.25)^2 \times 0.375 = 0.0184 \text{ m}^3$$

$$V_c = 0.00263 \text{ m}^3$$

$$r_c = \frac{V_c + V_s}{V_c} = \frac{0.00263 + 0.0184}{0.00263} = 8$$

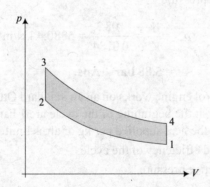

(i) Air standard efficiency,

$$\eta = 1 - \frac{1}{(r_c)^{\gamma-1}} = 1 - \frac{1}{(8)^{0.4}} = 0.5647 = 56.47\%$$

$$T_1 = 50 + 273 = 323 \text{ K}$$

$$p_1 = 1 \text{ bar}$$

For isentropic compression process 1-2,

$$\frac{T_2}{T_1} = (r_c)^{\gamma-1} = (8)^{0.4} = 2.2974$$

$$T_2 = 2.2947 \times 323 = 742 \text{ K}$$

$$p_1 V_1^\gamma = p_2 V_2^\gamma$$

$$p_2 = p_1 \left(\frac{V_1}{V_2} \right)^\gamma = p_1 (r_c)^\gamma = 1 \times 8^{1.4} = 18.38 \text{ bar}$$

$$p_3 = 25 \text{ bar}$$

For constant volume process 2-3,

$$\frac{p_3}{T_3} = \frac{p_2}{T_2}$$

$$\therefore \qquad T_3 = \frac{p_3}{p_2} \cdot T_2 = \frac{25}{18.38} \times 742 = 1009.25 \text{ K}$$

Heat absorbed, $Q_{2-3} = m \, Cv \, (T_3 - T_2) = 1 \times 0.717 \, (1009.25 - 742)$

$$= 191.62 \text{ kJ}.$$

Work done during the cycle,

$$W = \eta \times Q_{2-3} = 0.5647 \times 191.62 = 108.2 \text{ kJ}$$

Mean effective pressure,

$$p_m = \frac{W}{V_s} = \frac{108.2}{0.0184} = 5880.4 \text{ kN/m}^2$$

$$= \textbf{5.88 bar} \quad \textbf{Ans.}$$

Example 4.9: A petrol engine works on an air standard Otto cycle. The pressure and temperature of air at the beginning of the cycle are 1 bar and 15°C. The compression ratio is 8 and the heat supplied per kg of air is limited to 2000 kJ. Find:

   (i) The air standard efficiency of the cycle.

   (ii) The mean effective pressure.

Solution:

   (i) The efficiency of air standard Otto cycle.

$$\eta = 1 - \frac{1}{(r_c)^{\gamma-1}} = 1 - \frac{1}{(8)^{1.4-1}}$$

$$= 0.5647 = 56.47\%$$

   (ii) Heat supplied, $Q_{2-3} = 2000 \text{ kJ}$

   Work done, $\qquad W = Q_{2-3} \times \eta = 2000 \times 0.5647$

$$= 1229.4 \text{ kJ}$$

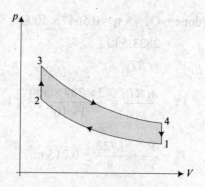

Now, $\qquad p_1 v_1 = m\, RT_1$

$\therefore \qquad V_1 = \dfrac{m\, RT_1}{p_1} = \dfrac{1 \times 287 \times 288}{1 \times 10^5} = 0.826 \text{ m}^3$

$$[\because R = 287 \text{ J/kg-K for air}]$$

$$V_2 = \dfrac{V_1}{r_c} = \dfrac{0.826}{8} = 0.10325 \text{ m}^3$$

Mean effective pressure

$$P_m = \dfrac{W}{V_s} = \dfrac{1229.4}{V_1 - V_2} = \dfrac{1229.4}{0.826 - 0.10325}$$

$$= 1562.64 \text{ kN/m}^2$$

$$= \mathbf{15.626\ bar} \quad \mathbf{Ans.}$$

Example 4.10: 2 kg of air at a pressure of 1 bar and 300 K is taken through an Otto cycle of compression ratio 8 involving 5000 kJ of heat addition. Determine the thermal efficiency, work done and mean effective pressure, if $\gamma = 1.4$ and $R = 287$ J/kg-K.

Solution:

$$m = 2 \text{ kg}$$

$$p_1 = 1 \text{ bar}$$

$$T_1 = 300 \text{ K}$$

$$r_c = 8$$

$$Q_{2-3} = 5000 \text{ kJ}$$

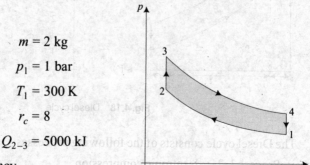

(i) Thermal efficiency,

$$\eta = 1 - \dfrac{1}{(r_c)^{\gamma-1}} = 1 - \dfrac{1}{(8)^{0.4}} = 0.5647 = 56.47\%$$

$$\text{Work done} = Q_{2-3} \times \eta = 0.5647 \times 5000$$
$$= 2823.5 \text{ kJ.}$$

$$p_1 V_1 = m\, RT_1$$

$$V_1 = \frac{m\, RT_1}{p_1} = \frac{2 \times 287 \times 300}{1 \times 10^5} = 1.1722 \text{ m}^3$$

$$V_2 = \frac{V_1}{r_c} = \frac{1.722}{8} = 0.215 \text{ m}^3.$$

Mean effective pressure,

$$P_m = \frac{W}{V_1 - V_2} = \frac{2823.5}{1.722 - 0.215} = 1873.6 \text{ kN/m}^3$$

$$= \textbf{18.74 bar} \quad \textbf{Ans.}$$

4.14 DIESEL CYCLE

Diesel cycle is named after a German engineer Rudolf Diesel who invented compression ignition engine. The ideal Diesel cycle consists of two reversible adiabatic or isentropic, a constant pressure and a constant volume processes. Slow speed diesel engines work on Diesel cycle, however, modern high speed diesel engines work on dual cycle. The air standard Diesel cycle is composed of four internally reversible processes. The *p-V* and *T-s* diagrams are shown:

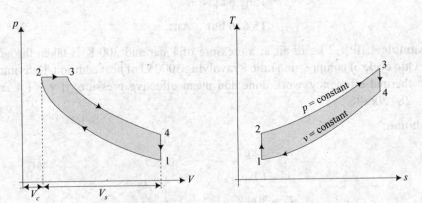

Fig. 4.18 Diesel cycle

The Diesel cycle consists of the following processes.

(i) Process 1-2 : Isentropic compression

(ii) Process 2-3 : Constant pressure heat addition

(iii) Process 3-4 : Isentropic expansion

(iv) Process 4-1 : Constant volume heat rejection

Let engine cylinder contain m kg of air at state 1 where air parameters are p_1, V_1, T_1.

1. *Process 1-2: Reversible adiabatic or isentropic compression*

   No heat is added or rejected through the process.

   $$Q_{1-2} = 0$$

   The law of process is $pV^\gamma = C.$

2. *Process 2-3: Constant pressure heat absorption*

   Heat is absorbed by air from a heat reservoir at constant pressure and temperature of air changes T_2 to T_3,

   Heat absorbed, $\quad Q_{2-3} = m\,Cp\,(T_3 - T_2)$

3. *Process 3-4: Reversible adiabatic or isentropic expansion*

   No heat is added or rejected during the process,

   $$Q_{3-4} = 0$$

   The law of process is $pV^\gamma = C.$

4. *Process 4-1: Constant volume heat rejection*

   Heat is rejected by air to a heat sink at constant volume and the temperature of air changes for T_4 to T_1.

   Heat rejected, $\quad Q_{4-1} = m\,Cv\,(T_4 - T_1)$

4.14.1 Cycle Analysis

Work done by the cycle = Heat absorbed – Heat rejected

$$W = m\,Cp\,(T_3 - T_2) - m\,Cv\,(T_4 - T_1)$$

$\therefore$ Air standard efficiency,

$$\eta = \frac{\text{Work done}}{\text{Heat absorbed}} = \frac{m\,Cp\,(T_3 - T_2) - m\,Cv\,(T_4 - T_1)}{m\,Cp\,(T_3 - T_2)}$$

$$= 1 - \frac{Cv}{Cp}\left[\frac{T_4 - T_1}{T_3 - T_2}\right] = 1 - \frac{1}{\gamma}\left[\frac{T_4 - T_1}{T_3 - T_2}\right]$$

The following ratios may be defined,

Compression ratio, $\quad r_c = \dfrac{V_1}{V_2}$

Expansion ratio, $\quad r_e = \dfrac{V_4}{V_3}$

Cut off ratio, $\qquad \rho = \dfrac{V_3}{V_2}$

Now expansion ratio, $\qquad r_e = \dfrac{V_4}{V_3} = \dfrac{V_4}{V_2} \times \dfrac{V_2}{V_3}$

$$= r_c \times \dfrac{1}{\rho} = \dfrac{r_c}{\rho}$$

For process 2-3 (constant pressure)

$$\dfrac{V_2}{T_2} = \dfrac{V_3}{T_3} \qquad\qquad (\because\ p_2 = p_3)$$

$$\therefore \qquad T_3 = T_2 \cdot \dfrac{V_3}{V_2} = T_1 \times \rho \qquad\qquad\qquad ...(1)$$

For process 3-4 (isentropic expansion)

$$\dfrac{T_4}{T_3} = \left(\dfrac{V_3}{V_4}\right)^{\gamma-1} = \left(\dfrac{1}{r_e}\right)^{\gamma-1} = \left(\dfrac{\rho}{r_c}\right)^{\gamma-1}$$

$$\therefore \qquad T_4 = T_3 \left(\dfrac{\rho}{r_c}\right)^{\gamma-1} = T_1 \times \rho \left(\dfrac{\rho}{r_c}\right)^{\gamma-1} \qquad ...(2)$$

For process 1-2 (isentropic compression)

$$\dfrac{T_2}{T_1} = \left(\dfrac{V_1}{V_2}\right)^{\gamma-1} = (r_c)^{\gamma-1}$$

$$\therefore \qquad T_2 = T_1\,(r_c)^{\gamma-1} \qquad\qquad\qquad\qquad ...(3)$$

Substituting the value of T_2 from equation (3) into equations (1) and (2)

$$\therefore \qquad T_3 = T_1\,(r_c)^{\gamma-1} \times \rho$$

$$T_4 = T_1\,(r_c)^{\gamma-1} \times \rho \left(\dfrac{\rho}{r_c}\right)^{\gamma-1} = T_1\,\rho^{\gamma}$$

Substituting the values of T_2, T_3 and T_4 into expression for efficiency,

$$\eta = 1 - \dfrac{1}{r}\left[\dfrac{T_1\rho^{\gamma} - T_1}{T_1(r_c)^{\gamma-1}\rho - T_1(r_c)^{\gamma-1}}\right]$$

$$\therefore \qquad \eta = 1 - \frac{1}{(r_c)^{\gamma-1}}\left[\frac{\rho^{\gamma}-1}{\gamma(\rho-1)}\right]$$

4.14.2 Observations

1. The Diesel cycle efficiency depends upon compression ratio, r_c, γ and cut off ratio ρ.

2. The air standard efficiency of Diesel cycle is lower than Otto cycle for same compression ratio. However, the compression ratio used in diesel engine is more than petrol engine. Overall, the thermal efficiency of diesel engine is more than petrol engine.

3. The Diesel cycle efficiency increases with decrease in cut-off ratio and maximum value can be equal to Otto cycle efficiency.

4. The cut-off ratio in diesel engines is proportional to the load on the engine, for this reason, the efficiency of the Diesel cycle decreases with the load on the engine.

5. The Diesel cycle efficiency differs from Otto cycle by the bracketed term

$$\left[\frac{\rho^{\gamma}-1}{\gamma(\rho-1)}\right] \text{ which is always greater than unity.}$$

Example 4.11: The Mercedes Benz car has four cylinder in-line diesel engine with compression ratio 20:1 and expansion ratio 10:1. Calculate:

  (i) Cut-off ratio

 (ii) Air standard efficiency

Take $\gamma = 1.4$.

Solution: The cycle is shown on p-V diagram,

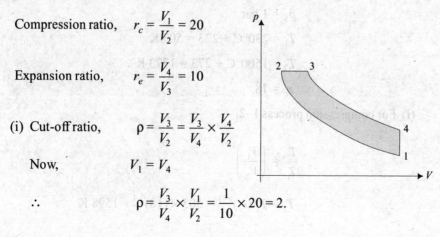

Compression ratio, $\quad r_c = \dfrac{V_1}{V_2} = 20$

Expansion ratio, $\quad r_e = \dfrac{V_4}{V_3} = 10$

(i) Cut-off ratio, $\quad \rho = \dfrac{V_3}{V_2} = \dfrac{V_3}{V_4} \times \dfrac{V_4}{V_2}$

Now, $\qquad\qquad V_1 = V_4$

$\therefore \qquad\qquad \rho = \dfrac{V_3}{V_4} \times \dfrac{V_1}{V_2} = \dfrac{1}{10} \times 20 = 2.$

(ii) Air standard efficiency,

$$\eta = 1 - \frac{1}{(r_c)^{\gamma-1}} \left[\frac{\rho^\gamma - 1}{\gamma(\rho - 1)} \right]$$

$$= 1 - \frac{1}{(20)^{0.4}} \left[\frac{(2)^{1.4} - 1}{1.4(2 - 1)} \right] = 1 - \frac{1}{3.314} \left[\frac{1.639}{1.4} \right] = 0.6467$$

$$= \mathbf{64.67\%} \quad \textbf{Ans.}$$

Example 4.12: Air enters at 1 bar and 230°C in an engine running on Diesel cycle whose compression ratio is 18. Maximum temperature of the cycle is limited to 1500°C. Calculate:

(i) Cut-off ratio

(ii) Heat supplied per kg of air

(iii) Cycle efficiency. [U.P.T.U. I Sem., 2005-06]

Solution:

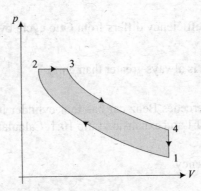

The *p-v* diagram of Diesel cycle is drawn.

$$p_1 = 1 \text{ bar}$$

$$T_1 = 230°C + 273 = 503 \text{ K}$$

$$T_3 = 1500°C + 273 = 1773 \text{ K}$$

$$r_c = 18.$$

(i) For compression process 1–2,

$$\frac{T_2}{T_1} = \left(\frac{V_1}{V_2} \right)^{\gamma-1}$$

∴

$$T_2 = T_1 (r_c)^{\gamma-1} = 503 \times (18)^{0.4} = 1598 \text{ K}$$

For constant pressure process 2-3,

Cut-off ratio, $\rho = \dfrac{V_3}{V_2} = \dfrac{T_3}{T_2} = \dfrac{1773}{1598}$ $\qquad (\because \quad p_2 = p_3)$

$$= 1.1 \quad \textbf{Ans.}$$

(ii) Heat supplied per kg of air,

$$Q_{2-3} = Cp\,(T_3 - T_2) = 1.005\,(1773 - 1598)$$

$$= \textbf{175.875 kJ/kg} \quad \textbf{Ans.}$$

(iii) Cycle efficiency,

$$\eta = 1 - \frac{1}{(r_c)^{\gamma-1}}\left[\frac{\rho^\gamma - 1}{\gamma\,(\rho - 1)}\right]$$

$$= 1 - \frac{1}{(18)^{0.4}}\left[\frac{(1.1)^{1.4} - 1}{1.4\,(1.1 - 1)}\right] = 1 - \frac{1}{3.17767}\left[\frac{0.142746}{0.14}\right] = 0.679$$

$$= \textbf{67.9\%} \quad \textbf{Ans.}$$

Example 4.13: A diesel engine operating on air standard Diesel cycle operates on 1 kg of air with an initial pressure of 98 kPa and a temperature of 36°C. The pressure at the end of compression is 35 bar and cut off is 6% of stroke. Determine:

(a) Thermal efficiency

(b) Mean effective pressure. [U.P.T.U. II Sem., 2004-05]

Solution:

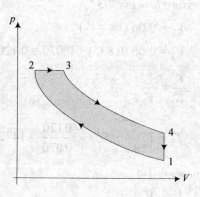

The *p-V* diagram of Diesel cycle is drawn,

$$p_1 = 98 \text{ kPa} = 0.98 \text{ bar}$$

$$T_1 = 36°C + 273 = 309 \text{ K}$$

$$m = 1 \text{ kg}$$

$$p_2 = 35 \text{ bar}$$

$$(V_3 - V_2) = 0.06 \, (V_1 - V_2)$$

(i) *At state 1*

$$p_1 V_1 = m \, R T_1$$

$$\therefore \qquad V_1 = \frac{m \, R T_1}{p_1} = \frac{1 \times 0.287 \times 309}{98} = 0.905 \text{ m}^3$$

(ii) *For compression process 1-2*

$$T_2 = T_1 \left(\frac{p_2}{p_1}\right)^{\frac{\gamma-1}{\gamma}} = 309 \left(\frac{35}{0.98}\right)^{\frac{1.4-1}{1.4}} = 858 \text{ K}$$

Also, $\qquad p_1 V_1^\gamma = p_2 V_2^\gamma$

$$\therefore \qquad \frac{V_2}{V_1} = \left(\frac{p_1}{p_2}\right)^{\frac{1}{\gamma}}$$

$$\therefore \qquad V_2 = V_1 \left(\frac{p_1}{p_2}\right)^{\frac{1}{\gamma}} = 0.905 \left(\frac{0.98}{35}\right)^{\frac{1}{1.4}} = 0.070 \text{ m}^3$$

Stroke volume, $\quad V_s = V_1 - V_2 = 0.905 - 0.070 = 0.835 \text{ m}^3$

(iii) *For constant pressure process 2-3*

$$V_3 - V_2 = 0.06 \, (V_1 - V_2)$$

$$V_3 = 0.06 \, (0.835) + 0.070 = 0.120 \text{ m}^3$$

$$\frac{T_3}{T_2} = \frac{V_3}{V_2} \qquad\qquad\qquad (\because \ p_2 = p_3)$$

$$\therefore \qquad T_3 = T_2 \frac{V_3}{V_2} = 858 \times \frac{0.120}{0.070} = 1472 \text{ K}$$

(iv) *For expansion process 3-4*

$$V_4 = V_1$$

$$\frac{T_4}{T_3} = \left(\frac{V_3}{V_4}\right)^{\gamma-1}$$

$$\therefore \qquad T_4 = T_3 \left(\frac{V_3}{V_4}\right)^{\gamma-1} = 1472 \left(\frac{0.120}{0.905}\right)^{0.4} = 472 \text{ K}$$

Heat supplied, $Q_{2-3} = m\, Cp\, (T_3 - T_2) = 1 \times 1.005\, (1472 - 858) = 617$ kJ

Heat rejected, $Q_{4-1} = m\, Cv\, (T_4 - T_1) = 1 \times 0.718\, (1472 - 309) = 117$ kJ

Work done, $W = Q_{2-3} - Q_{4-1} = 617 - 117 = 500$ kJ

(a) Efficiency, $\quad \eta = \dfrac{W}{Q_{2-3}} = \dfrac{500}{617} = 0.81 = 81\%$

(b) Mean effective pressure,

$$P_m = \frac{W}{V_s} = \frac{500}{0.835} = \textbf{598.8 kN/m}^3 = \textbf{5.99 bar} \quad \textbf{Ans.}$$

QUESTION BANK NO. 4

1. How can you define I.C. engines and how they are classified?
2. Differentiate between SI and CI engines.
3. Explain the working of a 4-stroke SI engine with the help of neat sketch.
4. With the help of neat sketches, explain the working of 2-stroke CI engine.
5. With the help of neat sketch, explain the working of 2-stroke SI engine
6. Derive a relation for air standard efficiency of Diesel cycle. Also draw the cycle on p-V and T-s diagrams.
7. What are the advantages and disadvantages of 2-stroke IC engines over 4-stroke IC engines?
8. Compare the working of 4-stroke and 2-stroke cycles of internal combustion engines.
9. Define: Bore, stroke, compression ratio, clearance ratio and mean effective pressure.
10. Explain the working of any air standard cycle (by drawing it on p-V diagram) known to you. Why is it known as 'Air standard cycle'?
11. Write notes on:
    (a) Air standard cycles
    (b) Mean effective pressure
    (c) Indicator diagram
    (d) Indicated power
12. Show Otto and Diesel cycles on p-V and T-s diagram.
13. Stating the assumptions made, describe air standard Otto cycle.

PART II:
FORCE AND STRUCTURE ANALYSIS
(ENGINEERING MECHANICS)

Force and Equilibrium

5.1 INTRODUCTION

Mechanics deals with dynamic behaviour of rigid bodies under the action of forces. Engineering mechanics deals with the laws and principles of mechanics and their applications to engineering problems. Engineering mechanics is also called structural mechanics or rigid-body mechanics and the body is idealized as a continuum (continuous distribution of matter) that undergoes theoretically no deformation whatsoever. In actual practice, no solid body is perfectly rigid and every body undergoes some deformation (change of shape and size) under the action of forces. However, the deformation may by very small and may not be considered.

In the study of dynamic behaviour of a body, the latter may be at rest and in equilibrium or it may be in motion. Statics deals with the behaviour of bodies at rest and dynamics deals with the bodies in motion under the action of forces.

Force is an agent or disturbance which changes or tends to change the state of rest or of uniform motion of a body in a straight line. Force may cause mechanical disturbance of a body on another body in direct contact or without any contact where force may be created by gravitational action.

5.2 LAWS OF MECHANICS

The laws of motion and gravity were established by Isaac Newton (1642–1727) which are the foundations of engineering mechanics. The Newton's laws of motion form a complete definition of force. The following basic laws and principles are considered to be the foundation of mechanics:

1. Newton's first and second laws of motion
2. Newton's third law
3. Newton's law of gravitation
4. The parallelogram law of forces
5. Principle of transmissibility of forces.

5.2.1 Newton's Laws of Motion

Newton's laws of motion establish relationship between force and motion. There are three laws of motion.

1. *Newton's First Law of Motion*

It states, "Every body continues in its state of rest or of uniform motion in a straight line unless it is acted upon by some external force to change that state".

As per Newton's first law of motion the action of a force is to change the velocity of a body, i.e., to produce an acceleration on the body on which it acts. This is the functional definition of force.

The first law of motion is also called the *law of inertia*. Inertia is the property of a body to remain in its present state of rest or of uniform motion.

2. *Newton's Second Law of Motion*

It states, "The rate of change of momentum of a body is directly proportional to the impressed force and takes place in the direction of the straight line in which the force acts."

The quantity of motion possessed by a body moving in a straight line is known as *momentum*. It is the product of mass of a body and its velocity.

If m = mass of a body, kg

u = initial velocity, m/s

v = final velocity, m/s

a = acceleration, m/s^2

t = time required to change the velocity of a body from u to v.

F = force required to change the above velocity [N]

Initial momentum = mu

Final momentum = mv

Change of momentum = $mv - mu$

Rate of change of momentum

$$= \frac{mv - mu}{t} = \frac{m(v - u)}{t}$$

$$= ma$$

As per Newton's second law of motion, force is the product of mass and acceleration produced on the mass by the force. Alternatively, the acceleration of a body is inversely proportional to its mass for a given force.

$$a \propto \frac{1}{m}$$

or,
$$a = \frac{F}{m}$$

or,
$$F = ma$$

where, F is the constant of proportionality and is called force.

Unit of Force

The unit of force in SI system is *Newton*. This is the force which acting on a mass of kilogram produces an acceleration of one meter per second, per second.

$$1 \text{ N} = 1 \text{ kg} \times 1 \text{ m/s}^2$$

$$= \frac{1 \text{ kg m}}{\text{s}^2}$$

$$[F] = [M] \frac{[L]}{[T]^2}$$

The second law of motion is the *quantitative* and *dimensional* definition of force.

5.2.2 Newton's Third Law of Motion

It states, "To every action there is an equal and opposite reaction".

Action is the force exerted by a body on another body and reaction is the equal and opposite force exerted by another body on the first body.

Example: A body C of weight W is resting on a plane surface and is exerting a force equal to $mg (= W)$ downwards. As per Newton's third law, a reaction R_A equal and opposite to W will be acting upwards on the body C at the point A.

Similarly, a force produces motion of the body on which it acts. If the motion of the body is stopped by another body, as per Newton's third law, there will be equal and opposite reaction by the second body on the first.

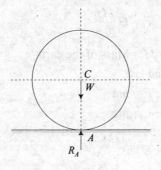

Fig. 5.1 Reaction

Therefore, as per Newton's third law of motion, force is an interaction and that is the nature of the force.

The Newton's three laws of motion form a complete definition of force.

The principle of reaction is used in ensuring equilibrium of bodies at rest, propulsion of jet aeroplanes and rockets, power production by steam, gas and water turbines, propulsion of ship, etc.

5.2.3 Newton's Law of Gravitation

Newton's law of gravitation states, "Every particle attracts any other particle with a *gravitational force* whose magnitude is given by the following equation."

$$F = G\frac{m_1 m_2}{r^2}$$

where, m_1 and m_2 = mass of particles, [kg]

r = distance between particles [m]

G = gravitational constant

$= 6.67 \times 10^{-11}$ N.m^2/kg

$= 6.67 \times 10^{-11}$ m^3/kg.s^2

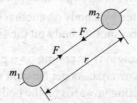

Fig. 5.2   Gravitation

Gravitational force near Earth's surface is given by

$$F = G\frac{Mm}{r^2}$$

where, M = Mass of the earth

From Newton's second law,

$$F = ma$$

$$= mg$$

$\therefore$ $$g = \frac{GM}{r^2}$$

$$= 9.81 \text{ m/s}^2$$

g = Acceleration due to gravity

The force can be applied directly or it can be a load of mass of body due to gravitational attraction called weight of the body. The weight of the body is found out from Newton's law of gravitation.

In SI system of units, a kilogram is the amount of mass that will accelerate 1 m/s² under the action of 1 Newton. One kilogram of force is the weight of 1 kilogram of mass at earth's surface, where acceleration of gravity is 9.81 m/s².

The weight,

$$W = mg \text{ [N]}$$

where,
$$m = \text{mass of body [kg]}$$
$$g = \text{acceleration due to gravity [m/s}^2\text{]}$$
$$= 9.81 \text{ m/s}^2.$$

5.2.4 Principle of Transmissibility

It states, "If a force acts at a point or a rigid body, it is assumed to act at any other point on the line of action of a force within the same body."

The effect of a force depends upon its following *characteristics:*

1. Magnitude
2. Direction or line of action with respect to a coordinate system
3. Sense of nature (push or pull)
4. Point of applications

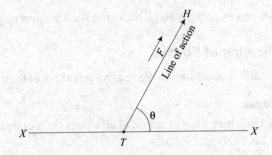

Fig. 5.3 Representation of a force

Force is a vector quantity as it possesses magnitude and direction. It can be represented as follows:

1. The *magnitude F* is indicated by the length of line *TH*, where *T* is the tail and *H* is the head of the line.
2. The *directon* is defined by angle θ with reference to axis *X-X*.
3. The *sense* is represented by arrowhead.

The sliding of a force along its line of action without altering the external static effect of the rigid body is called the principle of transmissibility.

Example:

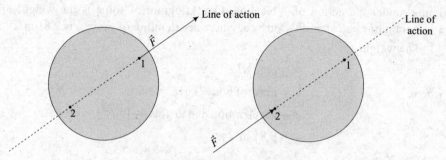

Fig. 5.4 Principle of transmissibility

The force $\vec{F}$ applied at point 1 of the rigid body can be slided along its line of action and applied at point 2. The magnitude, direction and line of action of the force do not change and, therefore, equilibrium conditions of rigid body remain unchanged.

Two forces having the same magnitude, direction and line of action but acting at different points 1 and 2 and producing the same external static effect on the rigid body are called *equivalent* forces.

5.3   SYSTEM OF FORCES

When two or more forces act on a body, they are called to form a system of forces.

5.3.1   Classification of Forces

The forces are classified according to the relative positions of their lines of action.

(a) Coplanar forces

A system of forces in which lines of action of all forces lie in one plane. The coplanar forces can be further classified as:

(i) *Collinear forces*:   The lines of action lie on the same straight line.

(ii) *Concurrent forces*:   The lines of action of all forces of the system meet at one point. The direction of forces are different and all forces lie in one plane.

(iii) *Non-concurrent forces*:   The lines of action of all forces of the system lie in one plane but do not pass through one point. The forces may be parallel or non-parallel.

Fig. 5.5 Collinear forces

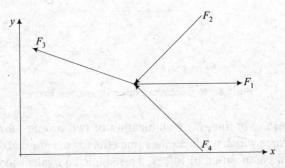

Fig. 5.6 Concurrent forces

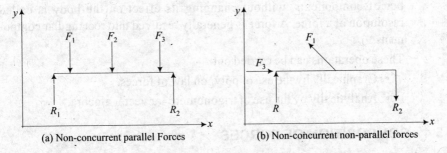

(a) Non-concurrent parallel Forces   (b) Non-concurrent non-parallel forces

Fig. 5.7 Non-concurrent forces

(b) Non-coplanar forces

The lines of action of forces do not lie in one plane. The non-coplanar forces can be
further classified as:

    (i) *Concurrent forces*:   The lines of action pass through a common point. For
         example, forces in the legs of tripod support of a camera.

    (ii) *Non-concurrent forces*:   The lines of action of all forces do not lie in the
         same plane nor do they meet in one point. See Fig. 5.8.

5.3.2 Basic Operations

In solving engineering problems, some basic operations have to be carried out with
the system of forces, i.e., addition of forces, subtraction of forces, composition of
forces, resolution of forces, etc.

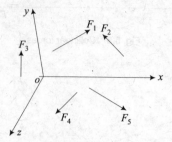

Fig. 5.8 Non-coplanar and non-concurrent forces

(i) *Composition of forces*: Combination of two or more forces to obtain a single force which produces the same effect as the original system of forces is called composition of forces. This single equivalent force is called the *resultant* of the original system of forces.

(ii) *Resolution of a force*: The process of splitting up a given force into a number of components, without changing its effect on the body is called resolution of a force. A force is generally resolved into rectangular components.

These operations can be carried out:

- Graphically by the use of polygon law of forces.
- Analytically by the use of trigonometry or vector algebra.

5.4 COMPOSITION OF FORCES

5.4.1 Polygon Law of Forces (Graphical Method)

The resultant of several forces acting at a point (concurrent forces) is found out graphically by the use of polygon law of forces. "If a number of coplanar forces are acting at a point such that they can be represented in magnitude and direction by the sides of a polygon taken in the same order, then their resultant is represented in magnitude and direction by the closing side of the polygon taken in the opposite direction".

Example:

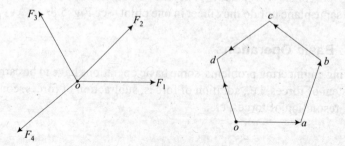

Fig. 5.9 Polygon law of forces

If the forces $F_1, F_2, F_3,$ and F_4 acting simultaneously on a particle O be represented in magnitude and direction by the sides oa, ab, bc and cd of a polygon respectively, their resultant is represented in magnitude and direction by the side od drawn in the opposite direction.

5.4.2 Triangle Law of Forces

If two forces acting simultaneously on a body are represented in magnitude and direction by the two sides of a triangle taken in order then their resultant may be represented in magnitude and direction by the third side taken in opposite direction.

Example: If F_1 and F_2 are two coplanar concurrent forces and are represented by sides Oa and ab of a triangle, their resultant is represented in magnitude and direction by the side Ob drawn in opposite direction.

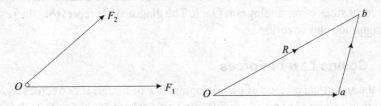

Fig. 5.10 Triangle law of forces

Tringle law of forces is useful in determining the condition of equilibrium.

"If three forces acting at a point be represented in magnitude and direction by the three sides of a triangle, taken in order, they will be in equilibrium."

Example: There are three forces F_1, F_2 and F_3. Draw Oa, ab and bO to represent the forces F_1, F_2 and F_3 in magnitude and direction. If the triangle Oab is closed, the forces are in equilibrium.

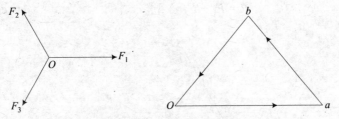

Fig. 5.11 Forces in equilibrium

5.4.3 Parallelogram Law of Forces

This law is also used to find the resultant of two concurrent forces.

"If two forces, acting at a point be represented in magnitude and direction by the two adjacent sides of a parallelogram, then their resultant is represented in mag-

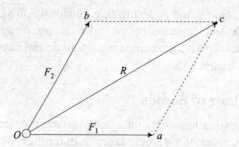

Fig. 5.12 Parallelogram law of forces

nitude and direction by the diagonal of the parallelogram passing through that point".

Example: Two forces F_1 and F_2 are acting at point O of a body. Oa and Ob are the adjacent sides of parallelogram $Oacb$. The diagonal Oc represents the resultant R in magnitude and direction.

5.4.4 Cosine Law of Forces

This is the analytical method of parallelogram law of forces. If two forces F_1 and F_2 are acting at a point O of a body making an angle α between them, then the magnitude of resultant R is given by:

$$R = \sqrt{F_1^2 + F_2^2 + 2F_1F_2 \cos \alpha}$$

The direction of resultant R with force F_1 is given by:

$$\theta = \tan^{-1}\left[\frac{F_2 \sin \alpha}{F_1 + F_2 \cos \alpha}\right]$$

Case 1. If $\alpha = 0$, forces F_1 and F_2 are collinear. Put $\alpha = 0$

$$\therefore \qquad\qquad R = F_1 + F_2 \qquad\qquad\qquad (\because \cos \theta° = 1)$$

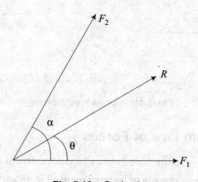

Fig. 5.13 Cosine law

Case 2. If $\alpha = 90°$, forces F_1 and F_2 are acting at right angles to each other.

$$\text{Cos } 90° = 0$$

$\therefore$

$$R = \sqrt{F_1^2 + F_2^2}$$

Case 3. If $\alpha = 180°$, the forces F_1 and F_2 are collinear but are acting in opposite direction to each other.

$$R = F_1 - F_2 \qquad\qquad (\because \cos 180° = -1)$$

5.4.5 Sine Law of Forces (Lami's Theorem)

"If three forces acting at a point are in equilibrium, each force will be proportional to the sine of the angle between the other two forces". This is also called *Lami's Theorem*.

$$\frac{F_1}{\sin \beta} = \frac{F_2}{\sin \gamma} = \frac{F_3}{\sin \alpha}$$

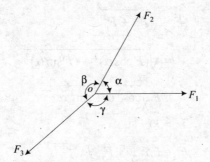

Fig. 5.14 Lami's theorem

5.5 RESOLUTION OF A FORCE

Resolution of a force is the reverse operation of composition of forces. In the operation of composition of forces, several forces acting at a particle are replaced by a resultant force which produces the same effect as original forces. In resolution of forces, a single force can be replaced by two forces in magnitude and direction different from the single force but producing the same effect as the given force. The given force is called the *resolved force* and parts are called *component forces* or the *resolutes*. A force is generally resolved into rectangular components.

The operation of resolution of forces can be carried out by the use of trigonometry or vector algebra.

A force F represented by a vector Oc inclined at an angle α with the horizontal axis has the following components.

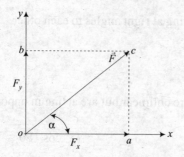

Fig. 5.15 Components of a force

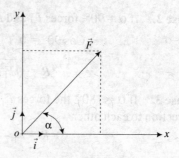

Fig. 5.16 Vector components of a force

Along x-axis: vector O-a

$$F_x = F \cos \alpha$$

Along y-axis: vector O-b

$$F_y = F \sin \alpha$$

The component forces satisfy the condition:

$$F = \sqrt{F_x^2 + F_y^2}$$

$$= \sqrt{(F \cos \alpha)^2 + (F \sin \alpha)^2}$$

$$= F \sqrt{\cos^2 \alpha + \sin^2 \alpha}$$

$$= F$$

Also $$\vec{F} = F \times \vec{i} + F_{yx} \vec{j}$$

where, $\vec{i}$ is unit vector in x-direction and $\vec{j}$ is the unit vector in y-direction.

5.5.1 Principle of Resolved Components

If a large number of forces are acting, the principle of resolved components states: "The sum of resolved components of forces acting at a point in any direction is equal to the resolved components of the resultant of forces in that direction".

If three forces F_1, F_2 and F_3 are acting at a point O making angles α_1, α_2 and α_3 with the x-axis respectively, the resolved components are:

On x-axis:

$F_1 \cos \alpha_1$, $F_2 \cos \alpha_2$ and $F_3 \cos \alpha_3$

On y-axis:

$F_1 \sin \alpha_1$, $F_2 \sin \alpha_2$ and $F_3 \sin \alpha_3$

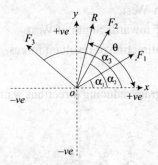

Fig. 5.17 Principle of resolved components

If R is resultant force making an angle θ with x-axis, its components are $R\cos\theta_1$ and $R\sin\theta_1$. According to the principle of resolved components:

$$R\cos\theta_1 = F_1\cos\alpha_1 + F_2\cos\alpha_2 + F_3\cos\alpha_3$$

$$= \Sigma F_x$$

$$R\sin\theta_1 = F_1\sin\alpha_1 + F_2\sin\alpha_2 + F_3\sin\alpha_3$$

$$= \Sigma F_y$$

$$R = \sqrt{(\Sigma F_x)^2 + (\Sigma F_y)^2}$$

and

$$\theta = \tan^{-1}\frac{\Sigma F_y}{\Sigma F_x}$$

5.5.2 Resultant of Coplanar Concurrent Forces

The application of laws and principles of mechanics to the solution of engineering problems is the main role of engineering mechanics. The engineering problems involve the analysis of static and dynamic stability of structures and machines against the action of forces. The analytical tools available are:

(i) Trigonometry

(ii) Vector algebra

(iii) Graphics

These tools have relative merits and demerits and can be used with advantage depending upon particular problem. The resultant of coplanar concurrent forces can best be solved by the use of principle of resolved components as discussed above.

Example 5.1: Following forces act at a point:

(i) 20 N inclined at 20° towards North of East.

(ii) 25 N towards North.

(iii) 30 N towards North-West.

(iv) 35 N inclined at 40° towards South of West.

Find the resultant and direction of resultant forces.

Solution:

1. Draw the force diagram reducing directions w.r.t. x-axis.

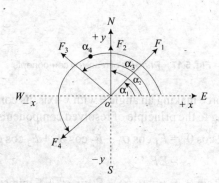

2. Find the component of forces along x-axis and y-axis.

3. Find the resultant forces in magnitude and direction employing the principle of resolved components.

$$F_1 = 20 \text{ N} \qquad \alpha_1 = 20° \text{ (North of East)}$$

$$F_2 = 25 \text{ N} \qquad \alpha_2 = 90° \text{ (North)}$$

$$F_3 = 30 \text{ N} \qquad \alpha_3 = 135° \text{ (North - West)}$$

$$F_4 = 35 \text{ N} \qquad \alpha_4 = 180° + 40° = 220° \text{ (40° South of West)}$$

$$\Sigma F_x = 20\cos 20° + 25\cos 90° + 30\cos 135° + 35\cos 220°$$

$$= 18.79 + 0 - 21.2 - 26.81 = -29.22 \text{ N}$$

$$\Sigma F_y = 20\sin 20° + 25\sin 90° + 30\sin 135° + 35\sin 220°$$

$$= 6.84 + 25 + 21.2 - 22.49 = 30.55 \text{ N}$$

The resultant, $\qquad R = \sqrt{(\Sigma F_x)^2 + (\Sigma F_y)^2}$

$$= \sqrt{(-29.22)^2 + (30.55)^2}$$

$$= 42.27 \text{ N}$$

The angle of resultant with x-axis (East)

$$\theta = \tan^{-1}\frac{\Sigma F_y}{\Sigma F_x} = \tan^{-1}\frac{30.55}{-29.22}$$

$$= -42.27°$$

Example 5.2:   A small block of weight 100 N is placed on an inclined plane which makes an angle, $\theta = 30°$ with the horizontal. What is the component of this weight parallel to inclined plane and perpendicular to inclind plane?

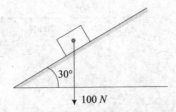

Solution:

1. Select the axis:

   x-axis parallel to inclined plane

   y-axis perpendicular to inclined plane

2. Draw the force diagram,

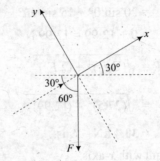

3. Find the force components,

$$F = 100 \text{ N}$$

$$\alpha = 60°$$

$$F_x = F\cos\alpha = 100\cos 60° = 50 \text{ N}$$

$$F_y = F\sin\alpha = 100\sin 60° = 86.6 \text{ N}$$

Example 5.3:   The angle between two forces of magnitude 20 kN and 15 kN is 60°, the force of 20 N being horizontal. Determine the resultant in magnitude and direction, if

(i) The forces are pulls.

(ii) The 15 kN force is a push and 20 kN force is a pull.

Solution:

Case 1.   Draw the force diagram,

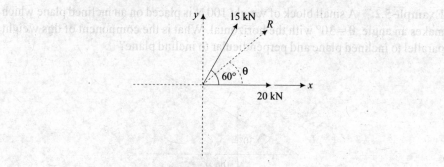

$$F_1 = 20 \text{ kN} \quad \alpha_1 = 0$$
$$F_2 = 15 \text{ kN} \quad \alpha_2 = 60°$$
$$\Sigma F_x = F_1 \cos \alpha_1 + F_2 \cos \alpha_2$$
$$= 20 \cos 0° + 15 \cos 60°$$
$$= 20 + 7.5 = 27.5 \text{ kN}$$
$$\Sigma F_y = F_1 \sin \alpha_1 + F_2 \sin \alpha_2$$
$$= 20 \sin 0° + 15 \sin 60°$$
$$= 0 + 12.99 = 12.99 \text{ N}$$

The resultant, $\qquad R = \sqrt{(\Sigma F_x)^2 + (\Sigma F_y)^2}$

$$= \sqrt{(27.5)^2 + (12.99)^2}$$

$$= \textbf{30.4 kN} \quad \textbf{Ans.}$$

The direction of resultant with x-axis,

$$\theta = \tan^{-1} \frac{\Sigma F_y}{\Sigma F_x} = \tan^{-1} \frac{12.99}{27.5} = \textbf{25.28°} \quad \textbf{Ans.}$$

Case 2. Draw the force diagram,

$$\Sigma F_x = F_1 \cos \alpha_1 + F_2 \cos \alpha_2$$
$$= 20 \cos 0° + 15 \cos 240°$$
$$= 20 - 7.5 = 12.5 \text{ kN}$$
$$\Sigma F_y = F_1 \sin \alpha_1 + F_2 \sin \alpha_2$$
$$= 20 \sin 0° + 15 \sin 240°$$
$$= 0 - 12.99 = -12.99 \text{ kN}$$

The resultant, $\qquad R = \sqrt{(\Sigma F_x)^2 + (\Sigma F_y)^2}$

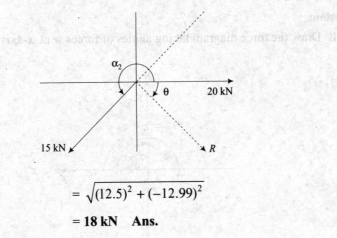

$$= \sqrt{(12.5)^2 + (-12.99)^2}$$

$$= 18 \text{ kN} \quad \textbf{Ans.}$$

The angle of resultant with x-axis,

$$\theta = \tan^{-1}\frac{\Sigma F_y}{\Sigma F_x} = \tan^{-1}\frac{-12.99}{12.5} = -46° \quad \textbf{Ans.}$$

5.5.3 Equilibrium of Concurrent Forces

The principle of equilibrium states, "A stationary body subjected to a system of concurrent forces will be in equilibrium if the resultant of the forces is zero".

The forces are generally resolved into components along Cartesian coordinates. Therefore,

$$\Sigma F_x = 0$$
$$\Sigma F_y = 0.$$

Example 5.4: The five forces F_1, F_2, F_3, F_4 and F_5 are acting at a point on a body as shown and the body is in equilibrium. If $F_1 = 18$ N, $F_2 = 22.5$ N, $F_3 = 15$ N and $F_4 = 30$ N, determine the force F_5 in magnitude and direction.

[U.P.T.U. I Sem., 2004-05]

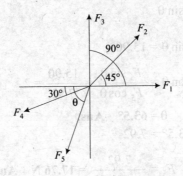

Solution:

1. Draw the force diagram taking angles of forces w.r.t. x-axis.

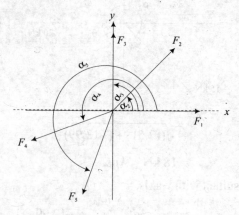

$$F_1 = 18 \text{ N} \qquad \alpha_1 = 0°$$
$$F_2 = 22.5 \text{ N} \qquad \alpha_2 = 45°$$
$$F_3 = 15 \text{ N} \qquad \alpha_3 = 90°$$
$$F_4 = 30 \text{ N} \qquad \alpha_4 = 210°$$
$$F_5 = ? \qquad \alpha_5 = (180° + \theta)$$

2. Calculate the components of forces

$$\Sigma F_x = F_1 \cos \alpha_1 + F_2 \cos \alpha_2 + F_3 \cos \alpha_3 + F_4 \cos \alpha_4 + F_5 \cos \alpha_5$$
$$= 18 \cos 0° + 22.5 \cos 45° + 15 \cos 90° + 30 \cos 210° + F_5 \cos (180° + \theta)$$
$$= 18 + 15.90 + 0 - 25.98 - F_5 \cos \theta$$
$$= 7.92 - F_5 \cos \theta°$$
$$= 0$$
$$\therefore \quad F_5 \cos \theta = 7.92$$
$$\Sigma F_y = F_1 \sin \alpha_1 + F_2 \sin \alpha_2 + F_3 \sin \alpha_3 + F_4 \sin \alpha_4 + F_5 \sin \alpha_5$$
$$= 18 \sin 0° + 22.5 \cos 45° + 15 \sin 90° + 30 \sin 210° + F_5 \sin (180° + \theta)$$
$$= 0 + 15.90 + 15 - 15 - F_5 \sin \theta$$
$$= 15.90 - F_5 \sin \theta$$
$$= 0$$
$$\therefore \qquad F_5 \sin \theta = 15.90°$$

$$\tan \theta = \frac{F_5 \sin \theta}{F_5 \cos \theta} = \frac{15.90}{7.92} = 2$$

$$\therefore \qquad \theta = 63.5° \quad \textbf{Ans.}$$
$$F_5 \cos 63.5° = 7.92$$

$$\therefore \qquad F_5 = \frac{7.92}{\cos 63.5°} = \textbf{17.76 N} \quad \textbf{Ans.}$$

Example 5.5: Determine the value of F and θ so that particle A is in equilibrium.

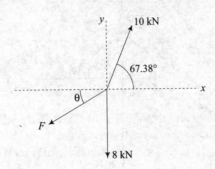

Solution:

1.
$$\Sigma F_x = 10 \cos 67.38° + F \cos (180° + \theta) + 8 \cos 270°$$
$$= 3.846 - F \cos \theta - 0$$
$$= 0$$

$\therefore$ $F \cos \theta = 3.846$

2.
$$\Sigma F_y = 10 \sin 67.38° + F \sin (180° + \theta) + 8 \sin 270°$$
$$= 9.23 - F \sin \theta - 0$$
$$= 0$$

$\therefore$ $F \sin \theta = 1.238$

3.
$$\tan \theta = \frac{F \sin \theta}{F \cos \theta} = \frac{1.23}{3.846} = 0.3198$$

$\therefore$ $\theta = 17.73°$

$F \cos \theta = 3.846$

$\therefore$ $F = \dfrac{3.846}{\cos 17.73°} = \mathbf{4\ kN}$ **Ans.**

Example 5.6: Four coplanar forces are acting at a point. Three forces have magnitude of 20 N, 50 N and 20 N at angles of 45°, 200° and 270° respectively. Fourth force is unknown. Resultant force has a magnitude of 50 N and acts along x-axis at an angle of 0°. Determine the unknown force and its direction from x-axis.

Solution:

1. Draw the force diagram.
2. Component forces,

$$\Sigma F_x = 20 \cos 45° + 50 \cos 200° + 20 \cos 270° + F \cos \theta$$
$$= 14.14 - 46.98 + 0 + F \cos \theta$$
$$= F \cos \theta - 32.84$$

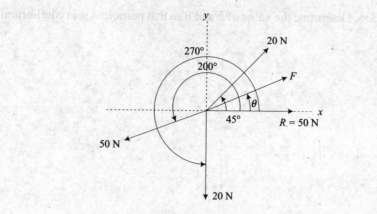

$$\Sigma F_y = 20 \sin 45° + 50 \sin 200° + 20 \sin 270° + F \sin \theta$$
$$= 14.14 - 17.1 - 20 + F \sin \theta$$
$$= -22.96 + F \cos \theta$$

3. Components of resultant,

$$R_x = R = 50°$$
$$= \Sigma F_x = F \cos \theta - 32.84$$

∴ $$F \cos \theta = 50 + 32.84 = 82.84$$

$$R_y = 0$$
$$= \Sigma F_y = F \sin \theta - 22.96$$

∴ $$F \sin \theta = 22.96$$

$$\tan \theta = \frac{F \sin \theta}{F \cos \theta} = \frac{22.96}{82.84} = 0.277$$

∴ $$\theta = 15.49°$$
$$F \sin \theta = 22.96$$

∴ $$F = \frac{22.96}{\sin 15.49°} = \textbf{85.96 N} \quad \textbf{Ans.}$$

Example 5.7: A lamp of mass 1 kg is hung from the ceiling by a chain and is pulled aside by a horizontal chord until the chain makes an angle of 60° with the ceiling. Find the tension in the chain and chord.

Solution:

1. Draw force diagram.
2. Apply equilibrium conditions.

(i) $$\Sigma F_x = 0$$
$$F_2 \cos 0° + F_1 \cos 120° + 9.8 \cos 270° = 0$$

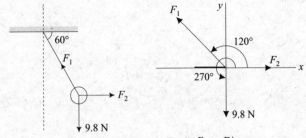

Force Diagram

$\therefore$ $F_2 - 0.5\,F_1 + 0 = 0$

     or $F_1 = 2F_2$

  (ii) $\Sigma F_y = 0$

        $F_2 \sin 0° + F_1 \sin 120° + 9.8 \sin 270° = 0$

$\therefore$ $0 + 0.866\,F_1 - 9.8 = 0$

$\therefore$ $F_1 = \dfrac{9.8}{0.866} = 11.3 \text{ N}$

$\therefore$ $F_2 = \dfrac{F_1}{2} = \dfrac{11.3}{2} = \textbf{5.658 N} \quad \textbf{Ans.}$

5.6   MOMENT OF A FORCE

A force tends to move a body in the direction of its applications. It also tends to rotate a body about an axis. The rotational tendency of the force is called moment. The magnitude of the moment is given by the product of the force and the perpendicular distance of the axis of rotation and line of action of force. The unit of moment is Nm. The moment can be clockwise or anticlockwise. The moment is, therefore, a vector quantity as it has magnitude as well as direction.

Example: A force $\vec{F}$ is acting at point A of a rigid body making an angle θ with the horizontal axis passing through the axis O of the body. The perpendicular distance between line of action and point O is given by OB.

        $OB = d = r \sin \theta$

where, $r = OA$.

    The moment of force $\vec{F}$ about point O can be expressed as

              $\vec{M}_o = \vec{r} \times \vec{F}$ (Vector equation)

                $\vec{r}$ = displacement vector of point A w.r.t. point O.

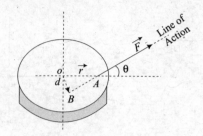

Fig. 5.18 Moment

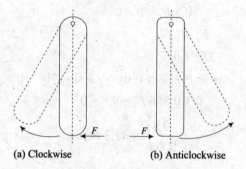

(a) Clockwise   (b) Anticlockwise

Fig. 5.19 Direction of moment

$$M_o = rF \sin \theta$$
$$= F \times d \quad \text{(Scalar equation)}$$

If a force tries to rotate a body in the clockwise direction, the moment is called clockwise moment.

Similarly, a body is rotated in anticlockwise direction by an anticlockwise moment.

5.6.1 Principle of Moments (Varignon's Theorem)

The principle of moments may be stated as "When a body acted upon by several forces is in rotational equilibrium, the sum of the clockwise moments of the forces about any point is equal to the sum of the anticlockwise moments of the forces about the same point".

There are four coplanar forces F_1, F_2, F_3 and F_4 acting on a body keeping the body in rotational equilibrium. The respective perpendicular distances from z-axis (axis of rotation) are d_1, d_2, d_3 and d_4. The moment of the forces about point O are:

$$M_1 = F_1 \times d_1 \text{ (clockwise)}, -ve$$
$$M_2 = F_2 \times d_2 \text{ (anticlockwise)}, +ve$$
$$M_3 = F_3 \times d_3 \text{ (clockwise)}, -ve$$
$$M_4 = F_4 \times d_4 \text{ (anticlockwise)}, +ve$$

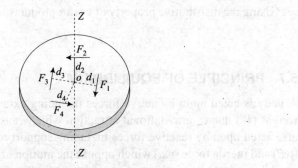

Fig. 5.20 Principle of moments

The resultant moment = $-F_1 d_1 + F_2 d_2 - F_3 d_3 + F_4 d_4 = 0$

$\therefore \qquad\qquad F_1 d_1 + F_3 d_3 = F_2 d_2 + F_4 d_4$

Sum of clockwise moments = sum of anticlockwise moment.

Thus principle of moments is verified.

Varignon's theorem states, "The moment of the resultant of several concurrent forces about a given point is equal to the sum of the moments of individual forces about the same point".

The theorem can be best expressed in vector algebra.

$\vec{F_1}$, $\vec{F_2}$ and $\vec{F_3}$ are three concurrent forces acting at point O. The resultant,

$$\vec{R} = \vec{F_1} + \vec{F_2} + \vec{F_3} + \dots$$

The moment of resultant

$$= \vec{r} \times \vec{R}$$

The sum of moment of original forces

$$= \vec{r}\, \vec{F_1} + \vec{r}\, \vec{F_2} + \vec{r}\, \vec{F_3} + \dots$$

$$\vec{r} \times (\vec{R}) = \vec{r} \times (\vec{F_1} + \vec{F_2} + \vec{F_3} + \dots)$$

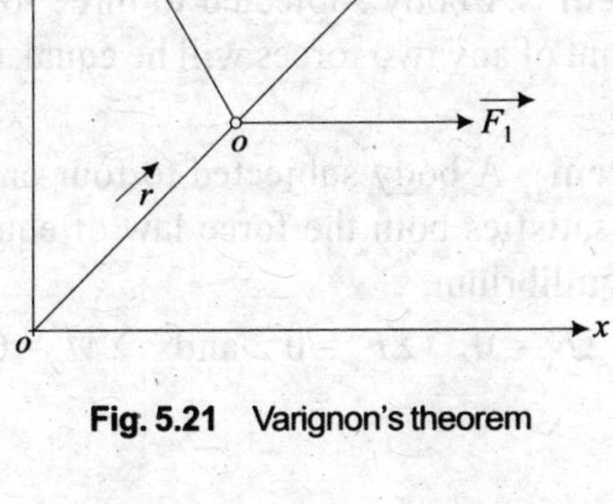

Fig. 5.21 Varignon's theorem

Using the distributive property of vector products,

$$\vec{r} \times (\vec{F_1} + \vec{F_2} + \vec{F_3} + ...) = \vec{r}\,\vec{F_1} + \vec{r}\,\vec{F_2} + \vec{r}\,\vec{F_3} + ...$$

5.7 PRINCIPLE OF EQUILIBRIUM

A body is acted upon by active forces including external applied forces (F) and weight (W) due to gravitational attraction which cause motion of the body. It is also acted upon by reactive forces including support reaction (R), frictional force (μR) and inertia force (ma) which oppose the motion of the body. A body will be in equilibrium if there is no linear motion due to combined action of active and reactive forces and also there is no rotation due to moments of these forces.

The principle of equilibrium states, "A stationary body subjected to a system of forces will be in equilibrium if the resultant of the forces and resultant moment are zero".

If F includes all active and reactive forces

$$\Sigma F = 0$$

$$\Sigma M = 0$$

The forces are generally resolved into components along Cartesian coordinates. Therefore,

$$\Sigma F_x = 0 \quad \text{and} \quad \Sigma M_x = 0$$
$$\Sigma F_y = 0 \quad \text{and} \quad \Sigma M_y = 0$$
$$\Sigma F_z = 0 \quad \text{and} \quad \Sigma M_z = 0$$

For coplanar forces, the principle of equilibrium can be simplified as:

$$\Sigma F_x = 0, \qquad \Sigma F_y = 0 \quad \text{and} \quad \Sigma M_o = 0$$

where 0 is any point.

5.7.1 Force Law of Equilibrium

The force law of equilibrium can be applied to different force systems.

(a) **Twoforce system**   A body subjected to two forces will be in equilibrium if the two forces are collinear, equal and opposite.

(b) **Threeforce system**   A body subjected to three forces will be in equilibrium if the resultant of any two forces will be equal and opposite to the third force.

(c) **Multi-force system**   A body subjected to four or more forces will be in equilibrium if it satisfies both the force law of equilibrium as well as the moment law of equilibrium.

$$\Sigma F_x = 0, \quad \Sigma F_y = 0 \quad \text{and} \quad \Sigma M_o = 0.$$

5.7.2 Coplanar Concurrent Forces

When several concurrent forces act on a body, it will be in equilibrium if there is no unbalanced force acting on it. The resultant of all forces should be zero. The following equilibrium conditions should be satisfied.

(a) **Analytical conditions**

The sum of all the components in the x-direction as well as y-direction must be zero,

$$\Sigma F_x = 0 \quad \text{and} \quad \Sigma F_y = 0$$

(b) **Graphical condition**

The force polygon must close.

5.7.3 Coplanar Non-concurrent Forces

When a body is under the action of coplanar non-concurrent force system, it may move linearly in horizontal or vertical direction due to unbalanced force components in respective directions. It may also rotate due to resultant moment of force system. The body will be in equilibrium if the algebraic sum of all external forces and their moments about any point in their plane are nil.

The conditions of equilibrium may be expressed mathematically as follows:

1. Sum of all component forces in x-direction must be zero,

$$\Sigma F_x = 0$$

2. Sum of all component forces in y-direction must be zero,

$$\Sigma F_y = 0$$

3. Sum of moments of all forces about any point O must be zero,

$$\Sigma M_o = 0$$

The *resultant* of coplanar non-concurrent non-parallel force system can be found out as follows:

(a) **Magnitude**

The magnitude of the resultant can be found out analytically as follows:

The resultant, $\quad R = \sqrt{(\Sigma F_x)^2 + (\Sigma F_y)^2}$

where, ΣF_x = algebraic sum of component forces in x-direction

ΣF_y = algebraic sum of component forces in y-direction

(b) **Direction**

The direction of the resultant with respect to x-axis can be found out as follows:

$$\theta = \tan^{-1} \frac{\Sigma F_y}{\Sigma F_x}$$

(c) Position

The position of resultant along x-axis can be found out by taking the moment of component force of resultant in y-direction about a point O and equating the same to the sum of moments of force components in y-direction about point O.

The sum of moments of component forces in y-direction about point O,

$$= \Sigma F_y.x_i$$

where, x_i = perpendicular distance of component forces in x-direction.

The moment of y-component of resultant about point O,

$$= Ry.x$$

where, x = perpendicular distance of R_y from point O along x-direction.

$\therefore$ $\qquad\qquad Ry.x = \Sigma Fy.x_i$

$\therefore$ $\qquad\qquad x = \dfrac{\Sigma F_y.x_i}{R_y}$

5.7.4 Coplanar Parallel Forces

There is no component force say in x-direction. The equilibrium conditions, therefore, reduce to the following:

1. The sum of parallel forces in y-direction must be zero,

$$\Sigma F_y = 0$$

2. The sum of moment of parallel forces about any point O must be zero,

$$\Sigma M_o = 0$$

The *resultant* of parallel forces be found out as follows:

(a) Magnitude

The magnitude of resultant will be the algebraic sum of parallel forces,

$$R = \Sigma F_y$$

(b) Direction

The direction of resultant will be parallel to the original forces.

(c) Position

The position of the resultant from a reference point O will be

$$x = \frac{\Sigma F_y.x_i}{R}$$

where, $\Sigma F_y x_i$ = algebraic sum of moment of parallel forces about reference point

R = resultant of parallel forces.

5.8 EQUIVALENT SYSTEMS

A force acting on a rigid body has two effects:

1. Tendency to push or pull the body in its direction.
2. Tendency to rotate the body about any axis which does not intersect the line of force.

The above dual effect of a force can be replaced by the following equivalent system.

1. An equal parallel force acting at point O, and
2. A couple to compensate for the change in moment of force.

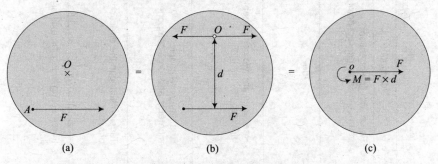

(a)　　　　　　　　(b)　　　　　　　　(c)

Fig. 5.22　Equivalent system

Example:　A force F is acting on a rigid body at point A tending to move the body towards right and rotate the body in anticlockwise direction about point O which is at a distance d from the line of action of force. The force can be shifted to point O by applying two equal and opposite forces F. The new system is equal to a direct force F acting towards right and a couple with a moment $F \times d$.

Therefore the equivalent system shown in Fig. 5.21(c) is a direct force F acting at point O and an anticlockwise moment,

$$M = F \times d$$

This method of movement of a force to parallel position is a powerful tool used in solving many problems in engineering mechanics.

5.9 SUPPORT REACTIONS

A rigid body to be in equilibrium requires both the balance of forces to prevent translation and balance of moments to prevent rotation. The balancing forces are provided by the support reactions as per Newton's third law of motion. Different types of supports used in mechanics and structures and their reactions are shown in Table 5.1.

Table 5.1 Support reactions

Sl. No.	Type of support/connections		Type of reaction offered		No. of unknown
	Description	Sketch	Description	Sketch	
1.	Roller		Force with line of action perpendicular to contact point		1 1
2.	Rocker		-do-		1
3.	Pin or Hinge support		Reactions are two components of force	F_x F_y	2
4.	Fixed support	a F	Reactions are couple moments and force components	F_x F_y $M = F \times a$	3
5.	Cable	F Cable	Tension acting in the direction of cable	T	1
6.	Link	Link	Force acting along the axis of link	F	1

5.10 FREE BODY DIAGRAM

A free body diagram is a sketch of the particle/body which represents it being isolated from its surrounding/system. All the forces that act on it are represented on the
sketch. The forces include active forces which cause motion of the particle/body and
reactive forces arising from supports or constraints. The reactions always act normal to the plane of contact as shown in Table 5.1.

Problems in static can be solved easily from free body diagram. Force system
acting on bodies in contact can be shown in the sketch. The solution of problem
becomes simple.

Example 1: A sphere resting on a horizontal surface,

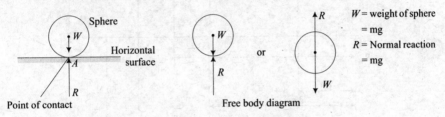

Fig. 5.23 FBD of a sphere on horizontal surface

Example 2: A sphere resting in a V-notch,

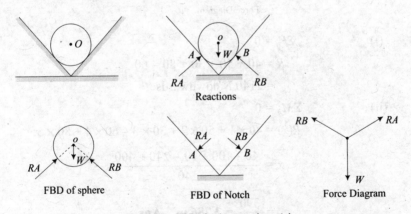

Fig. 5.24 FBD of sphere and a notch

Example 5.8: Determine the resultant R of five vertical forces acting on the beam
as shown.

Solution:

1. Draw force diagram.
2. Apply equilibrium conditions.

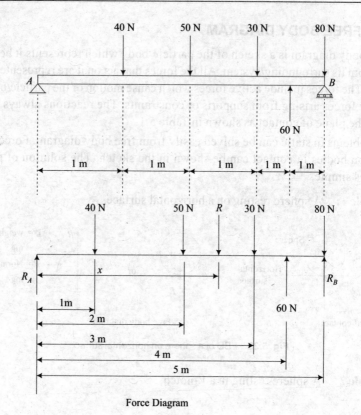

Force Diagram

(i) $\qquad \Sigma F_y = 0$

$$R = 40 + 50 + 30 + 80 - 60$$

$$= 1140 \text{ N downwards}$$

(ii) $\qquad \Sigma M_A = 0$

$$R(x) = 40 \times 1 + 50 \times 2 + 30 \times 3 - 60 \times 4 + 80 \times 5$$

∴ $\qquad x = \dfrac{40 + 100 + 90 - 240 + 400}{140}$

$$= \dfrac{390}{140} = \textbf{2.786 m} \quad \textbf{Ans.}$$

Example 5.9: A horizontal beam *PQRS* is 12 m long, where *PQ = QR = RS* = 4 m. Forces of 1000 N, 1500 N, 1000 N and 500 N act at point *P*, *Q*, *R* and *S* respectively with the downward direction. The lines of action of these forces make angles 90°, 60°, 45° and 30° respectively with *PS*. Find the magnitude, direction and position of resultant.

Solution:

1. Draw force diagram.

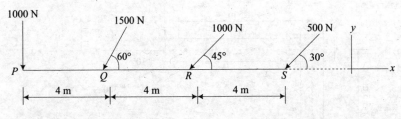

Force Diagram

2. Find the Σx and Σy

    (i) $\Sigma F_x = 1000 \cos 90° + 1500 \cos 60° + 1000 \cos 45° + 500 \cos 30°$

             $= 1000 \times 0 + 1500 \times 0.5 + 1000 \times 0.707 + 500 \times 0.866$

             $= 1890$ N

    (ii) $\Sigma Fy = 1000 \sin 90° + 1500 \sin 60° + 1000 \sin 45° + 500 \sin 30°$

             $= 1000 \times 1 + 1500 \times 0.866 + 1000 \times 0.707 + 500 \times 0.5$

             $= 3256$ N

    (iii) The resultant,

$$R = \sqrt{(\Sigma F_x)^2 + (\Sigma F_y)^2}$$

$$= \sqrt{(1890)^2 + (3256)^2}$$

$$= 3760 \text{ N}$$

$$\tan \theta = \frac{\Sigma F_y}{\Sigma F_x} = \frac{3256}{1890} = 1.865$$

$\therefore$ $\theta = 61°48'$

    (iv) $\Sigma M_p = 1000 \times 0 + 1500 \sin 60° \times 4 + 1000 \sin 45° \times 8$

                                   $+ 500 \sin 30° \times 12$

          $= 0 + 5196.15 + 5656.85 + 3000 = 13853$

          $= R_y \times x$

$\therefore$ $x = \dfrac{13853}{R \sin 61°48'} = \dfrac{13853}{3760 \sin 61°48'} =$ **4.18 m from P** **Ans.**

Example 5.10: Determine the moment of the force acting on the railway sign post about point O.

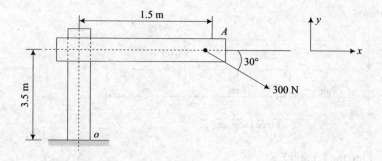

Solution:

1. The rectangular components of the force at A

$$F_x = F \cos \theta$$
$$= 300 \cos 30° = 259.8 \text{ N}$$
$$F_y = F \sin \theta = 300 \sin 30°$$
$$= 150 \text{ N}$$

2. Applying Varignon's theorem that the moment of resultant is equal to the sum of moments of component forces about the same point.

$$M_o = F_x (3.5) + F_y (1.5)$$
$$= 259.8 (3.5) + 150 (1.5)$$
$$= 1134.34 \text{ Nm (clockwise)}$$

Example 5.11: Replace the 20 kN force acting on steel column by a force-couple system at O.

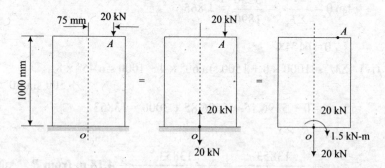

Solution: Step 1. Apply two concurrent forces of 20 kN each at O acting in opposite direction.

Step 2. Remove force of 20 kN at A and 20 kN upward force at O cancelling equal and opposite force.

Step 3. Apply a moment of couple of 20 kN acting at a distance of 75 mm clockwise.

$$M_o = 20 \times 0.075 = 1.5 \text{ kN-m}$$

Therefore, equivalent force system at O will consist of a vertical force of 20 kN acting downwards and a moment of 1.5 kN-m acting clockwise.

Example 5.12: What force moment is transmitted to the supporting wall at A in given cantilever beam. [U.P.T.U. II Sem., 2002-03]

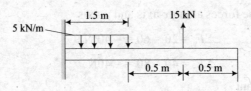

Solution:

1. Draw the force diagram,

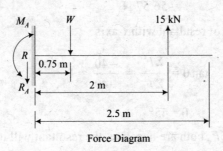

Force Diagram

The equivalent force W is equal to the area under the load curve and will act at c.g. of load curve.

$W = wL = 5 \times 1.5 = 7.5$ kN downwards at a distance of 0.75 m from A.

2. Apply equilibrium condition to find the resultant reaction.

$$R_A = 15 - 7.5 = 7.5 \text{ kN} \quad \text{downwards}$$

$$M_A = 15 \times 2 - 7.5 \times 0.75 = 30 - 5.625$$

$$= 24.375 \text{ kN-m} \quad \text{Anticlockwise}$$

Example 5.13: Four forces having magnitude of 20 N, 40 N, 60 N and 80 N respectively are acting along the fours sides (1 m each) of equate *ABCD*, taken in order, as shown. Determine the magnitude and direction of the resultant force.

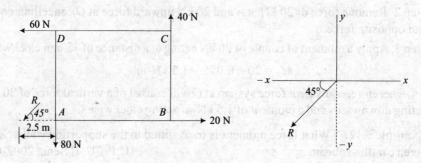

Solution:

1. Resolve the forces along x-axis and y-axis

$$\Sigma F_x = 20 - 60 = -40 \text{ N}$$

$$\Sigma F_y = 40 - 80 = -40\text{N}$$

2. The resultant

$$R = \sqrt{(\Sigma F_x)^2 + (\Sigma F_y)^2} = \sqrt{(-40)^2 + (-40)^2}$$

$$= 56.57 \text{ N}$$

3. The direction of resultant with x-axis

$$\tan\theta = \frac{\Sigma F_y}{\Sigma F_x} = \frac{-40}{-40} = 1$$

$\therefore$ $\theta = 45°$

As ΣF_x and ΣF_y both are negative, the resultant will make an angle of $45°$ with $-$ve x-axis.

4. *Position of resultant*

If x is the horizontal distance of resultant R from point A.

$$\Sigma M_A = 20 \times 0 + 40 \times 1 + 60 \times 1 + 80 \times 0$$

$$= 100 \text{ Nm}$$

$\therefore$ $$x = \frac{\Sigma M_A}{R_y} = \frac{100}{-40} = -2.5 \text{ m}$$

Resultant will be acting at a distance of 2.5 m to left of A.

Example 5.14:   A square block of wood of mass M is hinged at A and rests on roller at B. It is pulled by means of a string attached at D and inclined at an angle $30°$ with the horizontal. Determine the force P which should be applied to the string to just lift the block off the roller.   [U.P.T.U. I Sem., 2004-05]

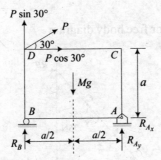

Solution:

$$R_B = 0$$

Apply equilibrium conditions,

(i) $$\Sigma F_x = 0$$

∴ $$P \cos 30° - R_{Ax} = 0$$

∴ $$R_{Ax} = P \cos 30°$$

(ii) $$\Sigma F_y = 0$$

∴ $$P \sin 30° - Mg + R_{Ay} = 0$$

(iii) $$\Sigma M_A = 0$$

$$P \cos 30° \times a - Mg \times \frac{a}{2} + P \sin 30° \times a = 0$$

∴ $$Pa(\cos 30° + \sin 30°) = Mg \frac{a}{2}.$$

∴ $$P = \frac{Mg \dfrac{a}{2}}{a(0.866 + 0.5)} = 0.366 \, Mg \quad \textbf{Ans.}$$

Example 5.15: A man weighing 75 N stands on the middle rung of a 25 N ladder resting on smooth floor and against a wall. The ladder is prevanted from slipping by a string *OD*. Find the tension in the string and reactions at *A* and *B*.

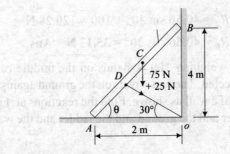

Solution:

1. Draw force diagram or free body diagram

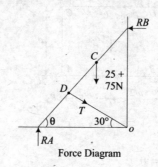

Force Diagram

$$\tan \theta = \frac{OB}{OA} = \frac{4}{2} = 2$$

∴ $\theta = 63.43°$

2. Apply equilibrium conditions.

(i) $\Sigma F_x = 0$

$$T \cos 30° - R_B = 0$$

$$R_B = T \cos 30°$$

(ii) $\Sigma F_y = 0$

$$R_A - (75 + 25) - T \sin 30° = 0$$

$$T \sin 30° = R_A - 100$$

(iii) $\Sigma M_o = 0$

$$R_B(4) + 100(1) - R_A(2) = 0$$

$$2R_A - 4R_B = 100$$

∴ $$2(T \sin 30° + 100) - 4T \cos 30° = 100$$

$$2T \times 0.5 - 3.464T = 100 - 200$$

∴ $$T = \frac{-100}{-2.464} = 40.56 \text{ N}$$

∴ $$R_A = 40.56 \sin 30° + 100 = 120.28 \text{ N}$$

$$R_B = 40.56 \cos 30° = \mathbf{35.13 \text{ N}} \quad \textbf{Ans.}$$

Example 5.16: A man weighing 100 N stands on the middle rung of a ladder whose weight can be neglected. The end A rests on the ground against a stop and the end B rests on the corner of a wall as shown. Find the reactions at A and B. Neglect friction between the ladder and the ground and the ladder and the wall.

Solution:
1. Draw the force diagram or free body diagram,

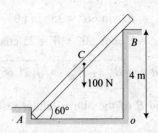

$$\sin 60° = \frac{OA}{AB}$$

∴

$$AB = \frac{4}{\sin 60°} = 4.62 \text{ m}$$

$$AD = AC \cos 60° = \frac{4.62}{2} \cos 60° = 1.155 \text{ m}$$

2. Apply equilibrium conditions,

   (i) $\Sigma F_x = 0$

                 $R_{Ax} - R_B \sin 60° = 0$

   (ii) $\Sigma F_y = 0$

                 $R_{Ay} - W - R_B \cos 60° = 0$

   (iii) $\Sigma M_A = 0$

                $R_B(AB) - 100(AD) = 0$

$$R_B = \frac{100(AD)}{AB}$$

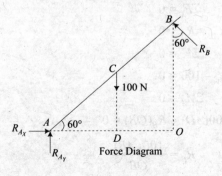

Force Diagram

$$= \frac{100 \times 1.155}{4.62} = 25 \text{ N}$$

$$R_{Ax} = R_B \sin 60° = 25 \sin 60° = 21.65 \text{ N}$$

$$R_{Ay} = R_B \cos 60° + W = 25 \cos 60° + 100 = 112.5 \text{ N}$$

$$R = \sqrt{R_{A_X}^2 + R_{A_Y}^2} = \sqrt{(21.65)^2 + (112.5)^2} = 114.56 \text{ N}$$

Example 5.17: If the end B of the above ladder rests against a wall, find the reactions at A and B.

Solution:

   1. Draw the force diagram

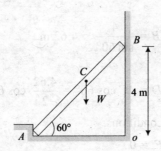

   2. Apply equilibrium condition

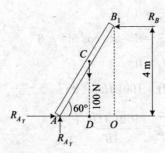

(i) $$\Sigma F_x = 0$$

$$R_{Ax} - R_B = 0$$

(ii) $$\Sigma F_y = 0$$

$$R_{Ay} - 100 = 0$$

(iii) $$\Sigma M_A = 0$$

$$100(AD) - R_B(OB) = 0$$

$$\therefore \qquad R_B = \frac{100(AB)}{OB}$$

$$R_B = \frac{100\,(1.155)}{4.62\,\cos 60°} = 50\ \text{N}$$

$$R_{Ax} = R_B = 50\ \text{N}$$

$$R_{Ay} = 100\ \text{N}$$

$$\therefore \qquad R = \sqrt{R_{AX}^2 + R_{AY}^2} = \sqrt{(50)^2 + (100)^2} = 111.8\ \text{N}$$

Example 5.18: Three cylinders are piled up in a rectangular channel as shown. Determine the reaction R_6 between cylinder A and vertical wall of the channel.

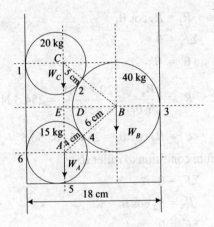

Solution:

1. Draw free body diagram of the three cylinders.

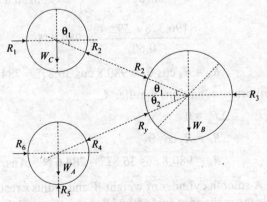

Free body diagram

2. Find the angles,

$$\cos \theta_1 = \frac{BD}{BC} = \frac{18 - 5 - 6}{6 + 5} = \frac{7}{11}$$

$$\therefore \qquad \theta_1 = 50.5°$$

$$\cos \theta_2 = \frac{BE}{AC} = \frac{18 - 4 - 6}{4 + 6} = \frac{8}{10}$$

$$\therefore \qquad \theta_2 = 36.87°$$

3. Apply equilibrium conditions to roller c,

    (i) $\qquad\qquad \Sigma F_x = 0$

$$R_1 = R_2 \cos \theta_1$$

    (ii) $\qquad\qquad \Sigma F_y = 0$

$$R_2 \sin \theta_1 = W_C$$

$$\therefore \qquad R_2 = \frac{W_C}{\sin \theta_1} = \frac{20 \times 9.81}{\sin 50.5°} = 254.5 \text{ N}$$

$$R_1 = R_2 \cos \theta_1 = 254.5 \cos 50.5° = 161.88 \text{ N}$$

4. Apply equilibrium condition to roller B.

    (i) $\qquad\qquad \Sigma F_x = 0$

$$R_3 = R_4 \cos \theta_2$$

    (ii) $\qquad\qquad \Sigma F_y = 0$

$$R_4 \sin \theta_2 - R_2 \sin \theta_1 - W_B = 0$$

$$\therefore \qquad R_4 = \frac{R_2 \sin \theta_1 + W_B}{\sin \theta_2} = \frac{254.5 \sin 50.5° + 40 \times 9.81}{\sin 36.87°}$$

$$= \frac{196.378 + 392.4}{0.60} = 980.8 \text{ N}$$

$$R_3 = R_4 \cos \theta_2 = 980.8 \cos 36.87° = 784.64 \text{ N}$$

5. Apply equilibrium conditions to roller A,

    (i) $\qquad\qquad \Sigma F_x = 0$

$$R_6 = R_4 \cos \theta_2$$

$$R_6 = 980.8 \cos 36.87° = \textbf{784.6 N} \quad \textbf{Ans.}$$

Example 5.19: A smooth cylinder of weight W and radius r rests in a V-shaped groove whose sides are inclined at angle α and β with horizontal. Find the reactions R_A and R_B at points of contact. Given $\alpha = 25°$, $\beta = 65°$, $W = 500$ N.

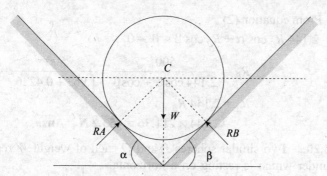

Solution:

1. Draw force diagram or free body diagram,

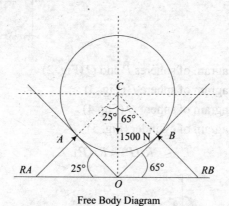

Free Body Diagram

Angle $ACO = \alpha = 25°$

Angle $BCO = \beta = 65°$

2. Apply equilibrium conditions to roller C.

(i) $\Sigma F_x = 0$

$R_A \sin \alpha - R_B \sin \beta = 0$ ...(1)

(ii) $\Sigma F_y = 0$

$R_A \cos \alpha + R_B \cos \beta - W = 0$ ...(2)

From equation (1)

$\therefore$ $$R_A = \frac{R_B \sin \beta}{\sin \alpha} = \frac{R_B \times 0.9063}{0.4226} = 2.144 \, R_B$$

From equation (2)

$$2.144 \, R_B \cos \alpha + R_B \cos \beta - W = 0$$

∴

$$R_B = \frac{500}{2.144 \cos\alpha + \cos\beta} = \frac{500}{1.963 + 0.4226} = 211.36 \text{ N}$$

∴

$$R_A = 2.144 \, R_B$$
$$= 2.144 \times 211.36 = \textbf{453.2 N} \quad \textbf{Ans.}$$

Example 5.20: Two similar spheres P and Q each of weight W rest inside a hollow cylinder which is resting on a horizontal plane. Draw the free-body diagram of:

(a) Both the spheres taken together
(b) The sphere P
(c) The sphere Q
(d) The cylinder.

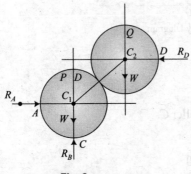

Fig. 1

Solution:

(a) Free body diagram of spheres P and Q (Fig. 2)
(b) Free body diagram of sphere P (Fig. 3)
(c) Free body diagram of sphere Q (Fig. 4)
(d) Free body diagram of cylinder (Fig. 5)

$$R_C(P) < R_C(P)$$

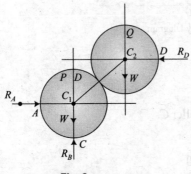

Fig. 2

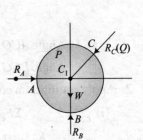

Fig. 3

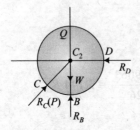

Fig. 4

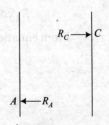

Fig. 5

Example 5.21: The figure shows a sphere resting in a smooth *V*-shaped groove and is subjected to a spring force. The spring is compressed to a length of 100 mm from its free length of 150 mm. If the stiffness of spring is 2 N/mm, determine the contact reactions at *A* and *B*. [U.P.T.U., May, 2003]

Solution:

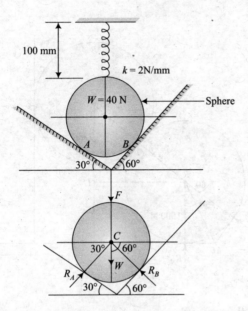

1. Draw the free-body diagram of sphere.

   Spring force, $F = k \times \delta = 2 \times (150 - 100) = 100$ N

2. Equilibrium of sphere *C*.

$$\Sigma F_x = 0; \quad R_A \sin 30° - R_B \sin 60°$$

$$\therefore \qquad R_A = \frac{\sin 60°}{\sin 30°} - R_B = 1.732 R_B$$

$$\Sigma F_y = 0; \quad R_A \cos 30° + R_B \cos 60° - W - F = 0$$

$$1.732\, R_B \cos 30° + R_B \cos 60° = 40 + 100 = 140$$

$$\therefore \qquad R_B = \frac{140}{1.732 \cos 30° + \cos 60°} = \textbf{70 N} \quad \textbf{Ans.}$$

$$R_A = 1.732 \times 70$$

$$= \textbf{1121.2 N} \quad \textbf{Ans.}$$

Example 5.22: Two cylinders A and B rest in a horizontal channel as shown. The cylinder A has a weight of 1000 N and radius of 9 cm. The cylinder B has a weigth of 400 N and a radius of 5 cm. The channel is 18 cm wide at the bottom, with one side vertical. The other side is inclined at 60° with the horizontal. Find the reactions.

Solution:

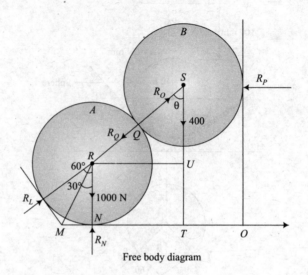

Free body diagram

1. Free body diagram is drawn.

2. $\sin \theta = \dfrac{RU}{RS} = \dfrac{18 - MN - OT}{9 + 5}$

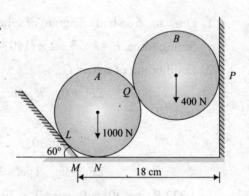

$\qquad = \dfrac{18 - RN \tan 30° - 5}{14}$

$\qquad = \dfrac{18 - 9 \tan 30° - 5}{14}$

$\qquad \theta = 33.86°$

3. Equilibrium of cylinder B.

$$\Sigma F_x = 0, \quad R_Q \sin(33° - 86°) - R_P = 0$$

$$\Sigma F_y = 0, \quad R_Q \cos 33.86° - 400 = 0$$

$\therefore \qquad R_Q = 481.9 \text{ N}$

$\qquad R_P = 481.9 \sin 33.86 = 268.5 \text{ N}.$

4. Equilibrium of cylinder A.

$$\Sigma F_x = 0; \quad R_L \sin 60° - R_Q \sin 33.86° = 0$$

$$\therefore \qquad R_L = \frac{481.9 \sin 33.86°}{\sin 60°} = 310 \text{ N}$$

$$\Sigma F_y = 0, \quad R_N - 1000 - R_Q \cos 33.86° + R_L \cos 60° = 0$$

$$\therefore \qquad R_N = 1248.2 \text{ N}.$$

Example 5.23: A roller of mass 150 kg has to be started rolling over the block A as shown. Find out the force T required.

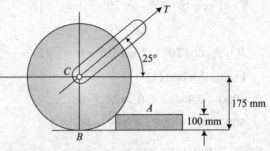

Solution: When the roller is about to be started, it will lose contact with ground at point B,
$\therefore R_B = 0$. As the roller is in equilibrium, there will be three concurrent forces acting at C. The free body diagram is drawn. For equilibrium of roller,

$$\Sigma F_x = 0, \quad T \cos 25° - R_A \cos \theta = 0$$
$$\Sigma F_y = 0, \quad T \sin 25° + R_A \sin \theta - W = 0$$

$$\theta = \tan^{-1} \frac{100}{175} = 53°$$

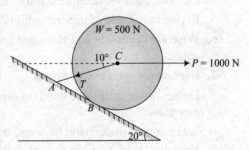

$$\therefore \qquad R_A = \frac{T \cos 25°}{\cos 53°} = \frac{0.906T}{0.602} = 1.5T$$

$$\therefore \qquad T \sin 25° + 1.5T \sin 53° = 150 \times 9.81$$

$$\therefore \qquad T = \frac{150 \times 9.81}{\sin 25° + 1.5 \sin 53°}$$

Free body diagram

$$= \frac{1471.5}{0.4226 + 1.19795} = \textbf{908 N} \quad \textbf{Ans.}$$

Example 5.24: A right circular roller of weight 5000 N rests on a smooth inclined plane and is held in position by a cord AC as shown in Fig. Find the tension in the cord if there is a horizontal force of magnitude 1000 N acting at C.

U.P.T.U. May, 2003]

Solution:

1. Draw free-body diagram of the roller.

2. For equilibrium

$$\Sigma F_x = 0$$

$$P - T\cos 10° + R_B \sin 20° = 0 \quad ...(1)$$

$$\Sigma F_y = 0$$

$$W - R_B \cos 20° + T \sin 10° = 0 \quad ...(2)$$

Free body diagram

From equation (1)

$$1000 - 0.9848T + 0.342R_B = 0$$

$$5000 - 0.93969R_B + 0.173648T = 0$$

$$\therefore \quad R_B = \frac{0.9848T - 1000}{0.342} = 2.8795T - 2923.9766$$

Substituting the value of R_B into equations (2).

$$5000 - 0.9369(2.8795T - 2923.9766) + 0.173648T = 0$$

$$5000 - 2.7T + 2747.63 + 0.173648T = 0$$

$$\therefore \quad T = \frac{5000 + 2747.63}{2.7 - 0.173648} = \frac{7747.63}{2.52635} = \textbf{3066.73 N} \quad \textbf{Ans.}$$

QUESTION BANK NO. 5

1. Enumerate different laws of motion, discussing the significance of each of them.
2. Discuss various laws and principles of engineering mechanics.
3. Explain various force systems.
4. What do you understand by resultant of a force system? Discuss various methods used for determination of the resultant of coplanar force systems.
5. Explain principle of transmissibility of a force.
6. What are necessary and sufficient conditions of equilibrium of a system of coplanar force system.
7. Explain polygon law of forces.
8. Define moment of force about a point. Also explain Varignon's theorem of moments.
9. What do you understand by transfer of force to a parallel position?
10. Explain free body diagram with suitable examples.

6

Friction

6.1 INTRODUCTION

Whenever the surfaces of two bodies are in contact, there is some resistance to sliding between them. The opposing force to the movement is called friction or force of friction. It is due to interlocking of surfaces as a result of the presence of some roughness and irregularities at the contact surfaces. The resisting force acts in the direction opposite to the movement. A force of friction comes into play whenever there is a relative motion between two parts. Some energy is wasted in order to overcome the resistance due to force of friction.

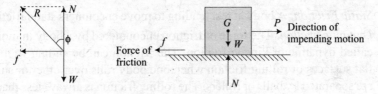

Fig. 6.1 Force of friction

Force of friction or frictional force is the opposing force to the movement of one body over the surface of another body.

In Fig. 6.1,

$$W = \text{Weight of the body (mg)}$$
$$N = \text{Normal reaction}$$
$$f = \text{Friction force}$$
$$P = \text{Force applied to the body}$$
$$R = \text{Total reaction}$$
$$\phi = \text{Angle of friction}$$

6.1.1 Engineering Applications of Friction

Friction is both desirable and undesirable in engineering applications.

1. *Friction devices:* There are devices in engineering which work due to friction. Examples are lifting machines such as screw jack, power transmission

by belts and ropes, friction clutches, brakes, nuts and bolts, jib and cotter joints, etc. The friction should be utilized by the use of proper materials and shape of components.

2. The vehicles can run on roads and inclined surfaces due to the friction only.

3. A man cannot walk if there is no friction.

However, there is resistance and wastage of energy in the bearings and sliding surfaces.

1. Friction causes loss of energy in moving parts of machines.

2. Friction causes heating of bodies and may affect the life of the machine.

3. The efficiency of machine decreases due to presence of friction.

Solid, liquid and gaseous lubricants are used to reduce friction by separating the moving surfaces.

The *characteristics* of the friction force are:

1. It always acts in a direction opposite to the motion and opposes the tractive effort.

2. It is a reactive or passive force and exists as long as tractive effort is present.

3. The friction force adjusts itself depending upon the magnitude of tractive effort or applied force.

6.2 TYPES OF FRICTION

1. *Static Friction:* A body at rest tending to move encounters static friction.

2. *Dynamic friction:* The force of friction encountered by a body in motion is called dynamic or kinetic friction. The friction can be sliding friction on flat surfaces or rolling friction when one body rolls over other or surfaces are separated by balls or rollers. The rolling friction is always less than sliding friction when surfaces slide one over the other. The friction that exists when no lubricant is used, is called *dry friction* or *solid friction*. Laws of static and dynamic friction are based on experiments conducted by Prof. Coulomb and are called laws of *Coulomb Friction*.

6.3 COULOMB LAWS OF FRICTION

6.3.1 Laws of Static Friction

The laws of static friction are as follows:

1. The force of friction always acts opposite to the direction of impending motion.

2. The magnitude of force of friction is equal to applied force tending to move the body.

$$f = P$$

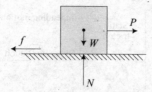

Fig. 6.2 Static friction

3. The force of friction is directly proportional to the normal reaction between the surfaces.

$$f \propto N$$

or

$$\frac{f}{N} = \text{constant} = \mu$$

μ is called coefficient of frictions and depends upon the nature of contact surfaces.

$$\therefore \qquad f = \mu N$$

4. The force of friction is independent of the area of contact of the two surfaces.
5. The force of friction depends upon the roughness of the surface.

6.3.2 Laws of Dynamic Friction

Limiting Friction

The friction force, f increases to oppose P until a maximum value of f_m is reached. If P is further increased, the body starts sliding as force of friction cannot balance the force P. The value of f is lowered from f_m, which is the *limiting* and maximum friction force that is exerted at the time when motion is about to begin.

The laws of dynamic friction are as follows:

1. The force of friction always opposes the motion of the body and acts in the direction opposite to the motion.
2. The frictional force is directly proportional to the normal reaction between the contacting surfaces in relative motion.
3. The force of friction although directly proportional to normal reaction but is slightly less than for static friction

$$f = \mu N$$

The coefficients of friction μ is less for dynamic friction than for limiting friction.

4. For low velocities, total amount of friction force that can be developed is practically independent of velocity.

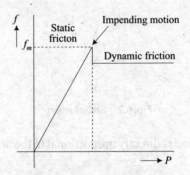

Fig. 6.3   Limiting friction

6.3.3   Important Definitions

(a) *Angle of Friction*

It is the angle made by the resultant (R) of the normal reaction (N) and limiting force of friction (f) and made with the direction of normal reaction.

R is the resultant of normal reaction N and force of friction f.

$$R = \sqrt{N^2 + f^2}$$

ϕ is the angle of friction

$$\tan \phi = \frac{f}{N}$$

or

$$\phi = \tan^{-1} \frac{f}{N}$$

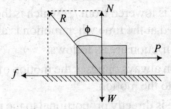

Fig. 6.4   Angle of friction

(b) *Coefficient of friction*

It is ratio of limiting frictional force and the normal reaction.

The coefficient of friction,

$$\mu = \frac{f}{N} = \tan \phi$$

$\therefore$

$$f = \mu N$$

μ depends upon the nature of contacting surface. Its value is very low for lubricated surfaces and high for dry friction.

(c) Angle of Repose

A body of weight W is lying on a rough plane inclined at an α with the horizontal. The body is in equilibrium under the action of following forces:

(i) *Weight of the body W*. It has two components: $W \sin \alpha$ parallel to inclined plane and $W \cos \alpha$ normal to the plane.

(ii) *Normal reaction*, N acting in a direction normal to inclined plane.

(iii) *Friction force, f* acting in a direction opposite to the motion.

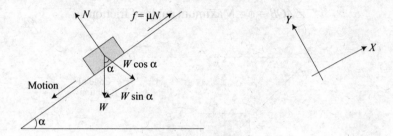

Fig. 6.5 Angle of repose

When the body tends to slide down the plane, the frictional force must act up the plane and when the body is being pulled up, the force of friction acts downwards to oppose the motion.

Selecting the reference coordinate system with X-axis in the direction of inclined plane and Y-axis perpendiculars to inclined plane,

Applying equilibrium conditions,

$$\Sigma F_X = 0$$

$$f = W \sin \alpha$$

$$\Sigma F_Y = 0$$

$$N = W \cos \alpha$$

$\therefore$
$$\frac{f}{N} = \frac{W \sin \alpha}{W \cos \alpha} = \tan \alpha$$

But
$$\frac{f}{N} = \mu = \tan \phi$$

where ϕ is called the angle of friction.

If angle α of inclined plane is altered so that the body is in the condition of limiting equilibrium, i.e., on the verge of sliding down, it is equal to angle of friction ϕ. This value of α is called *angle of repose*. Therefore, angle of repose may be defined as the angle of inclined plane on which body is in the condition of limiting equilibrium. The angle of repose is equal to angle of friction.

(d) Cone of Friction

In Fig. 6.6,

$$OA = N = \text{Normal reaction}$$
$$OC = f_{max} = \text{Limiting frictional force}$$
$$OR = \text{Resultant force}$$
$$\angle AOB = \phi = \text{Maximum angle of friction}$$

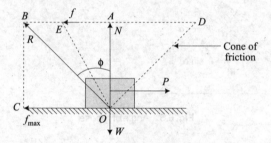

Fig. 6.6   Cone of friction

If line OB is revolved about vertical axis OA, the cone generated OBD is called cone of friction.

If the force of friction $f(AE)$ and resultant OC fall within the cone of friction, there will be no motion of the body. This principle is used in self-locking mechanism of lifting machines (i.e., screw jack) and taper pins.

6.4   EQUILIBRIUM OF BODIES INVOLVING DRY FRICTION

The force of dry friction is involved in the equilibrium and motion of bodies under following cases:

1. Rough Horizontal Plane
2. Rough Inclined Plane
3. Ladders
4. Wedges
5. Belt Drive
6. Lifting Machines

In drawing the free body diagrams, in addition to active and reactive forces, the force of friction is applied in a direction opposite to the motion. The forces are resolved into rectilinear components. The unknown forces are found from the equations of equilibrium.

6.4.1 Rough Horizontal Plane

There can be following cases of motion of body on rough horizontal plane. The equilibrium conditions for each case are discussed.

(a) No moving force

$$P = 0$$

∴ Friction force, $f = 0$

$$W = N$$

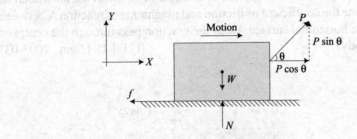

Fig. 6.7 Rough horizontal plane

(b) Body moving under pull or push

A force P is applied to the body. The force body diagram is shown.

Considering the equilibrium of the body.

$$\Sigma F_X = 0$$

$$P \cos \theta = f$$

But by definition, $f = \mu N$

∴ $$P \cos \theta = \mu N \qquad \qquad ...(1)$$

$$\Sigma F_Y = 0$$

∴ $$N = W \pm P \sin \theta$$

∴ $$W = N + P \sin \theta$$

or $$N = W - P \sin \theta \qquad \qquad ...(2)$$

From equation (1) and (2)

$$P \cos \theta = \mu(W - P \sin \theta)$$

$$P \cos \theta + \mu P \sin \theta = \mu W$$

$$P\left(\cos \theta + \frac{\sin \phi}{\cos \phi} \sin \theta\right) = \frac{\sin \phi}{\cos \phi} W \qquad \left[\because \mu = \tan \phi = \frac{\sin \phi}{\cos \phi}\right]$$

$$p(\cos \theta \cos \phi + \sin \theta \sin \phi) = W \sin \phi$$
$$P \cos(\theta - \phi) = W \sin \phi$$
$$P = \frac{W \sin \phi}{\cos(\theta - \phi)}$$

For P to be minimum, $\cos(\theta - \phi)$ should be maximum,

$$\therefore \qquad \cos(\theta - \phi) = 1$$
$$\theta = \phi$$

The angle of inclination of force P should be equal to the angle of friction, ϕ.

Example 6.1: The force required to pull the body of weight 50N on a rough horizontal surface is 20N where it is applied at an angle of 25° with the horizontal as shown. Determine the coefficient of friction and magnitude of reaction N between the body and the horizontal surface. Does the reaction pass through the centre of gravity of the body? [U.P.T.U. I Sem., 2002-03]

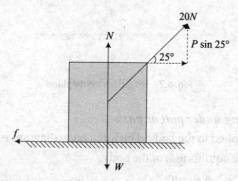

Solution: The body is in equilibrium

$$\therefore \qquad \Sigma F_y = 0$$
$$N = W - P \sin 25°$$
$$= 50 - 20 \times 0.42$$
$$= 41.55N$$
$$\Sigma F_x = 0$$
$$\therefore \qquad f = 20 \cos 25° = 18.13N$$

Now
$$\mu = \frac{f}{N} = \frac{18.13}{41.55} = 0.436 \quad \textbf{Ans.}$$

The reaction passes through the centre of gravity of the body as it is equal and opposite to weight of body W.

Example 6.2: A block of weight 5 kN is pulled by a force *P* as shown. The coefficient of friction between the contact surface is 0.35. Find the direction θ for which *P* is minimum and find the corresponding value of *P*.

[U.P.T.U. II Sem., 2003-04]

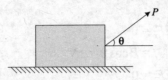

Solution:

  (i) Draw free body diagram of the block.

  (ii) Apply equilibrium conditions.

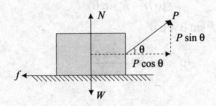

$$\Sigma F_x = 0$$
$$f = P \cos \theta$$
$$\Sigma F_y = 0$$
$$W = N + P \sin \theta$$

  (iii) The angle of inclination of force *P* will be equal to the angle of friction for minimum value of *P*.

∴
$$\theta = \phi = \tan^{-1} \mu = \tan^{-1} 0.35$$
$$= 20.48°$$

$$P = \frac{W \sin \phi}{\cos (\theta - \phi)}$$

For *P* to be minimum, $\cos (\theta - \phi) = 1$

∴
$$P = W \sin \phi = 5 \sin 20.48° = \textbf{1.75 kN} \quad \textbf{Ans.}$$

Example 6.3: Obtain the expression for minimum force required to drag a body on a rough horizontal plane.

Solution:

  (i) Draw free body diagram.

  (ii) Apply equilibrium conditions.

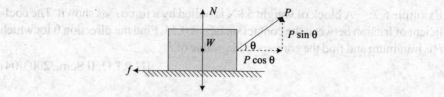

$$\Sigma F_x = 0$$

$\therefore$
$$f = P\cos\theta = \mu N \qquad\qquad\qquad ...(1)$$

$$\Sigma F_y = 0$$

$$N = W + P\cos\theta \qquad\qquad\qquad ...(2)$$

From equations (1) and (2)

$$P\cos\theta = \mu N$$

$$= \mu(W - P\sin\theta)$$

$$P\cos\theta + \mu P\sin\theta = \mu W$$

Now, $$\mu = \tan\phi = \frac{\sin\phi}{\cos\phi}$$

$$P\left(\cos\theta + \frac{\sin\phi}{\cos\phi}\sin\theta\right) = \frac{\sin\phi}{\cos\phi}.W$$

$$P(\cos\theta\cos\phi + \sin\theta\sin\phi) = W\sin\phi$$

$$P\cos(\theta - \phi) = W\sin\phi$$

$$P = \frac{w\sin\phi}{\cos(\theta - \phi)}$$

For P to be minimum, $\cos(\theta - \phi)$ should be maximum

$\therefore$
$$\cos(\theta - \phi) = 1$$

$$\theta = \phi$$

Minimum value of $P = W\sin\phi$

or $$= W\sin\theta$$

6.4.2 Rough Inclined Plane

(a) Equilibrium Condition for Different Angle of Inclination

(i) Angle of inclination less than angle of friction

1. Draw free body diagram.
2. Apply equilibrium conditions.

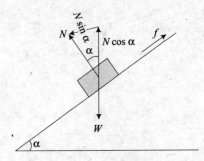 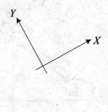

Fig. 6.8 Rough inclined plane ($\alpha < \phi$)

$$\Sigma F_x = 0$$
$$f = \mu N = N \sin \alpha$$
$$\Sigma F_y = 0$$
$$W = N \cos \alpha$$

The body is in equilibrium.

(i) **Angle of inclination more than angle of friction.**

The body will slide down and an upward force P is required to restrict the body from moving down. The restricting force can be applied in different ways.

1. *Along the inclined plane*

   Draw free body diagram.

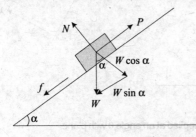 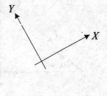

Fig. 6.9 Force parallel to inclined plane

Equilibrium conditions.

$$\Sigma F_x = 0$$
$$f = P - W \sin \alpha$$
$$\Sigma F_y = 0$$
$$N = W \cos \alpha$$

2. *The restricting force applied horizontally*

   Draw free body diagram.

   Apply equilibrium condition.

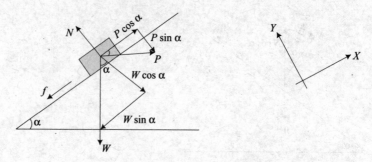

Fig. 6.10 Force applied horizontally

$$\Sigma F_x = 0$$
$$f = P \cos \alpha - W \sin \alpha$$
$$\Sigma F_y = 0$$
$$N = W \cos \alpha + P \sin \alpha$$

3. *Force applied at an angle* θ *with the inclined plane*
   Draw free body diagram.
   Apply equilibrium conditions.

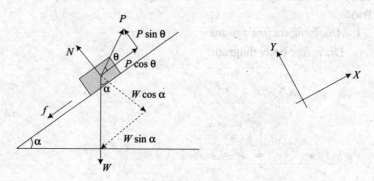

Fig. 6.11 Force applied at an angle θ with plane

$$\Sigma F_x = 0$$
$$f = P \cos \theta - W \sin \alpha$$
$$\Sigma F_y = 0$$
$$N = W \cos \alpha - P \sin \theta$$

(b) Body sliding downwards

A body of weight W is on the verge of sliding downwards. A minimum force P is required to restrict the motion.

   (i) Draw free body diagram.
   (ii) Apply equilibrium conditions.

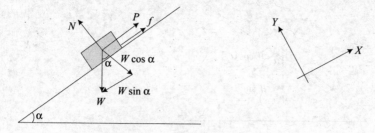

Fig. 6.12 Body sliding downwards

$$\Sigma F_x = 0$$

$$f = \mu N = W \sin \alpha - P$$

$\therefore$ $$P = W \sin \alpha - \mu N \qquad \qquad ...(1)$$

$$\Sigma F_y = 0$$

$$N = W \cos \alpha \qquad \qquad ...(2)$$

From equations (1) and (2)

$$P = W \sin \alpha - \mu (W \cos \alpha)$$

$$= W(\sin \alpha - \mu \cos \alpha)$$

$$= W\left(\sin \alpha - \frac{\sin \phi}{\sin \theta} \cos \alpha\right)$$

$\because$ $$\mu = \tan \phi = \frac{\sin \phi}{\cos \phi}$$

$$P \cos \phi = W(\sin \alpha \cos \phi - \cos \alpha \cos \phi)$$

$\therefore$ $$P = \frac{W \sin(\alpha - \phi)}{\cos \phi}$$

(c) Body up an inclined plane

Minimum force required to keep the body in equilibrium.

   (i) Draw free body diagram.

   (ii) Apply equilibrium conditions.

$$\Sigma F_x = 0$$

$$f = P - W \sin \alpha$$

$\therefore$ $$P = W \sin \alpha + f = W \sin \alpha + \mu N \qquad \qquad ...(1)$$

$$\Sigma F_y = 0$$

$$N = W \cos \alpha \qquad \qquad ...(2)$$

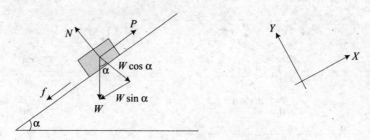

Fig. 6.13 Body up an inclined plane

From equations (1) and (2)

$$P = W \sin \alpha + \mu W \cos \alpha$$

$$= W \sin \alpha + \frac{\sin \phi}{\cos \phi} W \cos \alpha \qquad \left[\mu = \tan \phi = \frac{\sin \phi}{\cos \phi} \right]$$

$$P \cos \phi = W \sin \alpha \cos \phi + W \sin \cos \alpha$$

$$= W \sin (\alpha + \phi)$$

$$\therefore \qquad P = \frac{W \sin (a + \phi)}{\cos \phi}$$

Example 6.4: A body of weight 500 N is pulled up along an inclined plane having an inclination of 30° with the horizontal. If the coefficient of friction between the body and the plane is 0.3 and the force is applied parallel to inclined plane, determine the force required. [U.P.T.U. II Sem., 2005-06]

Solution:

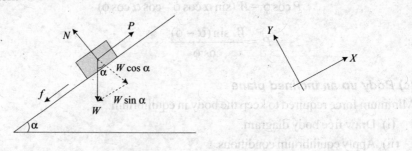

$$W = 500 \text{ N}$$

$$\alpha = 30°$$

$$\mu = 0.3$$

Angle of friction,

$$\phi = \tan^{-1} \mu = \tan^{-1} 0.3 = 16.7°$$

The force of friction will be acting downwards as the body is tending to move upwards.

Draw free body diagram and apply equilibrium conditions.

$$\Sigma F_x = 0$$
$$P - \sin \alpha = f = \mu N$$

∵
$$P = W \sin \alpha + \mu N \qquad ...(1)$$
$$\Sigma F_y = 0$$
$$N = W \cos \alpha \qquad ...(2)$$

From equations (1) and (2)

$$P = W \sin \alpha + \mu W \cos \alpha$$
$$= 500 \sin 30° + 0.3 \times 500 \cos 30°$$
$$= 250 \times 0.5 + 0.3 \times 500 \times 0.866$$
$$= 250 + 129.9 = \mathbf{379.9\ N} \quad \mathbf{Ans.}$$

Example 6.5: A body of weight W is placed on a rough inclined plane having inclination α to the horizontal. The force P is applied horizontal to drag the body. If the body is on the point of motion up the plane, prove that P is given by:

$$P = W \tan (\alpha + \phi)$$

where, ϕ = angle of friction.

Solution: Draw free body diagram and apply equilibrium conditions.

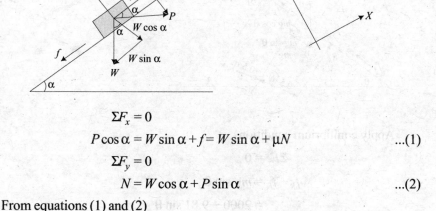

$$\Sigma F_x = 0$$
$$P \cos \alpha = W \sin \alpha + f = W \sin \alpha + \mu N \qquad ...(1)$$
$$\Sigma F_y = 0$$
$$N = W \cos \alpha + P \sin \alpha \qquad ...(2)$$

From equations (1) and (2)

$$P \cos \alpha = W \sin \alpha + \mu (W \cos \alpha + P \sin \alpha)$$
$$= W \sin \alpha + \mu W \cos \alpha + \mu P \sin \alpha$$
$$P \cos \alpha - \mu P \sin \alpha = W \sin \alpha + \mu W \cos \alpha$$

∴ $$P = W \frac{\sin \alpha + \mu \cos \alpha}{\cos \alpha - \mu \cos \alpha}$$

But $$\mu = \tan \phi \quad \text{(By definition)}$$

∴ $$P = W \frac{\sin \alpha + \tan \phi \cos \alpha}{\cos \alpha - \tan \phi \sin \alpha}$$

$$= W \frac{\sin \alpha \cos \phi + \sin \phi \cos \alpha}{\cos \alpha \cos \phi - \sin \alpha \sin \phi}$$

$$= W \frac{\sin (\alpha + \phi)}{\cos (\alpha + \phi)}$$

∴ $$P = W \tan (\alpha + \phi) \quad \textbf{Proved.}$$

Example 6.6: A four wheel drive car as shown has a mass of 2000 kg with passengers. The roadway is inclined at an angle θ with the horizontal. If the coefficient of friction between tyers and road is 0.3, what is the maximum inclination θ that can be climbed?

Solution: Draw free body diagram as shown. The wheels are rotating clockwise to climb the roadway, the forces of friction on the front wheel and rear wheel f_F and f_R will be acting upwards to oppose the direction of motion of the wheel.

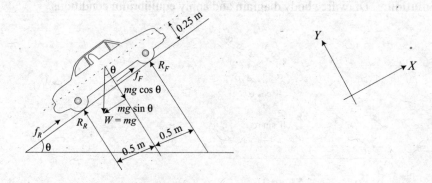

Apply equilibrium conditions,

$$\Sigma F_x = 0$$

∴ $$f_R + f_F = mg \sin \theta$$

$$= 2000 + 9.81 \sin \theta$$

$$= 19620 \sin \theta \ (N)$$

$$\Sigma F_y = 0$$
$$R_R + R_F = mg \cos \theta$$
$$= 2000 \times 9.81 \cos \theta$$
$$= 19620 \cos \theta \, (N)$$

But $\qquad f_R = \mu R_R$

and $\qquad f_F = \mu R_F$

$\therefore \qquad f_R + f_F = \mu (R_R + R_F)$
$$19620 \sin \theta = \mu (R_R + R_F)$$

$\therefore \qquad R_F + R_R = \dfrac{19620 \sin \theta}{0.3} = 65400 \sin \theta$
$$= 19620 \cos \theta$$

$\therefore \qquad \dfrac{\sin \theta}{\cos \theta} = \dfrac{19620}{65400} = 0.3$

$\therefore \qquad \theta = \mathbf{16.7°}$

6.4.3 Ladder Friction

A ladder is placed against a rough wall and a rough floor with coefficient of friction μ_2 and μ_1 respectively.

Example: A ladder, AB of length L and of weight W is placed against a rough wall and rough floor. The reactions R_A and R_B will be acting perpendicular to the point of supports at floor and wall. The ladder has tendency to slide down. Therefore, the forces of friction f_A and f_B will be acting to oppose the motion.

Draw free body diagram of ladder and apply conditions of equilibrium.

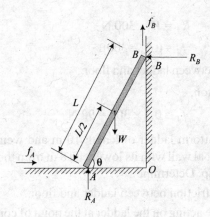

Fig. 6.14 Ladder friction

$$\Sigma F_x = 0 \quad \therefore f_A = R_B$$
$$\Sigma F_y = 0 \quad W = R_A + f_B$$

The number of unknowns are four. Therefore, take moments about point O.

$$\Sigma M_O = 0$$

Example 6.7: A uniform ladder of weight 300 N rests against a smooth vertical wall a rough horizontal floor making an angle 60° with the horizontal. Find the force of friction at floor. [U.P.T.U. I/II Sem., 2001-02]

Solution: Draw free body diagram of ladder AB and apply condition of equilibrium.

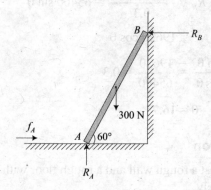

As the wall is smooth, $f_B = 0$.

$$\Sigma F_x = 0$$

$\therefore$
$$f_A = R_B.$$
$$\mu R_A = R_B$$

$$\Sigma F_y = 0$$

$$R_A = W = 300 \text{ N}$$

$\therefore$
$$f_A = \mu R_A = 300\mu$$

Assume $\mu = 0.3$ between ladder and floor.

The force of friction,

$$f_A = 0.3 \times 300 = \mathbf{90 \ N}$$

Example 6.8: A uniform ladder of length 10 m and weighing 20 N is placed against a smooth vertical wall with its lower end 8 m from the wall. In this position the ladder is just to slip. Determine:

(i) Coefficient of friction between ladder and floor.

(ii) Frictional force acting on the ladder at the point of contact between the ladder and floor. [U.P.T.U. I Sem., 2004-05]

Solution: Draw free body diagram of ladder.

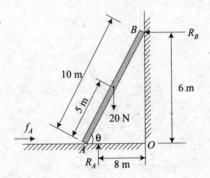

Angle of inclination,

$$\cos \theta = \frac{8}{10} = 0.8$$

$$\therefore \qquad \theta = 36.87°$$

$$\sin \theta = 0.6$$

Applying equilibrium conditions,

$$\Sigma F_x = 0$$

$$\therefore \qquad f_A = R_B$$

$$\Sigma F_y = 0$$

$$R_A = W = 20 \ N \quad (\because f_B = 0)$$

$$\Sigma M_O = 0$$

$$R_A \times 8 = 20 \times 4 + R_B \times 6$$

$$R_B = \frac{R_A \times 8 - 20 \times 4}{6} = \frac{20 \times 8 - 20 \times 4}{8}$$

$$= 10 \ N$$

Force of friction, $\quad f_A = R_B = 10 \ N$

$$\mu = \frac{R_B}{N} = \frac{10}{20} = 0.5 \quad \textbf{Ans.}$$

Example 6.9: A ladder of length 'l' rests against a wall, the angle of inclination being 45°. If the coefficient of friction between the ladder and the ground and that between the ladder and the wall is 0.5 each, what will be the maximum distance on ladder to which a man whose weight is 1.5 times the weight of the ladder may ascend before the ladder begins to slide? [U.P.T.U. I Sem., 2005-06]

Solution: Draw free body diagram as shown and apply equilibrium conditions.

(i) $\quad \Sigma F_x = 0$

$$f_A = R_B = \mu N = \mu R_A$$
$$= 0.5 R_A$$

$\therefore \qquad R_A = 2 R_B$

(ii) $\quad \Sigma F_y = 0$

$$R_A + f_B = W + 1.5 W = 2W$$
$$R_A = 2 R_B$$
$$f_B = \mu R_B = 0.5 R_B$$

$\therefore \qquad 2 R_B + 0.5 R_B = 2W$

$$R_B = \frac{2}{2.5} W = 0.8 W$$

$\therefore \qquad f_B = 0.5 R_B = 0.4 W$

(iii) $\quad \Sigma M_A = 0$

$$W \times \frac{l}{2} \cos 45° + 1.5 W\, x\cos 45° = R_B \times l \sin 45° + f_B\, l \cos 45°$$

$$= 0.8 W\, l \sin 45° + 0.4 W\, l \cos 45°$$

$$0.353\ Wl + 1.06\ Wx = 0.5656\, Wl + 0.2828\, Wl$$

$$1.06 x = (0.5656 + 0.2828 - 0.353) l = 0.49544 l$$

$\therefore \qquad \dfrac{x}{l} = \dfrac{0.49544}{1.06} = 0.467$

The man can ascend 46.7% of ladder length.

Example 6.10: A uniform ladder of weight 800 N and length 7 m rests on a horizontal ground and leans against a smooth vertical wall. The angle made by the ladder with the horizontal is 60°. When a man of weight 600 N stands on the ladder 4 m from the top of the ladder, the ladder is at the point of slipping. Determine the coefficient of friction between the ladder and the floor.

Solution: Draw free body diagram of the ladder and apply equilibrium conditions.

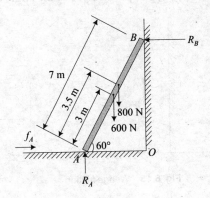

(i) $\Sigma F_x = 0$

$$f_A = R_B$$

(ii) $\Sigma F_y = 0$

$$R_A = 800 + 600 = 1400 \text{ N}$$

$\therefore$ $$f_A = R_B = \mu N = \mu R_A$$
$$= 1400\mu$$

(iii) $\Sigma M_B = 0$

$$R_A \times 7 \cos 60° = f_A \times 7 \sin 60° + 800 \times 3.5 \cos 60° + 600 \times 3 \cos 60°$$
$$1400 \times 7 \cos 60° = 1400\mu \times 7 \sin 60° + (800 \times 3.5 + 600 \times 3) \cos 60°$$

$\therefore$ $$\mu = \frac{1400 \times 7 \cos 60° - 4600 \cos 60°}{1400 \times 7 \sin 60°}$$

$$= \frac{4900 - 2300}{8487} = 0.3 \quad \textbf{Ans.}$$

6.4.4 Wedge Friction

A wedge is used to lift loads vertically by applying horizontal force. These are made in triangular or trapezoidal shape. It works on the principle of inclined plane.

A wedge of angle α is pushed under a block of weight W by applying a force P horizontally. The solution consists of finding the value P.

1. Draw free body diagram of block (Fig. 6.16). The frictional force f_1, between block and wall will act downwards to oppose the motion of block moving upwards. The frictional force f_2 between block and wedge will act downwords of inclined plane of block to oppose the motion. The reactions N_1 and N_2 will act normal to the supporting surface. R_1 and R_2 are the resultants of N_1, f_1 and N_2, f_2 respectively. ϕ is the angle of limiting friction.

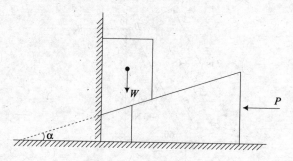

Fig. 6.15   Wedge friction

2. Apply equilibrium conditions.

(i) $\Sigma F_x = 0$

$$R_1 \cos \phi - R_2 \sin (\alpha + \phi) = 0 \qquad \qquad ...(1)$$

(ii) $\Sigma F_y = 0$

$$R_1 \sin \phi + W - R_2 \cos (\alpha + \phi) = 0 \qquad \qquad ...(2)$$

Calculate R_2 from equations (1) and (2),

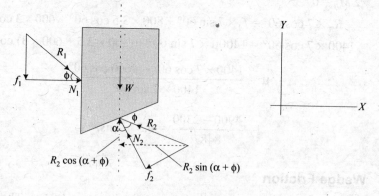

Fig. 6.16   FBD of block

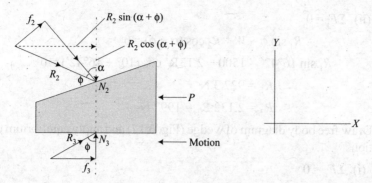

Fig. 6.17 FBD of wedge

3. Draw free body diagram of wedge (Fig. 6.17). The reactions $N_2' = -N_2$ and frictional force f_2 will be equal to frictional force acting on block but in opposite direction. Reactions N_2' and N_3 will be acting normal to supporting surfaces.

4. Apply equilibrium conditions to the wedge.

(i) $\Sigma F_x = 0$

$$R_2 \sin(\alpha + \phi) + R_3 \sin \phi - P = 0 \qquad \qquad ...(3)$$

(ii) $\Sigma F_y = 0$

$$R_3 \cos \phi - R_2 \cos(\alpha + \phi) = 0 \qquad \qquad ...(4)$$

Solve equations (3) and (4) simultaneously to find P.

Example. 6.11: A block weighing 1500 N, over laying a 10° wedge on a horizontal floor leaning against a vertical wall, is to be raised by applying a horizontal force to the wedge. Assuming the coefficient of friction between all surfaces in contact to be 0.3, determine the minimum horizontal force.

Solution:

1. Draw free body diagram for block (Fig. 6.16), and apply equilibrium conditions

(i) $\qquad \Sigma F_x = 0$

$$R_1 \cos \phi - R_2 \sin(\alpha + \phi) = 0$$

$$W = 1500 \text{ N}$$

$$\alpha = 10°$$

$$\mu = \tan \phi = 0.3$$

$$\phi = 16° \; 42'$$

$\therefore \qquad R_1 \cos 16°42' - R_2 \sin(10° + 16°42') = 0$

$\therefore \qquad R_2 = 2.132 R_1$

(ii) $\Sigma F_y = 0$

$$-R_1 \sin \phi - W + R_2 \cos (\alpha + \phi) = 0$$

$$-R_1 \sin 16°42' - 1500 + 2.13 R^1 \cos (10° + 16°42') = 0$$

$\therefore$
$$R_1 = 927.3 \text{ N}$$

$$R_2 = 2.132 \, R_1 = 1997 \text{ N}$$

2. Draw free body diagram of wedge (Fig. 6.17) and apply equilibrium conditions

(i) $\Sigma F_x = 0$

$$R_2 \sin (\alpha + \phi) + R_3 \sin \phi - P = 0$$

(ii) $\Sigma F_y = 0$

$$R_3 \cos \phi - R_2 \cos (\alpha + \phi) = 0$$

$$R_3 \cos 16°42' - 1977 \cos (10° + 16°42') = 0$$

$\therefore$
$$R_3 = 1844 \text{ N}$$

$\therefore$
$$P = R_2 \sin (\alpha + \phi) + R_3 \sin \phi$$

$$= 1977 \sin (10° + 16°42') + 1844 \sin 16°42'$$

$$= \mathbf{1418.3 \text{ N} \quad Ans.}$$

6.4.5 Belt Friction

Power is transmitted from one shaft to another by friction between the belt and the pulleys on which it passes. The driving pulley pulls the belt from one side and delivers the same to the other. The tension T_1 on tight side is more than T_2 on slack side.

If T_1 = Tension in belt on tight side, [N]

T_2 = Tension in belt on slack side, [N]

v = Velocity of the belt, [m/s]

r = Radius of the pulley, [m]

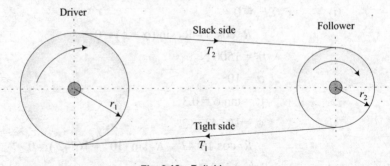

Fig. 6.18 Belt drive

Torque transmitted $= (T_1 - T_2)r \ [\text{N} - \text{m}]$

Work done $= \text{Force} \times \text{Distance} = (T_1 - T_2)v \ [\text{Nm/s}]$

Power transmitted $= (T_1 - T_2)v \ [\text{W}]$

Example 6.12: Find expression for the following due to belt friction is a pulley drive.

(i) Tension in the belt on tight and slack side.

(ii) Torque transmitted.

(iii) Reaction on the bearing of the pulley.

Solution: The belt and pulley system is shown in Fig. 6.19. The pulley is rotating in clockwise direction with angular velocity ω.

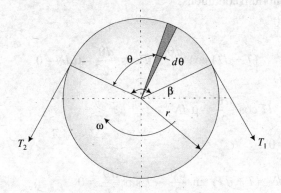

Fig. 6.19 Belt-pulley system

$$T_1 > T_2$$

$r = $ radius of pulley [m]

$\omega = $ angular velocity [rad/s]

$\beta = $ Angle of wrap of the belt over pulley.

Consider a small segment of belt of subtended angle $d\theta$ at an angle θ.

Draw free body diagram of segment of belt as shown in Fig. 6.20.

$T = $ Tension in the belt segment on slack side

$T + dT = $ Tension in belt segment on tight side

$dN = $ Normal reaction at the pulley rim.

$d\theta = $ Angle of embrace.

$f = $ frictional force

$ = \mu \, dN$

$\mu = $ Coefficient of friction between belt and pulley.

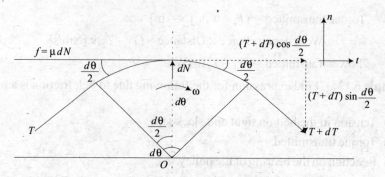

Fig. 6.20 Free body diagram

Apply equilibrium conditions,

(i) $\Sigma F_t = 0$

$$(T + dT) \cos \frac{d\theta}{2} - T \cos \frac{d\theta}{2} - \mu dN = 0$$

$\therefore$
$$dT \cos \frac{d\theta}{2} = \mu dN \qquad \qquad ...(1)$$

(ii) $\Sigma F_n = 0$

$$dN - (T + dT) \sin \frac{d\theta}{2} - T \sin \frac{d\theta}{2} = 0$$

$$dN - 2T \sin \frac{d\theta}{2} - dT \sin \frac{d\theta}{2} = 0$$

If $d\theta$ is very small, $\sin \dfrac{d\theta}{2} = \dfrac{d\theta}{2}$

and
$$\cos \frac{d\theta}{2} = \cos 0° = 1$$

$\therefore$
$$dN - Td\theta - dT \frac{d\theta}{2} = 0 \qquad \qquad ...(2)$$

From equation (1)

$$dT = \mu dN$$

$$dN = \frac{dT}{\mu}$$

From equation (2),

$$dN = -Td\theta$$

$$\left[\because dT.\frac{d\theta}{2} \approx 0, \text{product of two small quantities may be neglected}\right]$$

$$\therefore \qquad Td\theta = -dN = -\frac{dT}{\mu}$$

$$\therefore \qquad \frac{dT}{T} = -\mu d\theta$$

Integrating from $\qquad \theta = 0$ to $\theta = \beta$

$$\int_{T_1}^{T_2} \frac{dT}{T} = -\int_{\beta}^{0} \mu d\theta = \int_{0}^{\beta} \mu d\theta$$

$$\therefore \qquad \ln \frac{T_2}{T_1} = \mu\beta$$

$$\therefore \qquad \frac{T_1}{T_2} = e^{\mu\beta}$$

Torque transmitted $= (T_1 - T_2)r$ [N – m]

Reaction on pulley bearing $= (T_1 + T_2)$ [N]

Example 6.13: A flat belt is used to transmit torque from pulley A to pulley B. The radius of each pulley is 50 cm and coefficient of friction is 0.3. Determine the largest torque that can be transmitted if allowable belt tension is 3 kN.

Solution: The maximum allowable tension will be T_1

$$\therefore \qquad T_1 = 3 \text{ kN}$$

$$\mu = 0.3$$

$$\beta = 180° = \pi \text{ rad, as the pulley diameters are equal.}$$

$$r = 50 \text{ cm} = 0.5 \text{ m}$$

The belt tension ratio,

$$\frac{T_1}{T_2} = e^{\mu\beta}$$

$$\ln \frac{T_1}{T_2} = \mu\beta = 0.3\pi$$

$$\therefore \qquad \frac{T_1}{T_2} = 2.566$$

$$\therefore \qquad T_2 = \frac{T_1}{2.566} = \frac{3}{2.566} = 1.169 \text{ N}$$

$$\text{Torque} = (T_1 - T_2)r$$

$$= (3 - 1.169)0.5 = \mathbf{0.9155 \text{ N-m}} \quad \mathbf{Ans.}$$

Example 6.14: A belt drives a pulley of 200 mm diameter such that the ratio of tension in tight side and slack side is 1.2. If the maximum tension in the belt is not to exceed 240 kN. Find the safe power transmitted by the pulley at 60 rpm.

Solution: The maximum tension in the belt, $T_1 = 240$ kN.

$$\frac{T_1}{T_2} = 1.2$$

$$\therefore \qquad T_2 = \frac{240}{1.2} = 200 \text{ kN}$$

Speed of belt, $\qquad V = \dfrac{\pi d N}{60} = \pi \times \left(\dfrac{200}{1000} \right) \times \dfrac{60}{60} = 0.628 \text{ m/s}$

$\therefore$ Power transmitted, $\quad P = (T_1 - T_2)V = (240 - 200)0.628$

$$= \mathbf{25 \text{ kW}} \quad \mathbf{Ans.}$$

Example 6.15: A belt 100 mm wide and 8.0 mm thick is transmitting power at a belt speed of 160 m/min. The angle of lap of smaller pulley is 165° and coefficient of friction is 0.3. The maximum permissible stress in the belt is 2 MN/m^2 and mass of belt is 0.9 kg/m. Find the power transmitted and the initial tension in the belt.

Solution: The data given is:

$$b = 100 \text{ mm} = 0.1 \text{ m}$$

$$t = 8 \text{ mm} = 0.008 \text{ m}$$

$$V = 160 \text{ m/min} = \frac{160}{60} = 2.667 \text{ m/sec}$$

$$\beta = 165° \times \frac{\pi}{180} = 2.88 \text{ rad}$$

$$\mu = 0.3$$

$$m = 0.9 \text{ kg/m}$$

$$\sigma = 2 \text{ MN/m}^2 = 2 \text{ N/mm}^2$$

$$T_{max} = \sigma bt = 2 \times 100 \times 8 = 1600 \text{ N}$$

Centrifugal tension, $T_C = mv^2 = 0.9 (2.667)^2 = 6.4 \text{ N}$

$$T_1 = T_{max} - T_C = 1600 - 6.4$$

$$= 1593.6 \text{ N}$$

$$\frac{T_1}{T_2} = e^{\mu\beta} = e^{0.3 \times 2.88} = 2.3726$$

$$\therefore \qquad T_2 = \frac{1593.6}{2.3726} = 671.66 \text{ N}$$

Power transmitted,

$$P = (T_1 - T_2)V = (1593.6 - 671.66) \times 2.667 = 2.46 \text{ kW}$$

Initial tension,

$$T_o = \frac{T_1 + T_2}{2} + T_C = \frac{1593.6 + 671.66}{2} + 6.4 = \mathbf{1139 \text{ N}} \quad \textbf{Ans.}$$

Example 6.16: A belt is stretched over two identical pulleys of diameter D meter. The initial tension in the belt throughout is 2.4 kN, when the pulleys are at rest. In using these pulleys and belt to transmit torque, it is found that the increase in tension on one side is equal to the decrease on the other side.

Find the maximum torque that can be transmitted by the belt drive, given that the coefficient of friction between belt and pulley is 0.30.

[U.P.T.U. I Sem., 2002-03]

Solution:

$$\beta = \pi \text{ as the pulleys are identical}$$

$$\mu = 0.30$$

$$\frac{T_1}{T_2} = e^{\mu\beta} = e^{0.3\pi} = 2.566$$

$$T_o = \text{Initial tension}$$

$$= \frac{T_1 + T_2}{2} \qquad\qquad [\because T_1 - T_0 = T_0 - T_2]$$

$$= 2.4 \text{ kN}$$

$$T_1 + T_2 = 4.8$$

$$\therefore \qquad T_1 + \frac{T_1}{2.566} = 4.8$$

$$T_1 = 3.454 \text{ kN}$$
$$T_2 = 4.8 - 3.454 = 1.346 \text{ kN}$$

Max. torque $= (T_1 - T_2)\dfrac{D}{2} = (3.454 - 1.346)\dfrac{D}{2} = \mathbf{1.054D \text{ kN.m}}$ **Ans.**

Example 6.17: A horizontal drum of a belt drive carries the belt over a semicircle around it. It is rotated anti-clockwise to transmit a torque of 300 N-m. If coefficient of friction between the belt and drum is 0.3, calculate the tension in the limbs 1 and 2 of the belt shown, and the reaction on the bearings. The drum has a mass of 20 kg and the belt is assumed to be massless.   [U.P.T.U. I Sem., 2001-02]

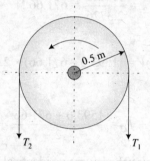

Solution: $T_2 > T_1$ given
$$\beta = \pi$$
$$\mu = 0.3$$
$$\therefore \quad \frac{T_2}{T_1} = e^{\mu\beta} = e^{0.3\pi} = 2.566$$

$$T_2 = 2.566 \, T_1$$
$$\text{Torque} = (T_2 - T_1)r$$
$$300 = (2.566 \, T_1 - T_1) \times 0.5$$
$$T_1 = 383 \text{ N}$$
$$T_2 = 2.566 \, T_1 = 983 \text{ N}$$

Reaction on the bearing $= T_1 + T_2 + mg = 383 + 983 + 20 \times 9.81$
$$= \mathbf{1562.2 \text{ N}} \quad \textbf{Ans.}$$

Example 6.18: A 100 N weight is on the verge of sliding downwards. It is prevented by a weight W as shown. The coefficient of friction between rope and the fixed pulley, and between other surfaces of contact $\mu = 0.3$. Determine the minimum weight W to prevent the downward motion of 100 N body.

[U.P.T.U. I Sem., 2001-02]

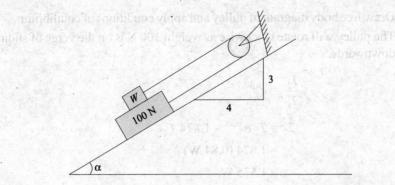

Solution:

$$\tan \alpha = \frac{3}{4}$$

$$\alpha = 36.87°$$

$$\cos \alpha = 0.8$$

$$\sin \alpha = 0.6$$

1. Draw free body diagram of weight W and apply conditions of equilibrium.

    (i) $\Sigma F_y = 0$

$$N_1 = W \cos \alpha = 0.8W$$

$$f_1 = \mu N_1 = 0.3 \times 0.8W$$

$$= 0.24W$$

    (ii) $\Sigma F_x = 0$

$$T_1 = f_1 + W \sin \alpha$$

$$= 0.24W + 0.6W = 0.84W$$

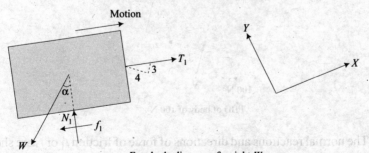

Free body diagram of weight W

2. Draw free body diagram of pulley and apply conditions of equilibrium.
   The pulley will rotate clockwise as weight 100 N is on the verge of sliding
   downwards.

$$\frac{T_2}{T_1} = e^{\mu\beta}$$

∴

$$T_2 = T_1\, e^{\,0.3\pi} = 1.874\, T_2$$
$$= 1.874\,(0.84\ W)$$
$$= 1.575\ W$$

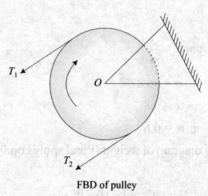

FBD of pulley

3. Draw free body diagram of body of 100 N and apply conditions of equilibrium.

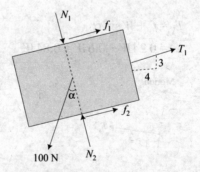

FBD of body of 100 N

The normal reactions and directions of force of friction f_1 or f_2 are shown to
oppose the motion.

(i) $\Sigma F_y = 0$

$$N_2 - N_1 - 100 \cos \alpha = 0$$

$$\therefore \qquad N_2 = N_1 + 100 \cos \alpha = 0.8W + 80$$

$$f_2 = \mu N_2 = 0.3(0.8W + 80)$$

$$= 0.24W + 24$$

(ii) $\Sigma F_x = 0$

$$f_1 + f_2 - 100 \sin \alpha + T_2 = 0$$

$$0.24W + (0.24W + 24) - 100 \times 0.6 + 1.575W = 0$$

$$2.55W = 36$$

$$\therefore \qquad W = 17.5 \text{ N} \quad \textbf{Ans.}$$

6.4.6 Screw Friction

The friction in the screw threads is widely used for fastening of components in machines and structures. The fastners such as studs, bolts, nuts and screws are very widely used. Screw friction is also used in the design of lifting machines. Square threads are used in screw jack which is a machine used for raising / lifting of heavy loads by applying a small effort. The screw jack works on the principle of inclined plane. The load W is placed on the head of screw jack which is rotated by application of effort P at the end of a lever for lifting or lowering of load. In one revolution the head moves through a distance, p which is the pitch of the screw. In case of multi-start screw, the distance moved is np where n is the number of starts. The inclination of the thread, α can be found out as:

$$\tan \alpha = \frac{p}{\pi d_m}$$

where, d_m = mean diameter of the thread

α = angle of helix.

The effort required can be found out from the principle of inclined plane.

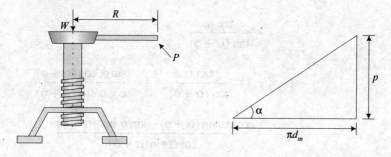

Fig. 6.21 Screw jack

The effort required to lift the load,

$$P = W \tan(\alpha + \phi)$$

where, ϕ = angle of friction.

The effort required to lever the load,

$$P = W \tan(\alpha - \phi) \qquad \text{(when } \alpha > \phi)$$

$$P = W \tan(\phi - \alpha) \qquad \text{(where } \alpha < \phi)$$

For ideal conditions, when there is no friction in the screw and nut,

$$P_o = W \tan \alpha$$

where, P_o = Ideal effort.

Example 6.19: Show that efficiency of a screw jack for raising a load W is given by,

$$\eta = \frac{\tan \alpha}{\tan(\alpha + \phi)}$$

where, α = angle of screw

ϕ = angle of friction

What is the condition for maximum efficiency?

Solution: The efficiency of a screw jack is defined as the ratio of ideal effort and actual effort.

The effort required to raise the load,

$$P = W \tan(\alpha + \phi)$$

The ideal effort required when there is no screw friction,

$$P_o = W \tan \alpha \qquad [\because \phi = 0]$$

$$\therefore \qquad \eta = \frac{\text{Ideal effort}}{\text{Actual effort}} = \frac{P_o}{P} = \frac{W \tan \alpha}{W \tan(\alpha + \phi)}$$

$$= \frac{\tan \alpha}{\tan(\alpha + \phi)}$$

$$1 - \eta = 1 - \frac{\tan \alpha}{\tan(\alpha + \phi)} = 1 - \frac{\sin \alpha \cos(\alpha + \phi)}{\cos \alpha \sin(\alpha + \phi)}$$

$$= \frac{\cos \alpha \sin(\alpha + \phi) - \sin \alpha \cos(\alpha + \phi)}{\cos \alpha \sin(\alpha + \phi)}$$

Let, $\qquad \alpha + \phi = A$ and $\alpha = \beta$

$\therefore \qquad \sin(A - B) = \sin A \cos B - \cos A \sin B$

$$1 - \eta = \frac{\sin(\alpha + \phi - \alpha)}{\cos\alpha \sin(\alpha + \phi)} = \frac{\sin\phi}{\cos\alpha \sin(\alpha + \phi)}$$

$$= \frac{2\sin\phi}{2\cos\alpha \sin(\alpha + \phi)}$$

But, $2\cos A \sin B = \sin(A + B) + \sin(A - B)$

$\therefore$

$$1 - \eta = \frac{2\sin\phi}{\sin(2\alpha + \phi) + \sin\phi}$$

For efficiency to be maximum, $1 - \eta$ should be minimum

$\therefore$ $2\sin(2\alpha + \phi)$ should be maximum.

$\therefore$ $(2\alpha + \phi) = 90°$

$$2\alpha = 90° - \phi$$

$$\alpha = \left(45° - \frac{\phi}{2}\right)$$

This is the condition for maximum efficiency of screw jack.

Example 6.20:   Show that maximum efficiency of a screw jack is given by

$$\eta_{max} = \frac{1 - \sin\phi}{1 + \sin\phi} \text{ at } \alpha = \left(45° - \frac{\phi}{2}\right)$$

where, α is the helix angle and ϕ is the angle of friction.

Solution:

$$\eta_{max} = \frac{P_o}{P} = \frac{\tan\alpha}{\tan(\alpha + \phi)}$$

For maximum efficiency,

$$\alpha = 45° - \frac{\phi}{2}$$

$\therefore$

$$\eta_{max} = \frac{\tan\left(45° - \frac{\phi}{2}\right)}{\tan\left(45° - \frac{\phi}{2} + \frac{\phi}{2}\right)} = \frac{\tan\left(45° - \frac{\phi}{2}\right)}{\tan\left(45° + \frac{\phi}{2}\right)}$$

$$= \frac{\dfrac{\tan 45° - \tan\phi/2}{1 + \tan 45° \tan\phi/2}}{\dfrac{\tan 45° + \tan\phi/2}{1 - \tan 45° \tan\phi/2}} = \frac{\dfrac{1 - \tan\phi/2}{1 + \tan\phi/2}}{\dfrac{1 + \tan\phi/2}{1 - \tan\phi/2}}$$

$$\eta_{max} = \frac{(1 - \tan\phi/2)^2}{(1 + \tan\phi/2)^2} = \frac{\left[1 - \dfrac{\sin\phi/2}{\cos\phi/2}\right]^2}{\left(1 + \dfrac{\sin\phi/2}{\cos\phi/2}\right)^2}$$

$$\therefore \qquad \eta_{max} = \frac{\left(\cos\dfrac{\phi}{2} - \sin\dfrac{\phi}{2}\right)^2}{\left(\cos\dfrac{\phi}{2} + \sin\dfrac{\phi}{2}\right)^2}$$

$$= \frac{\cos^2\dfrac{\phi}{2} + \sin^2\dfrac{\phi}{2} - 2\cos\dfrac{\phi}{2}\sin\dfrac{\phi}{2}}{\cos^2\dfrac{\phi}{2} + \sin^2\dfrac{\phi}{2} + 2\cos\dfrac{\phi}{2}\sin\dfrac{\phi}{2}}$$

$$= \frac{1 - 2\cos\dfrac{\phi}{2}\sin\dfrac{\phi}{2}}{1 + 2\cos\dfrac{\phi}{2}\sin\dfrac{\phi}{2}} = \frac{1 - \sin\phi}{1 + \sin\phi}$$

Example 6.21: A screw thread of a screw jack has a mean diameter of 10 cm and a pitch of 1.25 cm. The coefficient of friction between the screw and its nut housing is 0.25 cm. Determine the force F that must be applied at the end of a 50 cm lever arm to raise a mass of 5000 kg. Is the device self locking? Also determine its efficiency. [U.P.T.U. I Sem., 2001-02]

Solution:

Load,	$W = 5000 \times 9.81 = 49050$ N
Pitch,	$p = 12.5$ mm
Lever arm,	$R = 50$ cm $= 0.5$ m

Mean screw radius, $\quad r = \dfrac{10\text{ cm}}{2} = 50$ mm

$$\mu = 0.25$$

$$\phi = \tan^{-1}\mu = \tan^{-1}0.25 = 14°$$

$$\tan\alpha = \frac{p}{2\pi r} = \frac{12.5}{2\pi \times 50} = 0.0397887$$

$$\therefore \qquad \alpha = 2.2785°$$

Effort required to raise the load,

$$p = W \tan (\alpha + \phi)$$
$$= 49050 \tan (2.2785° + 14°)$$
$$= 14323.25 \text{ N}$$

Torque $T = Pr = 14323.25 \times 50 \times 10^{-3} \text{ N-m}$
$$= 716.16 \text{ N-m}$$

Force required at the end of the lever,

$$F = \frac{T}{R} = \frac{716.16}{0.5} = 1432.32 \text{ N}$$

Efficiency, $\eta = \dfrac{\tan \alpha}{\tan (\alpha + \phi)} = \dfrac{0.0397887}{0.2920} = 0.136$

$\therefore$ $\eta = 13.625\%$

Because η is less than 50%, it is self-locking device.

Example 6.22:   A screw jack has square threads of mean diameter of 10 cm and a pitch of 1.25 cm. Determine the force required at the end of 50 cm lever.

(i) to raise the load
(ii) to lower the load of 50 kN.

Find the efficiency of the jack. Is it self-locking?

Assume $\mu = 0.20$

Solution:

$$\mu = 0.20$$
$\therefore$ $\phi = \tan^{-1} \mu = \tan^{-1} 0.20 = 11.31°$

$$\tan \alpha = \frac{p}{2\pi r} = \frac{12.5}{2\pi \times \dfrac{100}{2}} = 0.0397887$$

$\therefore$ $\alpha = 2.279°$

(i) *Torque required to raise the load*

Effort $P = W \tan (\alpha + \phi)$
$$= 50 \tan (2.279° + 11.31°) = 12.1 \text{ kN}$$

Torque, $T = P \times r = 12.1 \times \dfrac{50}{1000} = 0.604 \text{ kN-m}$

Force to be applied at the end of the lever,

$$F \times R = T$$

$$\therefore \quad F = \frac{T}{R} = \frac{0.604}{0.50} = 1.208 \text{ kN}$$

(ii) *Torque required to lower the load*

Effort, $\qquad P = W \tan(\phi - \alpha)$

$$= 50 \tan(11.31 - 2.279)$$

$$= 7.947 \text{ kN}$$

Torque required, $\quad T = P \times r = 7.947 \dfrac{50}{1000} = 0.397 \text{ kN}$

Force to be applied at the end of the lever,

$$\therefore \quad F = \frac{T}{R} = \frac{0.0397}{0.50} = 1.7947 \text{ kN}$$

(iii) Efficiency, $\qquad \eta = \dfrac{\tan\phi}{\tan(\alpha + \phi)} = \dfrac{\tan 2.279°}{\tan(11.31° + 2.279°)} = 0.1646$

$$= 16.46\%$$

The efficiency is less then 5%. Therefore, it is self-locking.

QUESTION BANK NO. 6

1. What is friction? Give some useful applications of friction.
2. Explain laws of static and dynamic friction.
3. What are the characteristic of friction force? Explain the concept of equilibrium of bodies involving dry friction.
4. Write notes on:
   (i) Cone of friction
   (ii) Coefficient of friction
   (iii) Angle of friction
   (iv) Angle of repose
   (v) Limiting friction
5. Find expressions for the following due to belt friction in a pulley drive:
   (i) Tension in the belt on tight and slack side
   (ii) Torque transmitted
   (iii) Reaction on the bearing of pulley.

6. Draw free body diagram of the following systems:
   (i) A block pulled along a rough horizontal plane.
   (ii) A body sliding down a rough inclined plane.
   (iii) A body being pushed up a rough inclined plane.
   (iv) A wedge lifting a load.
   (v) A ladder resting on a rough floor and a wall.

7

Beams

7.1 INTRODUCTION

A beam is a structural member to support external loads. Normally a beam support transverse loads acting at right angle to its axis. The length of the beam is usually large as compared to its cross-section.

The load applied on the beam gets transferred to its supports. Any section of the beam experiences shearing off by the load and bending by a moment.

The beam supports different types of loads such as point loads, uniformly distributed loads, uniformly varying loads and the combination of the loads. The beams can be supported on roller supports, hinged supports or fixed supports.

7.1.1 Types of Loads

The various types of loads can be classified as:

(a) **Concentrated loads**

A concentrated load or a point load is considered to act at a point P_1, P_2 and P_3 are concentrated loads.

(b) **Uniformly distributed loads (UDL)**

A uniformly distributed load is considered distributed or spread over the length of the beam. In UDL, the intensity of load w is uniform over the length and is expressed as N/m or kN/m.

(c) **Uniformly varying loads**

If the spread or intensity of the load is uniformly varying from one end to another, it is called uniformly varying load. The triangular or trapezoidal distributed load is considered in this category. In trapezoidal distribution, intensity of load may vary linearly from w_1/m from left to w_2/m on right hand.

(d) **Externally applied moments**

External moments may also be applied to the beam. A moment M is applied to the beam at point C.

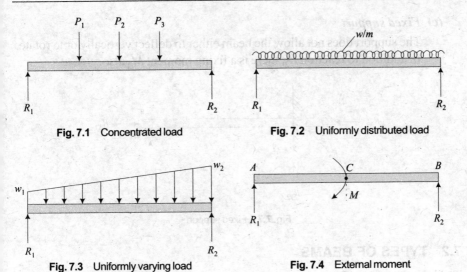

Fig. 7.1   Concentrated load

Fig. 7.2   Uniformly distributed load

Fig. 7.3   Uniformly varying load

Fig. 7.4   External moment

7.1.2   Types of Supports

The beams are supported on columns or walls by the following methods:

(a) Roller support or free support

The beam is supported on a ball support to allow for thermal expansion or thermal contraction and thus allow unhindered stretching or shrinking of the beam in axial direction. The reaction R at the roller support of the beam will act perpendicular to the roller bose.

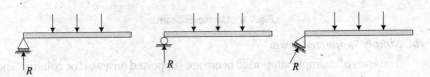

Fig. 7.5   Roller supports

(b) Pin support or hinged support

The beams with pin support are also called simply supported beams. The support does not allow vertical deflection or movement but rotation of beam to a slope angle is possible. The direction of reaction, R will depend upon the direction of loads on the beams.

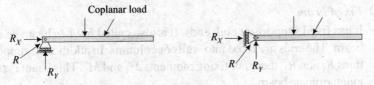

Fig. 7.6   Pin support

(c) Fixed support

The support does not allow the beam either to deflect vertically or to rotate. In addition to reaction R_2, there is a fixing moment M_2 also.

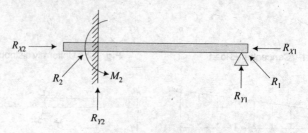

Fig. 7.7 Fixed support

7.2 TYPES OF BEAMS

The beams are classified as per type of support. Various types of beams are:

(a) Cantilever beam

It is fixed at one end and free at the other. In addition to reaction R, there is a fixing moment at the support.

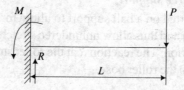

Fig. 7.8 Cantilever beam

(b) Simply supported beam

The ends of a simply supported beam are supported on a wall or column. The reactions at the support are vertical reactions.

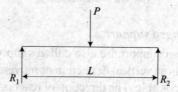

Fig. 7.9 Simply supported beam

(c) Fixed beam

It has fixed supports at both ends. It is also called *built-in* beam or *encastre* beam. The ends are fixed into walls or columns. In addition to support reactions R_1 and R_2, there are fixing moments M_1 and M_2. This beam is statically indeterminate beam.

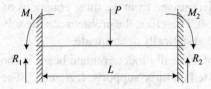

Fig. 7.10 Fixed beam

(d) Continuous beam

Such beams have more than two supports. Span between supports may be equal or unequal. This beam is also statically indeterminate beam.

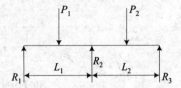

Fig. 7.11 Continuous beam

(e) Overhanging beam

The beam has one or both ends projecting beyond the supports.

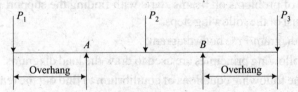

Fig. 7.12 Overhanging beam

7.3 STATICALLY DETERMINATE BEAM

If for a beam or a truss or any rigid body there are as many unknown supporting forces and couple-moments (including support reaction) as there are equations of equilibrium, and if the equations can be solved for these unknowns, the beam or truss or rigid body in called statically determinate.

1. The example of statically determinate beam is shown in Fig. 7.13. The loads F_1, F_2 and W are known. There are three unknown reactions R_A, R_{BY} and R_{BX}

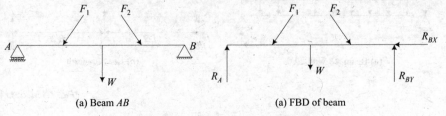

(a) Beam AB (a) FBD of beam

Fig. 7.13 Statically determinate beam

which can be determined from the three equations of equilibrium $\Sigma F_x = 0$; $\Sigma F_y = 0$; $\Sigma M_o = 0$. Therefore, the problem can the solved from statics alone, the beam is called statically determinate.

2. The example of statically indeterminate beam is shown in Fig. 7.14. The beam is supported on three supports A, B and C. There are four unknown, i.e., R_A, R_C, R_{BY} and R_{BX}. The beam becomes statically indeterminate as there are insufficient number of equilibrium equations (three) to calculate the four support reactions. In order to solve such problems, use of deflection of beam at various points has to be used.

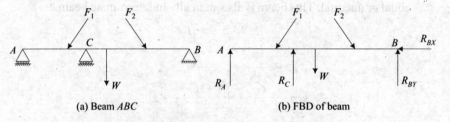

(a) Beam ABC (b) FBD of beam

Fig. 7.14   Statically indeterminate beam

7.4 SUPPORT REACTIONS

The solution of problems on beams starts with finding the support reactions. The solution consist of the following steps.

1. Load diagram/Free body diagram.

   The following principles are used to draw the load diagrams.

2. Use the following equations of equilibrium to find out the reactions.

$$\Sigma F_x = 0; \quad \Sigma F_y = 0; \quad \Sigma M_o = 0$$

where, 0 is any point (A or B).

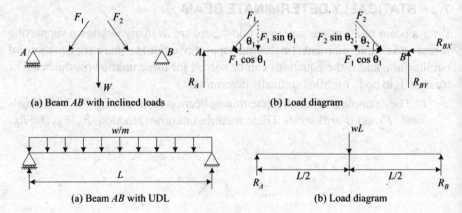

(a) Beam AB with inclined loads   (b) Load diagram

(a) Beam AB with UDL   (b) Load diagram

(Fig. 7.15 Contd.)

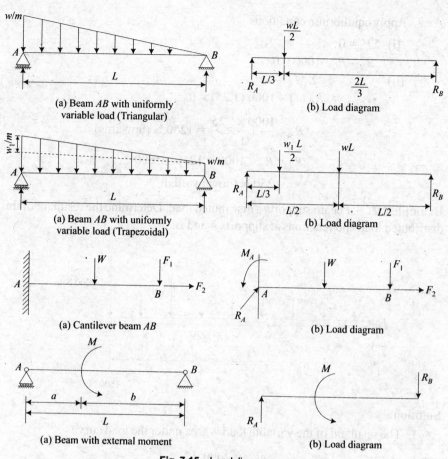

Fig. 7.15 Load diagrams

Example 7.1: Find the reactions of the beam *ABC*.

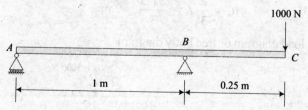

Solution:

1. Draw load diagram (or FBD)

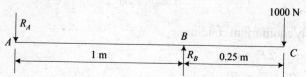

2. Apply equilibrium conditions

   (i) $\Sigma F_y = 0$

$$R_B - R_A - 1000 = 0$$

  (ii) $\qquad\qquad \Sigma M_A = 0$

$\therefore \qquad\qquad R_B (1) - 1000 (1.25) = 0$

$\therefore \qquad\qquad R_B = \dfrac{1000 \times 1.25}{1} = 1250 \text{ N (upwards)}$

$$R_A = R_B - 1000 = 1250 - 1000$$

$$= 250 \text{ N (downwards)}$$

Example 7.2: *A* beam supports a distributed load. Determine the resultant of this distributed load and reactions at supports *A* and *B*.

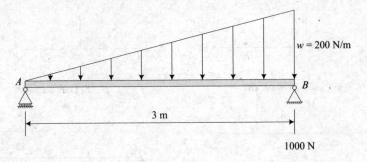

Solution:

1. The resultant of the variable load is area under the load curve

$$W = \frac{wL}{2} = \frac{2000 \times 3}{2} = 3000 \text{ N}$$

$$AC = \frac{2}{3} L = \frac{2 \times 3}{3} = 2 \text{ m}$$

2. Draw load diagram or free body diagram

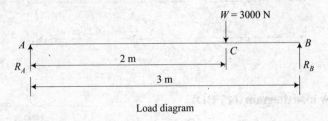

Load diagram

3. Apply equilibrium conditions

   (i) $\qquad \Sigma F_y = 0$

$\therefore \qquad\qquad R_A + R_B = W$

(ii) $\Sigma M_A = 0$

$\therefore$ $R_B \times 3 - W \times 2 = 0$

$$R_B = \frac{3000 \times 2}{3} = 2000 \ N.$$

$$R_A = W - R_B = 3000 - 2000 = 1000 \ N.$$

Example 7.3:   Two beams AB and CD are arranged and supported as shown. Find the reaction at D due to a force of 1000 N acting at B.

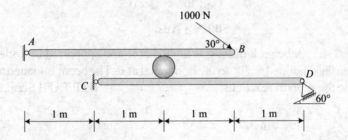

Solution:

  1. Draw free body diagram of beam AB.

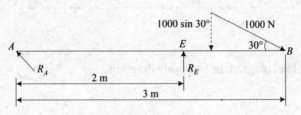

FBD of beam AB

  2. Apply equilibrium conditions

$\Sigma M_A = 0$

$\therefore$ $R_E \times 2 - 1000 \sin 30° \times 3 = 0$

$\therefore$ $R_E = \dfrac{3000}{2} = \sin 30°$

                $= 750 \ N$

  3. Draw free body diagram of beam CD.
  4. Apply equilibrium condition

$\Sigma M_C = 0$

$R_D \sin 30° \times 3 - R_E \times 1 = 0$

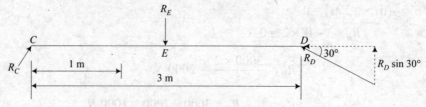

FBD of beam *CD*

$$\therefore \qquad R_D = \frac{R_E \times 1}{3 \sin 30°} = \frac{75° \times 1}{3 \times 0.5}$$

$$= 500 \text{ N} \quad \textbf{Ans.}$$

Example 7.4: A beam 8 m long is hinged at *A* and supported on rollers over a smooth surface inclined at 30° to the horizontal at *B*. The beam is loaded as shown. Determine the support reactions. [U.P.T.U II Sem., 2002-03]

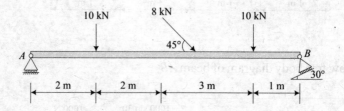

Solution:

1. Draw load diagram or free body diagram.

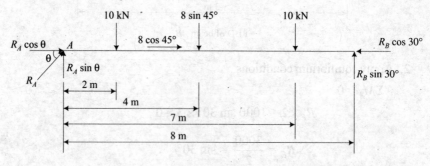

FBD of beam *AB*

2. Apply equilibrium condition

    (i) $\Sigma F_y = 0$

        $R_A \sin \theta - 10 - 8 \sin 45° - 10 + R_B \sin 30° = 0.$

(ii) $\qquad \Sigma F_x = 0$

$\qquad R_A \cos \theta + 8 \cos 45° - R_B \cos 30° = 0.$

(iii) $\qquad \Sigma M_A = 0$

$\qquad 10 \, (2) + 80 \sin 45° \, (4) + 10 \, (7) - R_B \sin 30° \, (8) = 0$

$\therefore \qquad R_B = \dfrac{20 + 22.63 + 70}{8 \times 0.5} = 28.16 \text{ kN}$

$\qquad R_A \sin \theta = 10 + 8 \sin 45° + 10 - 28.16 \sin 30°$

$\qquad\qquad = 11.66 \text{ kN}$

$\qquad R_A \cos \theta = R_B = \cos 30° - 8 \cos 45°$

$\qquad\qquad = 28.16 \cos 30° - 8 \cos 45°$

$\qquad\qquad = 18.73 \text{ kN}$

$\qquad R_A = \sqrt{(11.66)^2 + (18.73)^2}$

$\qquad\qquad = 22 \text{ kN}$

$\qquad \theta = \tan^{-1} \dfrac{11.66}{18.73} = 31.9°$

Example 7.5: Compute the simplest resultant force for the loads shown acting on the cantilever. What force and moment is transmitted by this force to supporting wall at A? [U.P.T.U. II Sem., 2004-05]

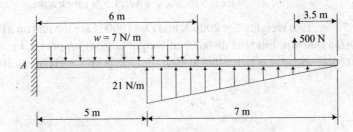

Solution:

1. Draw force diagram/free body diagram
2. Find resultant

(i) *Magnitude*

$\qquad R = \Sigma F_y$

$\qquad\qquad = -42 + 73.5 + 500$

$\qquad\qquad = 531.5 \, N$

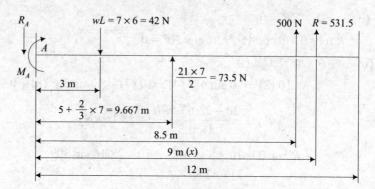

Free body diagram

(ii) *Direction*

Vertically upwards

(iii) *Position*

$$\Sigma M_A = 0$$
$$42 \times 3 - 73.5 \times 9.667 - 500 \times 8.5 - R_x = 0$$

$\therefore$
$$x = \frac{-126 + 7105 + 4250}{531.5} = 9 \text{ m from } A$$

3. Wall reactions

$$R_A = -R = 531.5 \text{ N downwards}$$
$$M_A = Rx = 531.5 \times 9 = 4783.5 \text{ N clockwise.}$$

Example 7.6: Two weights $C = 2000$ N and $D = 1000$ N are located on a horizontal beam AB as shown. Find the distance of weight C from support A, i.e. x so that the support reaction at A is twice that at B. [U.P.T.U. II Sem., 2000-01].

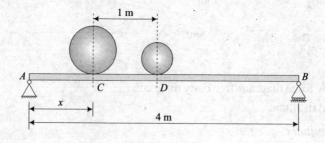

Solution:

1. Draw load diagram/free body diagram

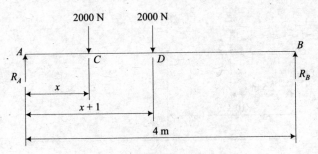

Free body diagram

2. Apply equilibrium conditions

(i) $\Sigma F_y = 0$

$$R_A + R_B = 2000 + 1000 = 3000 \text{ N}$$
$$R_A = 2\,R_B$$

∴
$$2R_B + R_B = 3000 \text{ N}$$
$$R_B = 1000 \text{ N}$$
$$R_A = 2000 \text{ N}$$

(ii) $\Sigma M_A = 0$

$$2000x + 1000(x+1) - R_B(4) = 0$$
$$2000x + 1000x + 1000 = 1000(4)$$

∴
$$x = \frac{4000 - 1000}{3000} = \textbf{1 m} \quad \textbf{Ans.}$$

Example 7.7: Determine the reactions A, B and D of the system.

[U.P.T.U. I Sem., 2001-02]

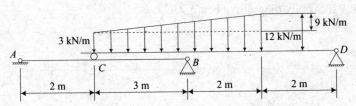

Solution:

1. Draw load diagram/free body diagram of beam CD. The equivalent load (area under the load curve),

$$W_1 = 3 \times 5 = 15 \text{ kN}$$

$$x_1 = \text{distance of load from } C$$

$$= \frac{5}{2} = 2.5 \text{ m}$$

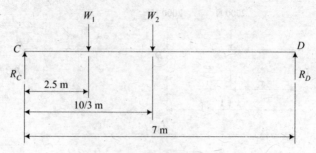

FBD of beam *CD*

$$W_2 = \frac{9 \times 5}{2} = 22.5 \text{ kN}$$

$$x_2 = \text{distance of load } W_2 \text{ from } C$$

$$= \frac{2L}{3} = \frac{2 \times 5}{3} = \frac{10}{3} \text{ m}$$

2. Apply equilibrium conditions

   (i) $\Sigma M_c = 0$

   $$15 \times 2.5 + 22.5 \times \frac{10}{3} = R_D \times 7$$

   ∴ $$R_D = 16 \text{ kN}$$
   $$\Sigma F_y = 0,$$

   (ii) $$R_c = W_1 + W_2 - R_D = 15 + 22.5 - 16 = 21.5 \text{ kN}$$

3. Draw free body diagram of beam *AB*.

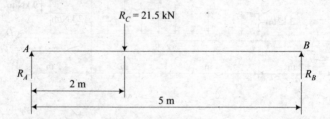

4. Apply equilibrium conditions

   (i) $\Sigma M_A = 0$

   $$21.5 (2) = R_B (5)$$

   ∴ $$R_B = \frac{21.5 \times 2}{5} = 8.6 = \text{kN}$$

   (ii) $\Sigma F_Y = 0$

   ∴ $$R_A = 21.5 - R_B = 21.5 - 8.6 = 12.9 \text{ kN}.$$

Example 7.8:   Find the support reactions in the beam shown below.

[U.P.T.U. I Sem., 2000-01]

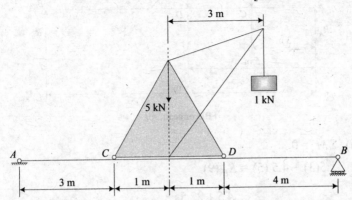

Solution:

1. Draw free body diagram of crane.

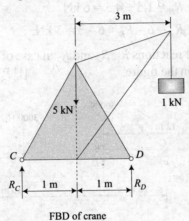

FBD of crane

2. Apply equilibrium conditions

(i) $\Sigma F_y = 0$

$$R_C + R_D = 5 + 1 = 6 \text{ kN}$$

(ii) $\Sigma M_C = 0$

$$R_D \times 2 = 5 \times 1 + 1 \times 4$$

$$R_D = 4.5 \text{ kN}$$

$$R_C = 6 - 4.5 = 1.5 \text{ kN}$$

3. Draw free body diagram of beam *AB*.

4. Apply equilibrium conditions

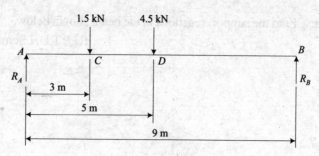

FBD of beam *AB*

(i) $\Sigma M_A = 0$

$$1.5(3) + 4.5\,(5) = R_B\,(9)$$

∴
$$R_B = \frac{4.5 + 22.5}{9} = 3 \text{ kN}$$

(ii) $\Sigma F_y = 0$

$$R_A + R_B = 1.5 + 4.5 = 6 \text{ kN}$$
$$R_A = 6 - R_B = 6 - 3 = 3 \text{ kN}.$$

Example 7.9: Find the reactions R_1, R_2 and R_3 in case of two beams placed one over the other as shown in the figure. [U.P.T.U. II Sem., 2000-01]

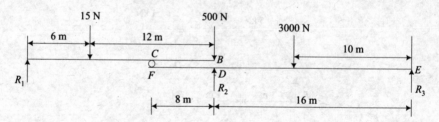

Solution:

1. Draw free body diagram of beam *AB*.

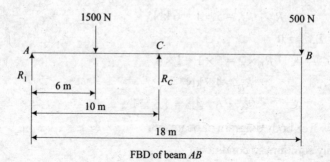

FBD of beam *AB*

2. Apply equilibrium conditions

   (i) $\Sigma M_A = 0$

$$1500\,(6) - R_C\,(10) + 500\,(18) = 0$$

$$\therefore \quad R_C = \frac{1500 \times 6 + 500 \times 18}{10} = 1800$$

$$R_1 - 1500 + 800 - 500 = 0$$

$$R_1 = 1500 + 800 - 500 = 0$$

$$R_1 = 200\ N$$

3. Draw free body diagram of beam *FE*.

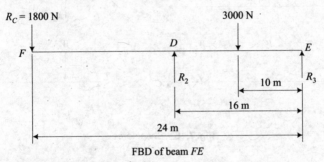

FBD of beam *FE*

4. Apply equilibrium conditions

   (i) $\Sigma F_y = 0$

$$-1800 + R_2 - 3000 + R_3 = 0$$

$$R_2 + R_3 = 4800\ N$$

   (ii) $\Sigma M_E = 0$

$$1800\,(24) - R_2(16) + 3000(10) = 0$$

$$\therefore \quad R_2 = \frac{1800 \times 24 \times 10}{16} = 2700\ N$$

$$\therefore \quad R_3 = 4800 - 2700 = 2100\ N$$

7.5 SHEAR FORCE AND BENDING MOMENT

The load applied on a beam gets transferred to its supports. Any section of the beam will experience shearing off by vertical forces and bending by moments.

7.5.1 Shear Force

A beam *AB* is supporting vertical loads W_1, W_2 and W_3 and the beam is supported at *A* and *B*. The supports *A* and *B* are providing reactions R_1 and R_2. The beam is in equilibrium under five forces R_1, W_1, W_2, W_3 and R_2. If the beam is cut at section

$X-X$, a force F_X must he applied at the section to maintain equilibrium of left portion and right. This force F_X is called shear force. This force was supported by left portion to the right portion. Similarly right portion was supported by left portion. Hence, there will be a shearing force at the action.

Share force is defined as the force that tries to shear off the section of the beam and is numerically equal to the algebraic sum of all the forces acting normal to the axis of the beam either to the left or to the right of the section. In the given example,

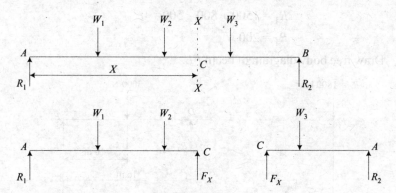

Fig. 7.16 Shear force

$$F_x = R_1 - W_1 - W_2$$
or
$$F_x = R_2 - W_3.$$

The shear force that tends to move the left portion upwards relative to right portion is taken as positive shear force.

The sign convention of shear force is illustrated in Fig. 4.17.

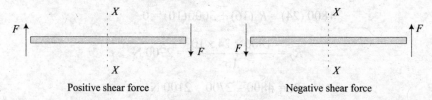

Positive shear force   Negative shear force

Fig. 7.17 Sign convention of shear force

7.5.2 Bending Moment

Banding moment at any section is the moment of forces which tries to bend it.

A beam AB is in equilibrium and there is no resultant moment. If the beam is cut at X-X there should be a balancing moment M_X at point C to keep separately the right portion and left protion in equilibrium. This moment is provided by the internal forces acting in the beam at section $X-X$. M_X is called the bending moment at the section.

The bending moment is numerically equal to the algebraic sum of moments of all the forces about the section, either acting to the left or to the right of the section. In the given example,

$$M_X = R_1 (x) - W_1 (x - l_1) - W_2 (x - l_2)$$

or

$$M_X = R_2 (l - x) - W_3 (x - l_3)$$

The bending moment which tends to sag the beam is taken as positive. The bending moment which tries to hog the beam is taken as negative.

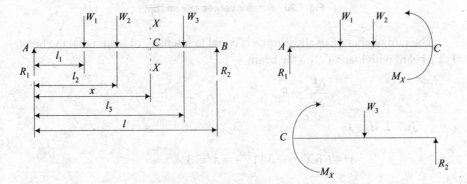

Fig. 7.18 Bending moment

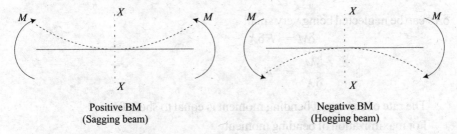

Positive BM
(Sagging beam)

Negative BM
(Hogging beam)

Fig. 7.19 Sign convention of bending moment

7.5.3 Relation between Shear Force and Bending Moment

There is a relationship between load on the beam, the shearing force and bending moment.

Consider the equilibrium of a small element of length δx in a beam AB of span l and under *udl* of w/m. The values of shear force and bending moment at left section be F and M which change to $F + \delta F$ and $M + \delta M$ over the elemental length δX.

Apply equilibrium conditions to small length δx

(i) $\Sigma F_y = 0$

$$F + w\,\delta x = F + \delta F$$

$$\therefore \qquad \frac{\delta F}{\delta x} = w$$

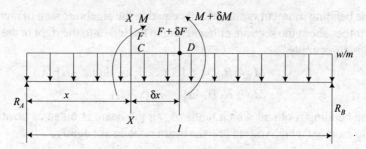

Fig. 7.20 Relation between SF and BM

The rate of change of shear force is equal to intensity of loading on the beam. For a beam with load of P on the beam

$$\frac{\delta F}{\delta x} = P$$

(ii) $\Sigma M_D = 0$

$$M - F\delta X - (w\,\delta X)\frac{\delta x}{2} = M + \delta M$$

$$-F\delta X - \frac{w}{2}\,\delta X^2 = \delta M$$

δX^2 can be neglected being very small

$$\delta M = -F\delta X$$

$$\frac{\delta M}{\delta X} = -F$$

The rate of change of bending moment is equal to shear force.

For maximization of bending moment

$$\frac{\delta M}{\delta X} = 0$$

$$F = 0$$

The bending moment is maximum or minimum where shear force is zero.

7.6 SHEAR FORCE AND BENDING MOMENT DIAGRAMS

The shear force and bending moment in a beam vary from section to section along its length.

7.6.1 Shear Force Diagram (SFD)

The variation of shear force along the length of the beam can be represented graphically in which ordinate represents the shear force. Such a graph is called Shear

Force Diagram (SFD). Positive shear force is represented above the beam line and negative force below the beam. Salient values of shear force are written on the diagram.

The section at which shear force changes its sign is the point or section where maximum bending moment occurs.

7.6.2 Bending Moment Diagram (BMD)

The variation of bending moment along the length of the beam can be represented graphically. Such a graph is called Bending Moment Diagram (BMD). Positive bending moment is represented above the beam line and negative bending moment below the line. Salient values of bending moment are written on the diagram.

The bending moment changes its sign at the point of contraflexure. The bending moment at that section is obviously zero.

7.6.3 Some Rules for SFD and BMD

(i) Whenever a concentrated load is acting (including support reactions) the shear force will suddenly change. The change is numerically equal to the load and in the direction of load.

(ii) If there is no loading between any two sections, the magnitude of shear force will not change between the given vertical loads and SFD will be a horizontal line.

(iii) The variation of shear force will be a horizontal line for vertical loads, it is an inclined straight line for uniformly distributed load and a parabola for uniformly varying load between the sections.

(iv) Bending moment is zero at supports of simply supported beam.

(v) Bending moment is zero at the free end of a cantilever or overhanging beam.

(vi) The bending moment is maximum or minimum where shear force is zero.

(vii) The bending moment curve will be one degree higher than that of shear force and two degree higher than load curve.

(viii) Since $\dfrac{dF}{dX} = P$ and $\dfrac{dM}{dX} = -F$, nature of variation of shear force and bending moment diagram with load will be as follows.

Load	SFD	BMD
1. Concentrated or no load.	Horizontal line	Inclined linear line
2. UDL (horizontal line)	Inclined linear line	Parabolic
3. Uniformly varying load (inclined line)	Parabolic	Cubic

(ix) If an external moment is acting at a point or a section of a beam, there will be sudden change in the value of bending moment, the change being numerically equal to external moment.

While proceeding from left, it will increase in value, if the moment is clockwise. There will be drop in bending moment value if the external moment is anticlockwise. Exactly opposite phenomenon will be observed while proceeding from right to left.

The external moment will affect the magnitude of support sections. However, there is no sudden change in the value of shear force at the section where external moment is acting.

(x) Inflection or contraflexure point occurs where bending moment is zero.

(xi) The location of points of maximum bending moment and contraflexure must be shown in the bending moment diagram.

7.7 SFD AND BMD FOR STANDARD CASES

The load diagram, S.F. diagrams and B.M. diagram for some standard cases are tabulated in Table 7.1 (See Page 361).

Example 7.10: Draw shear force and bending moment diagram for a cantilever of length L carrying a concentrated load W at the free end.

Solution:

1. *Draw the Load Diagram.*

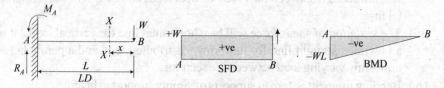

2. *Support reaction*

Applying equations of equilibrium, find the support reactions

(i) $\Sigma F_y = 0$

$\therefore$ $R_A = W$

(ii) $\Sigma M_A = 0$

$\therefore$ $- M_A = + WL = 0$

$M_A = WL$ (Anticlockwise)

3. *Shear Force*

Cut a section X-X at a distance X from end B.

$F_X = + W$ which is constant over the length

4. *Bending Moment*

$M_X = - Wx$ (Hogging)

when, $X = 0$ $M = 0$

when, $X = L$ $M = WL$

The B.M. changes linearly from B to A.

Table 7.1 SFD and BMD for standard case

Sl. No.	Type of beam	Load diagram	S.F. diagram	B.M. diagram
1.	Cantilever	$M = WL$, $R = W$; W, L	W, +ve	WL, −ve, Inclined line
		w/m, M, R, L	$R = wL$, +ve	$\dfrac{wL^2}{2}$, −ve, Parabola
2.	Simply supported beam	$\dfrac{W}{2}$, W, $\dfrac{W}{2}$, L	$\dfrac{W}{2}$ +ve, $-\dfrac{W}{2}$ −ve	$+\dfrac{WL}{4}$
		$\dfrac{wL}{2}$, w/m, $\dfrac{wL}{2}$, L	$\dfrac{wL}{2}$ +ve, $-\dfrac{wL}{2}$ −ve	$+\dfrac{wL^2}{4}$, Parabola
		R_A, F, d, a, b, R_B, L	$-R_A = \dfrac{Fd}{L}$, −ve	M, $\dfrac{Fda}{L}$, $\dfrac{Fdb}{L}$, +ve, −ve
		F, d, c, a, b, R_A, R_B, L; F, Fd, R_A, R_B	$\dfrac{F(a+b)}{L}$ +ve, B F, $\dfrac{-F(a+b)}{L}$ −ve	Fd, $\dfrac{F(a-b)b}{L}$, +ve

Example 7.11: Draw SFD and BMD for the cantilever shown carrying several concentrated loads.

Solution:
1. Draw the Load Diagram.

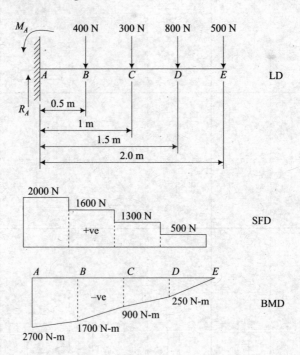

2. *Support Reactions,*

Apply equilibrium conditions

    (i) $\Sigma F_y = 0$

           $R_A = \Sigma W$

                $500 + 800 + 300 + 400 = 2000\text{N}$

    (ii) $\Sigma M_A = 0$

           $M_A = \Sigma Wx$

                $= 500 \times 2 + 800 \times 1.5 + 300 \times 1 + 400 \times 0.5$

                $= 2700 \text{ Nm}$

3. *Shear Force*

$$F_A = R_A = 2000 \text{ N}$$
$$F_B = F_A - W_B = 2000 - 400 = 1600 \text{ N}$$
$$F_C = F_B - W_C = 1600 - 300 = 1300 \text{ N}$$
$$F_D = F_C - W_D = 1300 - 800 = 500 \text{ N}$$
$$F_E = F_D - W_E = 500 - 500 = 0 \text{ N}$$

Draw SFD using horizontal lines between the loads.

4. *Bending Moment*

$$M_E = 0 \text{ (Free end)}$$

$$M_D = W_E (DE) = 500 \times 0.5 = 250 \text{ N-m} \qquad \text{(Beam hogging, –ve)}$$

$$M_C = W_E (CE) + W_D (CD) = 500 \times 1 + 800 \times 0.5$$

$$= 900 \text{ N-m (–ve)}$$

$$M_B = W_E (BE) + W_B (BD) + W_C (BC)$$

$$= 500 \times 1.5 + 800 \times 1 + 300 \times 0.5$$

$$= 1700 \text{ N-m (–ve)}$$

$$M_D = 2700 \text{ N-m (–ve)}$$

Draw BMD using linear inclined lines.

Example 7.12: Draw SFD and BMD of a cantilever of length L carrying a uniformly distributed load of w per unit run over whole length.

Solution:

1. Draw free body diagram,

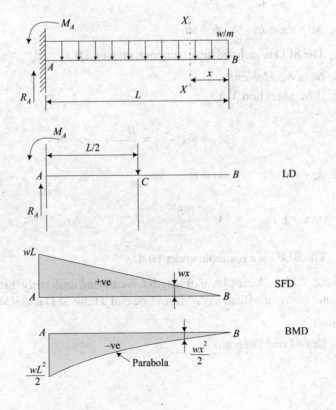

2. Draw equivalent load diagram,

$$W = wL \text{ (Area under the curve)}$$

Point of application $= L/2$

3. *Support Reactions*

Apply equilibrium conditions

(i) $\Sigma F_y = 0$

$\therefore \quad\quad R_A = W \text{ (Left up positive)}$

$\quad\quad\quad = WL$

(ii) $\Sigma M_A = 0$

$\therefore \quad\quad M_A = W \times \dfrac{L}{2} = WL\left(\dfrac{L}{2}\right) = \dfrac{WL^2}{2} \text{ (Hogging beam –ve)}$

4. *Shear Force*

Shear force at section X-X which is x distance from B

$$F_X = wx$$

At $\quad x = 0, \quad\quad F = 0$

At $\quad x = L, \quad\quad F = wL$

The SFD is inclined linear diagram under UDL.

5. *Bending Moment*

B.M. at section X-X,

$$M_X = wx\left(\dfrac{x}{2}\right) = \dfrac{Wx^2}{2} \text{ (–ve)}$$

At $x = 0, \quad\quad M = 0$

At $x = L, \quad\quad M = \dfrac{WL^2}{2}$

The BMD is a parabola under UDL.

Example. 7.13: A cantilever of length L is carrying uniformly distributed load of w per unit run over a distance 'a' from free end. Draw SFD and BMD.

Solution:

1. Draw Load Diagram.

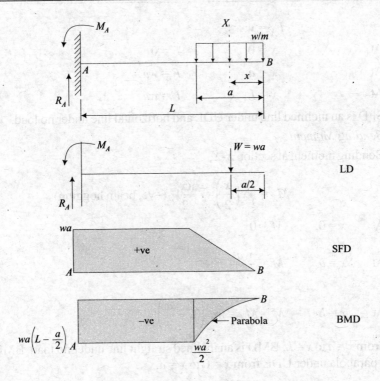

$$W = wa \text{ (Area under load curve)}$$

Point of application $= \dfrac{a}{2}$ from free end.

2. *Support Reactions*

   Apply equilibrium conditions

   (i) $\Sigma F_y = 0$

   $$R_a = W = wa$$

   (ii) $\Sigma M_A = 0$

   $$M_A = W\left(L - \frac{a}{2}\right)$$

   $$= wa\left(L - \frac{a}{2}\right) \text{ (Hogging } -ve)$$

3. *Shear Force*

   Shear force at section $X\text{-}X$

$$F_X = wx$$

At $\qquad x = 0, \qquad\qquad F = 0$

At $\qquad x = a, \qquad\qquad F = wa$

At $\qquad x = L, \qquad\qquad F = wa$

SFD is an inclined line under UDL and horizontal line under no load.

4. *Bending Moment*

Bending moment at section *X-X*,

$$M_X = wx\left(\frac{x}{2}\right) = \frac{wx^2}{2} \quad \text{(–ve, beam hogging)}$$

At $\quad x = 0, \qquad M = 0$

At $\quad x = a \qquad M = \dfrac{wa^2}{2}$

At $\quad x = L \qquad M = M_A = wa\left(L - \dfrac{a}{2}\right)$

From $x = a$ to $x = L$, BMD is an inclined straight line under no load. BMD is a parabola under UDL from $x = 0$ to $x = a$.

Example 7.14: Draw the Shear Force and Banding Moment diagrams for the cantilever beam shown in figure. [U.P.T.U. II Sem., 2005-06]

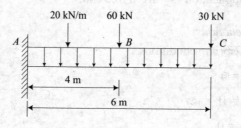

Solution:

1. Draw Equivalent Loading Diagram,

   A load of 120 kN (area under the load curve equal to 20×6 kN will be acting at 3 m from *A* i.e, c.g. of load curve).

2. *Support Reactions*

   Apply equilibrium conditions.

   (i) $\Sigma F_y = 0$

   $\therefore \qquad\qquad R_A = \Sigma W = 120 + 60 + 30 = 210 \text{ kN}$

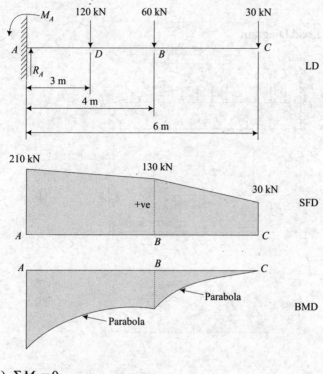

(ii) $\Sigma M_A = 0$

$\therefore$
$$M_A = 120\,(3) + 60\,(4) + 30\,(6)$$
$$= 360 + 240 + 180 = 780 \text{ kN-m}$$

3. *Shear Force*

$$F_C = 30 \text{ kN}$$
$$F_B = 30 + 60 + (20 \times 2) = 130 \text{ kN}$$
$$F_A = \Sigma W = 120 + 60 + 30 = 210 \text{ kN}$$

SFD will be an inclined straight line under UDL.

4. *Bending Moment*

$$M_C = 0 \text{ (Free end)}$$
$$M_B = 30 \times 2 + 20 \times 2 \times 1 = 60 + 40 = 100 \text{ kN-m}$$
$$M_A = 780 \text{ kN-m}$$

BMD will be parabolas under UDL.

Example. 7.15: A cantilever of length L carries a load whose intensity varies uniformly from zero at the free end to w per unit length at the fixed end. Draw SFD and BMD.

Solution:

1. Draw Load Diagram

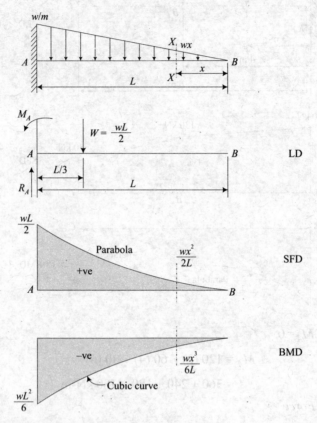

2. Draw Equivalent Load Diagram

   W is area under load curve

   $$W = \int_{o}^{L} w_x \, dx = \frac{WL}{2}$$

   where, $w_X = \dfrac{wx}{L}$

   Location W will be acting at a distance $L/3$ from A, i.e., c.g. of load curve.

3. *Support Reactions*

   Apply equilibrium conditions,

   (i) $\Sigma F_y = 0$

   $\therefore$

   $$R_A = W = \frac{wL}{2}$$

(ii) $\Sigma M_A = 0$

$$M_A = W \times \frac{L}{3} = \frac{wL}{2} \times \frac{L}{3} = \frac{wL^2}{6}$$

4. *Shear Force*

Shear force at section X–X,

$$F_X = w_x \cdot \frac{x}{2} \quad \text{(Area under load curve on the spare } X\text{-}B\text{)}$$

$$= \frac{wL}{2} \cdot \frac{x}{2} = \frac{wx^2}{2L}$$

At $x = 0$ $F = 0$

At $x = L$ $F = \frac{wL}{2}$

SFD is a parabola under uniformly varying load.

5. *Bending Moment*

Bending moment at section X-X is equal to area under the load curve X distance of centroid,

$$M_X = \frac{wx^2}{2L} \cdot \frac{x}{3} = \frac{wx^3}{6L}$$

At $x = 0$ $M = 0$

At $x = L$ $M = \frac{wL^2}{6}$ [Hogging beam –ve]

BMD is a cubic under a uniformly varying load.

Example 7.16:   A simply supported beam of span L carries a concentrated load W at the mid-span. Draw SFD and BMD.

Solution:

1. Draw Load Diagram

2. *Support Reactions*

The loading is symmetrical

∴

$$R_A = R_B = \frac{W}{2}$$

$$M_A = M_B = 0 \text{ at free supports.}$$

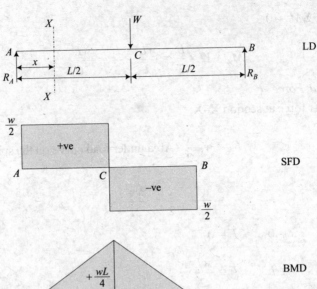

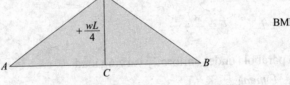

3. *Shear Force*

$$F_A = R_A = + \frac{W}{2}$$

$$F_C = F_A - W_C = \frac{W}{2} - W = - \frac{W}{2}$$

$$F_B = R_B = - \frac{W}{2}$$

SFD is horizontal straight lines under concentrated loads.

4. *Bending Moment*

The beam is sagging, bending moment is +ve.

Bending moment at section *X-X*,

$$M_X = R_A x = \frac{W}{2} x$$

At $\quad x = 0 \qquad M = 0$ \
At $\quad x = L \qquad M = 0$ $\Big\}$ Free supports

At $\quad x = \dfrac{L}{2} \qquad M = \dfrac{WL}{4}$

BMD is inclined straight line under concentrated load.

Example 7.17: The loading of a simply supported beam *AB* of span 8 m is shown. Draw SFD and BMD.

Solution:

1. Draw load diagram

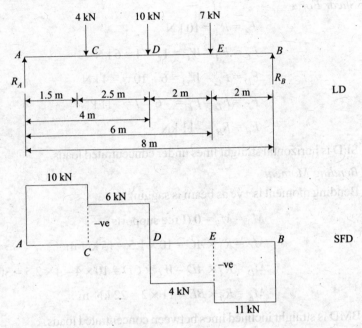

2. *Support Reactions*

Apply equilibrium conditions,

(i) $\Sigma M_A = 0$

$$R_B \times 8 = 4 \times 1.5 + 10 \times 4 + 7 \times 6$$

$$R_B = 11 \text{ kN}$$

(ii) $\Sigma F_y = 0$

$$R_A + R_B = 4 + 10 + 7 = 21 \text{ kN}$$

$\therefore$

$$R_A = 21 - R_B = 10 \text{ kN}$$

3. *Shear Force*

$$F_A = R_A = 10 \text{ kN}$$

$$F_C = F_A - W_C = 11 - 4 = 6 \text{ kN}$$

$$F_D = F_C - W_D = 6 - 10 = -4 \text{ kN}$$

$$F_E = F_D - F_E = -6 - 7 = -11 \text{ kN}$$

$$F_B = R_B = -11 \text{ kN}$$

SFD is horizontal straight lines under concentrated loads.

4. *Bending Moment*

Bending moment is +ve as beam is sagging.

$$M_A = M_B = 0 \text{ (Free supports)}$$

$$M_C = R_A \times AC = 10 \times 1.5 = 15 \text{ kN-m}$$

$$M_D = R_A \times AD - W_C \times CD = 10 \times 4 - 4 \times 2.5 = 30 \text{ kN-m}$$

$$M_E = R_B \times BE = 11 \times 2 = 22 \text{ kN-m}$$

BMD is straight inclined lines between concentrated loads.

Example 7.18: A simply supported beam is shown. Draw SFD and BMD.

[U.P.T.U. I Sem., 2004-05]

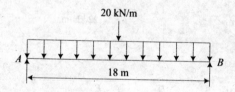

Solution:

1. Draw equivalent load diagram.

Equivalent load W is equal to area under the load curve and acts at c.g. of load curve.

$\therefore$

$$W = 20 \times 18 = 360 \text{ kN}$$

and acts at mid-span.

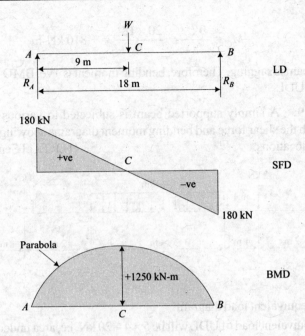

2. Support Reactions

The beam is symmetrically loaded,

$$\therefore \qquad R_A = R_B = \frac{W}{2} = 180 \text{ kN}$$

3. Shear Force

$$F_A = R_A = 180 \text{ kN.}$$

$$F_B = R_B = -180 \text{ kN}$$

$$F_C = F_A - \frac{wL}{2} = 0$$

SFD is a straight inclined line under UDL.

4. Bending Moment

$$M_A = M_B = 0 \text{ (Free supports)}$$

$$M_C = R_A \times \frac{L}{2} - \frac{WL}{2} \times \frac{L}{4}$$

$$= 180 \times 9 - 180 \times \frac{18}{4} = 1620 - 810$$

$$= 810 \text{ kN-m}$$

Check: $\qquad M_C = \dfrac{WL^2}{8} = \dfrac{20 \times 18^2}{8} = +810$ kN-m

The beam is sagging. Therefore, bending moment is +ve. BMD is a parabola under UDL.

Example 7.19: A simply supported beam is subjected to various loadings as shown. Sketch the shear force and bending moment diagrams showing their values at significant locations.   [U.P.T.U.I Sem., 2005-06]

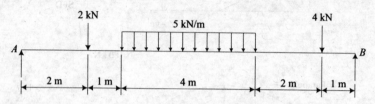

Solution:

1. Draw equivalent load diagram.

   The equivalent load of UDL will be $5 \times 4 = 20$ kN. i.e, area under UDL curve and will act at c.g of load curve.

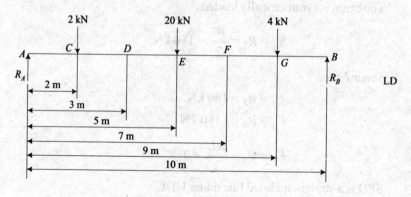

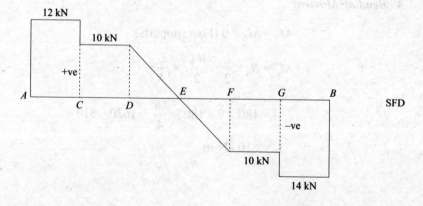

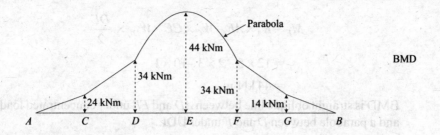

Parabola

44 kNm

34 kNm

34 kNm

24 kNm

14 kNm

BMD

A   C   D   E   F   G   B

2. *Support Reactions*

Apply equilibrium condition

(i) $\Sigma F_Y = 0$

$$R_A + R_B = 2 + 20 + 4 = 26 \text{ kN}$$

(ii) $\Sigma M_A = 0$

$$R_B \times 10 = 2 \times 2 + 20 \times 5 + 4 \times 9 = 4 + 100 + 36 = 140 \text{ kN-m}$$

$$R_B = 14 \text{ kN}$$

$$\therefore \quad R_A = 26 - 14 = 12 \text{ kN}$$

3. *Shear Force*

$$F_A = R_A = 12 \text{ kN}$$
$$F_C = F_A - W_C = 12 - 2 = 10 \text{ kN}$$
$$F_D = F_C = 10 \text{ kN}$$
$$F_B = R_B = -14 \text{ kN}$$
$$F_G = F_B - W_G = -14 + 4 = -10 \text{ kN}$$
$$F_F = F_G = -10 \text{ kN}$$
$$F_E = F_D - W_{DE} = 10 - 5 \times 2 = 0$$

SFD is horizontal straight line between *AD* and *FB* and straight inclined line under UDL between *D* and *F*.

4. *Bending Moment*

Bending moment is +ve for sagging beam.

$$M_A = M_B = 0 \text{ (Free supports)}$$
$$M_C = R_A \times AC = 12 \times 2 = 24 \text{ kN-m}$$
$$M_D = R_A \times AD - W_C \times CD = 12 \times 3 - 2 \times 1 = 34 \text{ kN-m}$$
$$M_G = R_B \times GB = 14 \times 1 = 14 \text{ kN-m}$$
$$M_F = R_B \times BF - WG \times GF = 14 \times 3 - 4 \times 2 = 34 \text{ kN-m}$$

$$M_E = R_A \times AE - W_C \times CE - W_{DE} \times \frac{DE}{2}$$

$$= 12 \times 5 - 2 \times 3 - 10 \times 1$$

$$= 44 \text{ kN-m}$$

BMD is straight oblique line between AD and FB under concentrated loads and a parabola between D and F under UDL.

Example 7.20: The intensity of loading on a simply supported beam of span 10 m increases uniformly from 10 kN/m at left support to 20 kN/m at the right. Find the position and magnitude of maximum bending moment.

[U.P.T.U. I Sem., 2004-05]

Solution:

1. Draw the load diagram

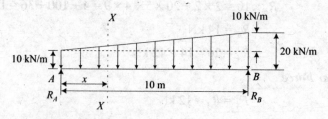

The load may be divided into two parts:

(i) A UDL of 10 kN/m

(ii) A uniformly increasing load which is zero at A and 10 kN/m at B.

2. *Support Reactions*

Apply equilibrium conditions

$$\Sigma M_A = 0$$

$$R_B \times 10 = 10 \times 10 \times \frac{10}{2} + \frac{10 \times 10}{2} \times \frac{2}{3} \times 10 = 500 + 333.34$$

$$= 833.34 \text{ kN-m}$$

∴ $$R_B = 83.334 \text{ kN}$$

$$R_A = \left(10 \times 10 \times \frac{10 \times 10}{2}\right) - 83.334 = 66.666 \text{ kN}$$

3. *Shear Force*

The rate of loading at section X-X,

$$w_X = 10 + \frac{10x}{10} = 10 + x$$

Shear force at section X-X

$$F_X = R_A - W_{AX} = 66.666 - [10x + \frac{1}{2}(10 + x)x]$$

$$= 66.666 - 15x - \frac{x^2}{2}$$

4. *Bending Moment*

Bending moment at section X-X,

$$M_X = R_A x - \frac{10x^2}{2} - \left(\frac{10x + x^2}{2}\right) \times \frac{2x}{3}$$

$$= 66.666x - 5x^2 - \frac{10x^2}{3} - \frac{x^2}{3}$$

For maximum bending moment, $F_X = 0$

$\therefore$ $\qquad 66.666 - 15x - \frac{x^2}{2} = 0$

$\therefore$ $\qquad x^2 + 15x - 133.332 = 0$

$$x = \frac{-15 + \sqrt{(15)^2 + 4 \times 133.332}}{2} = 6.269 \text{ m}$$

$$M_{max} = 66.666 \times 6.269 - 5(6.269)^2 - \frac{10}{3}(6.269)^2 - \frac{(6.269)^3}{3}$$

$$M_{max} = 417.93 - 196.5 - 131 - 82.12 = 8.31 \text{ kN-m}$$

Example 7.21: Draw SFD and BMD of the overhanging beam AE as shown and find out point of contraflexure.

Solution:

1. Draw the load diagram

2. *Support Reactions*

Apply equilibrium conditions

(i) $\Sigma F_Y = 0$

$$R_A + R_B = 2w - w + \frac{W}{2} = \frac{3W}{2}$$

(ii) $\Sigma M_A = 0$

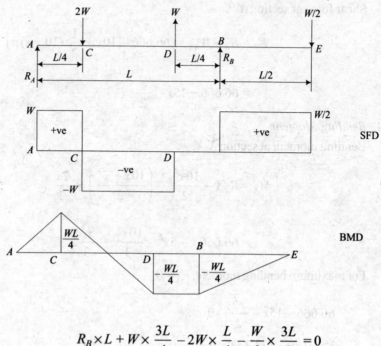

$$R_B \times L + W \times \frac{3L}{4} - 2W \times \frac{L}{4} - \frac{W}{2} \times \frac{3L}{2} = 0$$

$$R_B = \frac{W}{2}$$

$$R_A = \frac{3W}{2} - \frac{W}{2} = W$$

3. *Shear Force*

$$F_A = R_A = +W$$
$$F_C = R_A - W_C = W - 2W = -W$$
$$F_D = F_C - W_D = -W + W = 0$$

$$F_B = R_B = \frac{W}{2}$$

$$F_E = W_E = \frac{W}{2}$$

4. *Bending Moment*

$$M_A = 0 \qquad \text{Free support}$$
$$M_E = 0 \qquad \text{Free end}$$

$$M_C = R_A \times A_C = W \times \frac{L}{2} = + \frac{WL}{4} \text{ (sagging)}$$

$$M_D = R_A \times A_D - W_C (CD)$$

$$= W \times \frac{3L}{4} - 2W \times \frac{L}{2} = -\frac{WL}{4} \text{ (Hogging)}$$

$$M_B = W_E (BE) = \frac{W}{2} \times \frac{L}{2} = -\frac{WL}{4} \text{ (Hogging)}$$

The point of contraflexure is between C and D.

Example 7.22:   The bending moment diagram (BMD) of a simply supported beam is shown. Calculate the support reactions of the beam.

[U.P.T.U. I Sem., 2000-01]

Solution: The BMD is drawn.

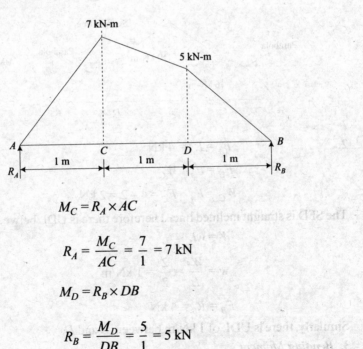

$$M_C = R_A \times AC$$

$$\therefore \qquad R_A = \frac{M_C}{AC} = \frac{7}{1} = 7 \text{ kN}$$

$$M_D = R_B \times DB$$

$$\therefore \qquad R_B = \frac{M_D}{DB} = \frac{5}{1} = 5 \text{ kN}$$

Example 7.23: The shear force diagram (SFD) of a simply supported beam is shown. Calculate the support reactions of the beam and also draw BMD.

Solution:

1. Draw load diagram.

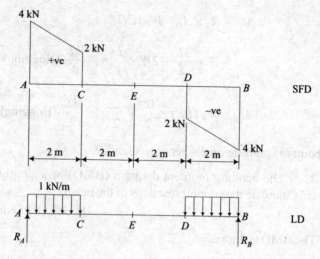

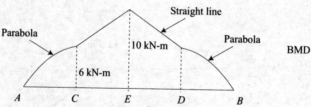

2.
$$F_A = R_A = 4 \text{ kN}$$
$$F_C = F_A - W_C$$
$$\therefore \qquad W_C = F_A - F_C = 4 - 2 = 2 \text{ kN}$$

The SFD is straight inclined line. Therefore there is UDL between *AC*.
$$W = wL$$

$$\therefore \qquad w = \frac{W}{L} = \frac{2}{2} = 1 \text{ kN/m}$$

$$F_B = R_B = 4 \text{ kN}$$

Similarly, there is UDL of 1 kN/m between. *B* and *D*.

3. *Bending Moment*
$$M_A = M_B = 0 \text{ (Free support)}$$
$$M_C = w\,(AC). \left(\frac{AC}{2}\right) - R_A\,(AC)$$

$$= -1 \times 2 \times 1 + 4 \times 2$$

$$= 6 \text{ kN-m (sagging)}$$

$$M_E = R_A(AE) = w(AC) \times \left(CE + \frac{AC}{2} \right)$$

$$= 4 \times 4 - 1 \times 2 \times 3 = 10 \text{ kN-m}$$

$$M_D = M_C = 6 \text{ kN-m.}$$

BMD is a parabola in the portion *AC* and *BD* and straight inclined lines in the portion *CE* and *ED*.

Example 7.24: The SFD of a simply supported beam is shown. Calculate the support reactions of the beam and draw BMD. [U.P.T.U. I Sem., 2001-02]

Solution:

1. *Loading Diagram*

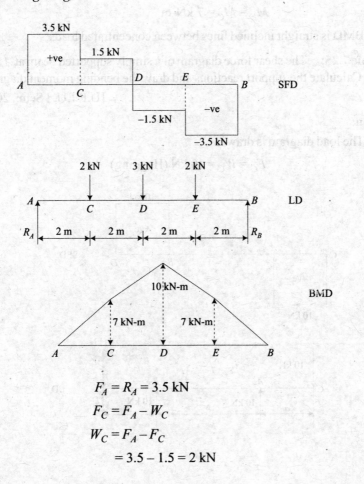

$$F_A = R_A = 3.5 \text{ kN}$$

$$F_C = F_A - W_C$$

$$\therefore \qquad W_C = F_A - F_C$$

$$= 3.5 - 1.5 = 2 \text{ kN}$$

$$F_D = F_C - W_D$$

$$\therefore \qquad W_D = F_D - F_C = 1.5 - (-1.5)$$

$$= 3.0 \text{ kN}$$

$$F_B = R_B = -3.5 \text{ kN}$$

$$F_E = F_D - W_E = -1.5 - (-3.5)$$

$$= 2 \text{ kN}$$

2. *Bending Moment*

$$M_A = M_B = 0 \text{ (Free supports)}$$

$$M_C = R_A (AC) = 3.5 \times 2 = 7 \text{ kN-m (+ve sagging)}$$

$$M_D = R_A (AD) - W_C (CD)$$

$$= 3.5 \times 4 - 2 \times 2 = 10 \text{ kN-m (+ve sagging)}$$

$$M_E = M_C = 7 \text{ kN-m}$$

BMD is straight inclined lines between concentrated loads.

Example 7.25: The shear force diagram of a simply supported beam at *A* and *B* is shown. Calculate the support reactions and draw the bending moment diagram.

[U.P.T.U. I Sem., 2002-03]

Solution:

1. The load diagram is drawn.

$$F_C = W_C = 10 \text{ kN (Hogging)}$$

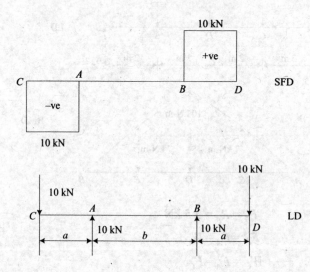

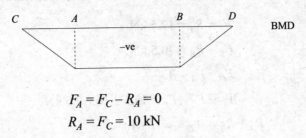

$$F_A = F_C - R_A = 0$$
$$\therefore \quad R_A = F_C = 10 \text{ kN}$$

By symmetry

$$W_D = 10 \text{ kN}$$
$$R_B = 10 \text{ kN}$$

2. *Bending Moment*

$$M_C = M_D = 0 \text{ (Free ends)}$$
$$M_A = R_A \times (a) = -10a \text{ kN-m (Hogging)}$$
$$M_B = M_A = -10a \text{ kN-m}$$

Example 7.26: Draw the load diagram for the SF diagram shown for a simply supported beam. Calculate the maximum bending moment and its location.

[U.P.T.U. I Sem., 2001-02]

Solution:

1. *Load Diagram*

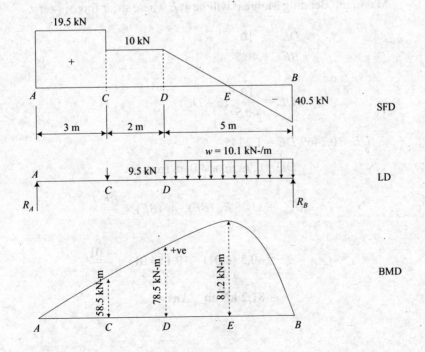

$$F_A = R_A = 19.5 \text{ kN}$$

$$F_B = R_B = 40.5 \text{ kN}$$

$$F_C = F_A - W_C$$

$$W_C = F_A - F_C = 19.5 - 10 = 9.5 \text{ kN}$$

$$R_A + R_B = W_C + w \ (DB)$$

$$\therefore \qquad w = \frac{R_A + R_B - W_C}{DB}$$

$$= \frac{19.5 + 40.5 - 9.5}{5} = 10.1 \text{ kN}$$

2. *Bending Moment*

$$M_A = M_B = 0 \text{ (Free supports)}$$

$$M_C = R_A \ (CA)$$

$$= 19.5 \times 3 = + 58.5 \text{ kN-m (sagging)}$$

$$M_D = R_A \ (AD) - W_C \ (CD)$$

$$= 19.5 \times 5 - 9.5 \times 2 = 78.5 \text{ kN-m (sagging)}$$

Maximum Bending Moment will be at E where shear force is zero.

$$\frac{DE}{BE} = \frac{10}{40.5}$$

$$DE = \frac{10}{40.5} \ (5 - DE)$$

$$\therefore \qquad DE + 0.2469 \ DE = 1.2345$$

$$\therefore \qquad DE = 0.99 \text{ m or } 4.01\text{m from } B$$

$$M_{max} = M_E = R_B \ (BE) - w \ (BE) \times \frac{BE}{2}$$

$$= 40.5 \ (4.01) - 10.1 \ (4.01) \times \frac{4.01}{2}$$

$$= \textbf{81.2 kN-m} \quad \textbf{Ans.}$$

1. Define a beam and classify different types of beams on the basis of:
   (i) Support conditions
   (ii) Loadings.
2. What do you understand by 'shear force' and 'bending moment' and what is their importance in beam design?
3. What do you understand by statically determinate beam?
4. How shearing force and bending moment diagrams are drawn for a beam?
5. What do you understand by the term 'point of contraflexure'.
6. Explain rules for shear force and bending moment diagram.
7. What is the relationship between shear force and bending moment?
8. State true or false:
   (i) The shear force changes suddenly at a section where, there is a vertical point load.
   (ii) The shear force between any two vertical loads remains constant.
   (iii) BM is maximum at a section where SF is zero.
   (iv) Where a beam is subjected to a couple at a section, BM changes suddenly at the section but SF remains unaltered.
   (v) At point of contraflexure, BM is zero.

Trusses

8.1 INTRODUCTION

A truss is a system of members that are fastened together at their ends to support stationary and moving loads. These are used in bridges, buildings, roofs of industrial sheds, railway platforms, godown, transmission towers, etc.

The main characteristics of a truss are as follows:

1. Each member of a truss is usually of uniform cross-section along its length.
2. Members of a truss have different cross-sections and cross-sectional areas as they must transmit different forces.
3. The common cross-sections of truss members are angles (L), I-section (I); channels ([), hollow tubes made from rolled steel.
4. The members of a truss are always connected at their ends. A truss is different than a frame. A frame may have some members connected along their length in addition to end connections.
5. In order to maximize the load-carrying capacity of a truss, the external loads must be applied at the joints, because the members of a truss are long and slender, the compression members may buckle under transverse loads.
6. The members of the trusses are fastened together by welding, riveting or bolting, through gusset plates in plane trusses.

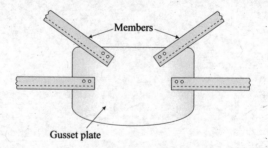

Fig. 8.1 Truss joint

7. For analysis of forces, the joint is taken as pin connection for coplanar concurrent members and a simple ball and socket connection for a space truss.

Pin joint   Ball and socket joint

Fig. 8.2   Different joints

8. The weight of the member of a truss is small as compared to external loads applied on a truss. The weight of the member is usually neglected. In case the weight of the member has to be considered, half load is considered acting at each joint.
9. Each member of a truss is two-force member, i.e., a tensile member or compression member. A compression member pushes the joint whereas a tensile member pulls the joint where it is connected.

Compression member   Tensile member

Fig. 8.3   Loading of truss members

8.2   TYPES OF TRUSSES

There are two categories of trusses according to *geometry*:

1. *Plane truss* consists of coplanar system of members.

   Examples: Roof truss, sides of a bridge.

2. *Space truss* consists of a three-dimensional system of members.

   Example: Electric power transmission tower.

8.2.1   A Simple Truss

A simple truss is just-rigid truss and removal of any of its members destroys its rigidity. If removing a member does not destroy rigidity, the truss is over-rigid.

The most elementary just-rigid truss has three members connected to form a triangle (*ABC*). The plane and space trusses are built-up from this triangle by adding for each new joint three new members. The trusses so built are called simple trusses.

A simple relationship exists between the number of joints, *j* and the number of members, *m* in a simple truss.

Simple Space Truss, $m = 3j - 6$

Simple Plane Truss, $m = 2j - 3$

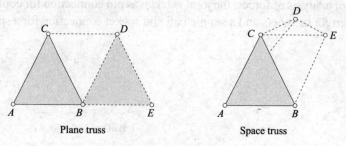

Plane truss   Space truss

Fig. 8.4 The simple truss

8.2.2 Classification of Simple Truss

A simple truss can be classified as follows:

(a) Perfect truss (stable truss)

In a perfect truss the above relationship holds good. The truss does not collapse under the loading.

Example: In a triangular truss ABC,

    No. of joints, $j = 3$

    No. of members, $m = 3$

    For plane truss,

$$m = 2j - 3$$
$$= 2 \times 3 - 3 = 3$$

A triangular truss satisfies the relationship and is a perfect truss.

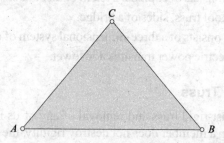

Fig. 8.5 Perfect truss

(b) Imperfect truss (unstable truss)

A truss which does not satisfy the above relationship is called imperfect truss. A truss which collapses when loaded is called unstable truss.

(c) Redundant truss (over-rigid truss)

In a redundant truss,

$$m > 2j - 3$$

Example: In Fig. 8.6 a redundant truss *ABCD* is shown.

No. of joints, $j = 4$

No. of members, $m = 6$

$$2j - 3 = 2 \times 4 - 3 = 5$$

Therefore, $m > 2j - 3$

One member is surplus and can be removed.

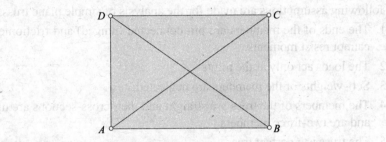

Fig. 8.6 Redundant truss

(d) Deficient truss

A truss in which $m < 2j - 3$ is called deficient truss and will collapse under loading.

Example: In Fig. 8.7, *ABCD* is a truss which has

Members, $m = 4$

Joints, $j = 4$

Now, $2j - 3 = 2 \times 4 - 3 = 5$

Therefore, one member is short.

To make the truss just-rigid or perfect, add a member *AC*.

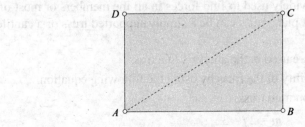

Fig. 8.7 Deficient truss

8.3 ANALYSIS OF SIMPLE PLANE TRUSS

When a truss is loaded, the loads are transferred to the supports and members are subjected to longitudinal forces (either tensile or compressive). The *solution* of simple truss consists of :

1. Computation of supporting reactions that must exist to keep the truss in equilibrium.
2. Computation of forces (in magnitude and direction) in each member of the truss so that every joint should be in equilibrium and the forces acting at every joint should form a system in equilibrium.

8.3.1 Assumptions

The following assumptions are made for the analysis of simple plane truss.

1. The ends of the members are pin-connected (hinged) and frictionless and cannot resist moments.
2. The loads act only at the joints.
3. Self-weights of the members are neglected.
4. The members of the truss are straight and their cross-sections are uniform and are two-force members.
5. The truss is a perfect truss.
6. The truss can be considered coplanar force system.
7. The truss is statically *determinate* and the equations of external loads, member forces and support reactions can be completely solved using equations of equilibrium only.

8.3.2 Methods of Analysis

Broadly there are two methods of analysis, namely, graphical as analytical. The analytical methods are:

(a) Method of Joints or Method of Resolution.
(b) Method of Sections.

(a) Method of Joints

This method is very widely used to find forces in all the members or most of the members of a truss. A plane truss can be a simply supported truss or a cantilever truss.

Following steps are used in the analysis of a truss.

(i) Check the stability of the truss by using the following equation.

For simply supported truss,

$$m = 2j - 3$$

For a cantilever truss,

$$m = 2j - 4$$

(ii) Draw free-body diagram of entire truss using the basic principles.

(iii) Use three equations of equilibrium to find out the reactions at the supports. Determination of support reactions may not be necessary in case of cantilever truss.

Example 8.1: Find out the support reactions of the truss shown.

Solution:

(a) Draw free-body diagram of the entire truss as shown.

(b) Apply equilibrium conditions.

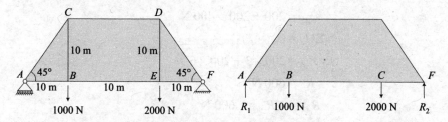

Fig. 8.8 Calculation of support reactions

A simply supported truss has one hinged support (F) and one roller support (A) to take care of thermal expansion and contraction.

$$\Sigma F_X = 0$$

$$\Sigma F_y = 0$$

∴ $$R_1 + R_2 = 1000 + 2000 = 3000$$

Taking moments about point A,

$$\Sigma M_A = 0$$

∴ $$R_1 \times 0 + R_2 \times 30 - 1000 \times 10 - 2000 \times 20 = 0$$

∴ $$R_1 = \frac{10,000 + 40,000}{30} = 1666 \text{ N}$$

and $$R_2 = 3000 - 1666 = 1334 \text{ N}$$

Example 8.2: Find the reactions of the cantilever truss shown.

Solution:

(a) Draw free-body diagram

(b) Apply equilibrium conditions

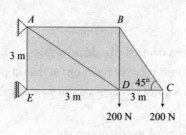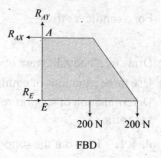

Fig. 8.9 Calculation of support reaction

$$\Sigma F_X = 0$$

$$\Sigma F_Y = 0$$

$$R_{AY} = 200 + 200 = 400 \text{ N}$$

$$\Sigma M_A = 0$$

$$R_E \times 3 = 200 \times 3 + 200 \times 6$$

$$\therefore \quad R_E = 600 \text{ N}$$

$$\therefore \quad R_{AX} = -R_E = -600 \text{ N}$$

(iv) Select a joint such that maximum two unknown forces exist.

(v) Draw FBD of selected joint by assuming forces in the members to be tensile.

(vi) The forces at the joint form a planar concurrent system in equilibrium. Apply equilibrium conditions:

$$\Sigma F_x = 0$$

$$\Sigma F_y = 0$$

(vii) Find out some forces by inspection.

(viii) Repeat the procedure by selecting different joints and find out forces in all members of the truss.

(ix) Tabulate the results in a Force Table using positive as tensile and negative as compressive.

Force Table

Sl. No.	Member	Magnitude	Nature

(x) Draw final FBD of truss showing magnitude and nature of all external and internal forces.

Important Rules

1. When two non-collinear members of a truss meet at a point and no external force or reaction act, then the force in both the members will be zero.

$$F_1 = F_2 = 0$$

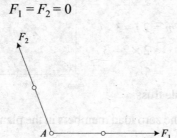

Fig. 8.10 Two-force joint

2. When two collinear members of a truss meet at a point, and a third member is connected at this point, then the force in the third member will be zero.

$$F_1 = F_2 \quad \text{and} \quad F_3 = 0$$

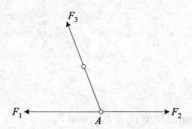

Fig. 8.11 Three-force joint

3. If a joint carries four forces with two pairs of collinear forces (cross joint), the collinear force must be equal and opposite.

$$F_1 = F_2 \quad \text{and} \quad F_3 = F_4$$

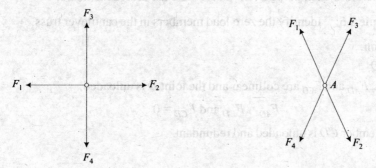

Fig. 8.12 Four-force joint

Example 8.3: Find whether the truss shown is a simple truss.

Solution:

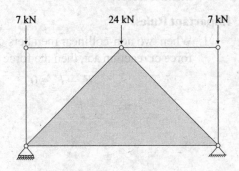

No. of joints, $j = 5$

No. of members, $m = 7$

For a simple truss,

$$m = 2j - 3$$
$$= 2 \times 2 - 3$$
$$= 7$$

Therefore, it is a simple truss.

Example 8.4: Identify the zero load members in the plane truss shown.

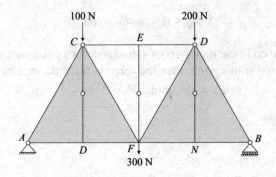

Solution:

Joint D

Three forces acting at joint D are F_{AD}, F_{FD} and F_{CD}.

Forces F_{AD} and F_{FD} are collinear and no external force or reaction is acting at joint D.

Therefore, $F_{AD} = F_{FD}$ and $F_{CD} = 0$

Similarly, forces F_{EF} and F_{GH} are zero.

Therefore, member, CD, EF and HG are zero force members.

Example 8.5: Identify the zero load members in the cantilever truss.

Solution:

Joint D

Forces, F_{AD} and F_{ED} are collinear and the joint D is unloaded.

$\therefore$ $F_{AD} = F_{ED}$ and $F_{CD} = 0$

Member CD is unloaded and redundant.

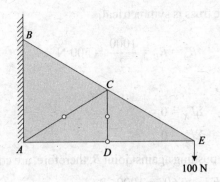

Joint C

Forces F_{BC} and F_{EC} are collinear and joint is unloaded ($F_{DC} = 0$)

$\therefore$ $$F_{AC} = 0$$

Member AC is unloaded.

Example 8.6: Find out forces in all the members of the truss ABC shown.

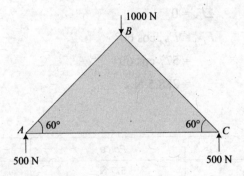

Solution:

(i) Check the stability of truss

No. of joints, $j = 3$

No. of members, $m = 3$

For a stable truss,

$$m = 2j - 3$$

$$= 2 \times 3 - 3 = 3$$

$\therefore$ The truss is stable.

(ii) *Reactions*. The truss is symmetrical,

$$\therefore \qquad R_A = R_B = \frac{1000}{2} = 500 \text{ N}$$

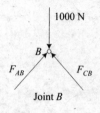

(iii) *FBD of joint B*

$$\Sigma F_X = 0$$
$$\Sigma F_Y = 0$$

The forces are pushing against joint *B*, therefore, are compressive.

$$F_{AB} \sin 60° + F_{CB} \sin 60° = 1000$$

But, $\qquad F_{AB} = F_{CB} = \dfrac{1000}{2 \sin 60°}$

$$= 577 \text{ N}$$

(iv) *FBD of joint A*

$$\Sigma F_X = 0$$
$$\Sigma F_Y = 0$$
$$F_{CA} = F_{BA} \cos 60°$$
$$= 577 \cos 60°$$
$$= 288.5 \text{ N}$$

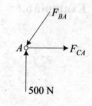

(v) *Force Table*

Member	Force	Nature
AB	577 N	C
BC	577 N	C
AC	288.5 N	T

(vi) *Force diagram*

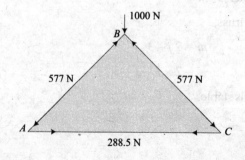

(b) Method of Section

This method is useful for calculating forces in some members of the truss and avoid laborious process of proceeding joint by joint reaching a joint on which the desired unknown force acts.

Following steps are used in the analysis of a truss.

(i) Check the stability of a truss.

(ii) Draw FBD and calculate reactions by applying three conditions of equilibrium.

(iii) Select a section, cutting maximum three members (maximum three unknown forces) including the member in which force is to be determined.

(iv) Draw FBD of any cut portion of truss and show forces in the cut members as towards the section (away from joints).

The FBD should include only external forces acting on that part and internal tensile forces in the cut members.

(v) Apply equations of equilibrium for a coplanar forces system.

Important Rules

1. A section can cut more than three members but the unknown forces should not exceed three.

2. In case of cantilever truss, draw FBD of part not containing the wall.

Example 8.7: Calculate the force in the member *CE* only.

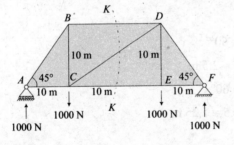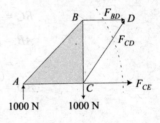

Solution:

1. Draw FBD of entire truss.

2. Reaction $R_A = R_B = 1000$ N by symmetry.

3. Cut section *K-K*.

4. Cut FBD of position *ABCD* showing the internal tensile forces F_{BD}, F_{CD} and F_{CE}.

5. Apply the condition,

$$\Sigma M_D = 0$$

Forces F_{BD} and F_{CD} pass through joint D.

$\therefore$
$$F_{CE} \times DE + 1000 \times CE - 1000 \times AE = 0$$
$$F_{CE} \times 10 + 1000 \times 10 - 1000 \times 20 = 0$$

$\therefore$
$$F_{CE} = 1000 \text{ N } (T).$$

Example 8.8: Find the forces in the members AB, AC and BC of the truss shown.

[U.P.T.U. II Sem., 2002-03]

Solution:

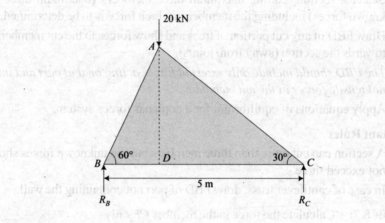

1. Dimensions

$$BC = 5 \text{ m}$$
$$AB = BC \cos 60°$$
$$= 5 \times 0.5 = 2.5 \text{ m}$$
$$BD = AB \cos 60°$$
$$= 2.5 \times 0.5 = 1.25 \text{ m}$$

2. Reactions

$$\Sigma F_y = 0$$
$$R_A + R_C = 20 \text{ kN}$$
$$\Sigma M_B = 0$$

$\therefore$
$$R_C \times 5 = 20 \times 1.25$$
$$R_C = \frac{20 \times 1.25}{5} = 5 \text{ kN}$$

$\therefore$
$$R_B = 20 - 5 = 15 \text{ kN}$$

3. *Consider Joint B*

  (i) Draw FBD

  (ii) Apply equilibrium conditions

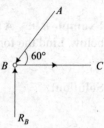

FBD of joint *B*

$$\Sigma F_X = 0$$

$$\therefore \quad F_{CB} - F_{AB} \cos 60° = 0$$

$$\Sigma F_Y = 0$$

$$\therefore \quad F_{AB} \sin 60° - R_B = 0$$

$$\therefore \qquad F_{AB} = \frac{R_B}{\sin 60°} = \frac{15}{0.866} = 17.32 \text{ kN } (C)$$

$$F_{CB} = F_{AB} \sin 60°$$

$$= 17.32 \times 0.5 = 8.66 \text{ kN } (T)$$

4. *Consider Joint C*

  (i) Draw FBD.

  (ii) Apply equilibrium conditions

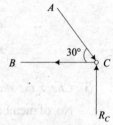

FBD of joint *C*

$$\Sigma F_X = 0$$

$$F_{AC} \cos 30° - F_{BC} = 0$$

$$\therefore \quad F_{AC} = \frac{F_{BC}}{\cos 30°} = \frac{8.66}{0.866} = 10 \text{ kN } (C)$$

Force Table

Member	Force
AB	17.32 kN (C)
AC	10.00 kN (C)
BC	8.66 kN (T)

Force Diagram

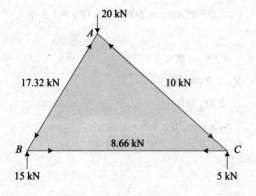

Example 8.9: A truss having a span of 6 m, carries a load of 30 kN and is shown below. Find the forces in members, *AB*, *AC*, *BC* and *AD*.

Solution:

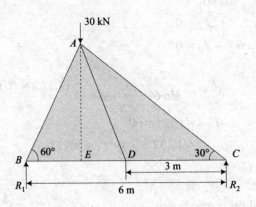

(i) *Check the stability of the truss,*

No. of members, $m = 5$

No. of joints, $j = 4$

For a stable truss,

$$m = 2j - 3$$
$$= 2 \times 4 - 3$$
$$= 5$$

Therefore, the truss is stable and just rigid.

(ii) *Dimensions*

Draw a vertical *AE*,

$$BD = 3 \text{ m and } BC = 6 \text{ m}$$
$$AB = BC \cos 60° = 6 \times 0.5 = 3 \text{ m}$$
$$BE = AB \cos 60° = 3 \times 0.5 = 1.5 \text{ m.}$$

(iii) *Reactions*

$$\Sigma F_Y = 0$$

∴

$$R_1 + R_2 = 30$$

$$\Sigma M_B = 0$$

∴

$$R_2 \times 6 = 30 \times 1.5$$

$$\therefore \quad R_2 = \frac{3.0 \times 1.5}{6} = 7.5 \text{ kN}$$

$$R_1 = 30 - 7.5 = 22.5 \text{ kN}$$

(iv) *Joint B*

Consider equilibrium of joint *B*. Draw its FBD,

$$\Sigma F_X = 0$$

$$\therefore \quad F_{DB} - F_{AB} \cos 60° = 0$$

$$\Sigma F_Y = 0$$

$$\therefore \quad F_{AB} \sin 60° = R_1 = 22.5$$

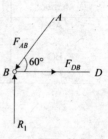

$$\therefore \quad F_{AB} = \frac{22.5}{\sin 60°} = \frac{22.5}{0.866} = 26 \text{ kN}$$

FBD of joint *B*

$$\therefore \quad F_{DB} = 26 \times 0.5 = 13 \text{ kN}$$

(v) *Joint C*

Consider equilibrium of joint *C*. Draw its FBD,

$$\Sigma F_X = 0$$

$$F_{DC} - F_{AC} \cos 30° = 0$$

$$\Sigma F_Y = 0$$

$$F_{AC} \sin 30° = R_2 = 7.5$$

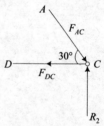

$$\therefore \quad F_{AC} = \frac{7.5}{0.5} = 15 \text{ kN}$$

FBD of joint *C*

$$\therefore \quad F_{DC} = 15 \times 0.866 = 13 \text{ kN}$$

(vi) *Joint A*

Consider equilibrium of joint *A* and *B* draw its FBD,

From geometry,

$$\angle BAD = 60°$$

$$\angle CAD = 30°$$

$$\therefore \quad \angle BAE = 30°$$

$$\angle CAE = 60°$$

$$\Sigma F_X = 0$$

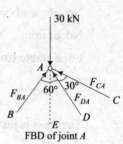

FBD of joint *A*

$$F_{BA} \cos 30° + F_{DA} \cos 30° + F_{CA} \cos 60° = 30$$

$$26 \times 0.866 + F_{DA} \times 0.866 + 15 \times 0.5 = 30$$

$$\therefore \quad F_{DA} = 30 - 22.5 - 7.50 = 0$$

(vii) The force table and force diagram are shown,

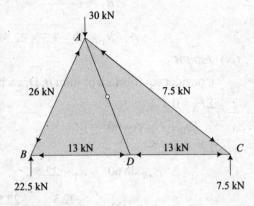

Member	Force
AB	26 kN (C)
AC	7.5 kN (C)
BC	13 kN (T)
AD	0

Example 8.10: Each member of the following truss is 2 m long. The truss is simply supported at the ends. Determine forces in all members clearly showing whether they are in tension or compression. [U.P.T.U. I Sem., 2005-06]

Solution:

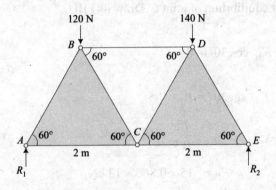

(i) *Check the stability of the truss,*

No. of members, $m = 7$

No. of joints, $j = 5$

For a stable truss,

$$m = 2j - 3$$
$$= 2 \times 5 - 3 = 7$$

∴ The truss is just-rigid and stable.

(ii) *Reactoins*

$$\Sigma F_Y = 0$$

∴ $$R_1 + R_2 = 120 + 140 = 260 \text{ kN.}$$

$\Sigma M_A = 0$

$$R_2 \times 4 = 120 \times 1 + 140 \times 3 = 540$$

$\therefore$ $\qquad R_2 = 135 \text{ kN}$

$$R_1 = 260 - 135 = 125 \text{ kN}$$

(iii) *Joint A*

Consider the equilibrium of joint A and draw its FBD.

$\Sigma F_Y = 0$

$$F_{BA} \sin 60° = R_1 = 125$$

$\therefore$ $\qquad F_{BA} = \dfrac{125}{0.866} = 144.3 \text{ kN } (C)$

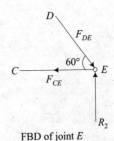

FBD of joint A

$\Sigma F_X = 0$

$$F_{CA} = F_{BA} \cos 60°$$
$$= 144.3 \times 0.5 = 72.15 \text{ kN } (T)$$

(iv) *Joint E*

Consider the equilibrium of joint E and draw its FBD.

$\Sigma F_Y = 0$

$$F_{DE} \sin 60° = R_1 = 135$$

$\therefore$ $\qquad F_{DE} = \dfrac{135}{\sin 60°} = 155.88 \text{ kN } (C)$

FBD of joint E

$\Sigma F_X = 0$

$$F_{CE} = F_{DE} \cos 60°$$
$$= 155.88 \times 0.5 = 77.94 \text{ kN } (T)$$

(v) *Joint C*

Consider the equilibrium of joint C and draw its FBD.

$$F_{AC} = 72.15 \text{ kN } (C)$$
$$F_{EC} = 77.94 \text{ kN } (T)$$

$\Sigma F_Y = 0$

$$F_{BC} \cos 30° + F_{DE} \cos 30° = 0$$
$$F_{BC} = -F_{DC}$$

$\Sigma F_X = 0$

$$-F_{AC} + F_{EC} = F_{BC} \cos 60° - F_{DC} \cos 60°$$

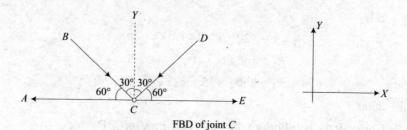

FBD of joint C

$$F_{BC} = \frac{F_{EC} - F_{AC}}{2 \cos 60°} = \frac{77.94 - 72.15}{2 \times 0.5} = 5.79 \text{ kN } (C)$$

$$\therefore \qquad F_{DC} = 5.79 \text{ kN } (T)$$

(vi) *Joint D*

Consider the equilibrium of joint D and draw its FBD,

$$F_{CD} = 5.79 \text{ kN } (T)$$

$$F_{ED} = 155.88 \text{ kN } (C)$$

$$\Sigma F_X = 0$$

$$F_{BD} = -F_{CD} \cos 60° - F_{ED} \cos 60°$$

$$= -5.79 \times 0.5 + 155.88 \times 0.5$$

$$= 75 \text{ kN } (C)$$

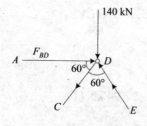
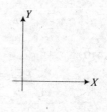

FBD of joint D

(vii) *Force Table and Diagram*

Member	Force
AB	144.3 kN (C)
BC	5.79 kN (C)
CD	5.79 kN (T)
DE	155.88 kN (C)
CE	77.94 kN (T)
AC	72.15 kN (T)

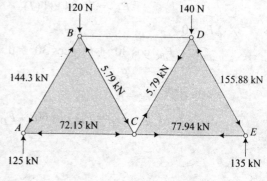

Example 8.11: Determine the forces in all the members of the cantilever truss shown.

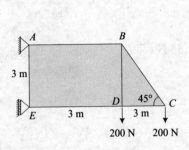
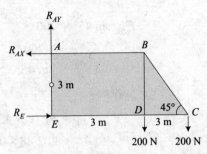

FBD of truss

Solution:

(i) Draw FBD of the truss.

(ii) Check the stability of the truss,

No. of members, $m = 6$

No. of joints, $j = 5$

For a stable truss,

$$m = 2j - 4 = 2 \times 5 - 4 = 6$$

The truss is stable.

(iii) *Reactions*

$$\Sigma F_Y = 0$$

$$\therefore \quad R_{AY} = 200 + 200 = 400 \text{ N}$$

$$\Sigma M_A = 0$$

$$R_E \times 3 = 200 \times 3 + 200 \times 6$$

$$R_E = 600 \text{ N}$$

$$R_{AX} = -R_E = -600 \text{ N}$$

(iv) *Joint E*

Consider the equilibrium of joint E and draw its FBD.

$$\Sigma F_X = 0$$

$$\therefore \quad F_{DE} = R_E = 600 \text{ N } (C)$$

$$F_{AE} = 0$$

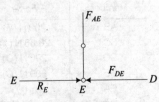

FBD of joint E

(v) *Joint C*

Consider the equilibrium of joint C and draw its FBD.

$$\Sigma F_Y = 0$$

$$F_{BC} \sin 45° = 200$$

$$F_{BC} = \frac{200}{\sin 45°} = 282.88 \text{ N } (T)$$

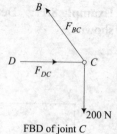

FBD of joint C

$$\Sigma F_X = 0$$

$$F_{DC} = F_{BC} \cos 45°$$

$$= 282.88 \times 0.707$$

$$= 200 \text{ N } (C)$$

(vi) *Joint B*

Consider the equilibrium of joint B and draw its FBD,

$$\Sigma F_Y = 0$$

$$F_{DB} = F_{CB} \cos 45°$$

$$= 282.88 \times 0.707$$

$$= 200 \text{ N } (C)$$

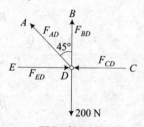

FBD of joint B

$$\Sigma F_X = 0$$

$$F_{AB} = F_{CB} \cos 45° = 200 \text{ N } (T)$$

(vii) *Joint D*

Consider the equilibrium of joint D and draw its FBD,

$$\Sigma F_X = 0$$

$$F_{AC} \sin 45° + F_{ED} - F_{CD} = 0$$

$$F_{AD} \sin 45° + 600 - 200 = 0$$

$$\therefore \quad F_{AC} = \frac{400}{\sin 45°} = 565.77 \ (T)$$

(vii) *Draw force table and diagram*

FBD of joint D

Member	Force
AB	200 N (T)
BC	28.88 N (T)
CD	200 N (C)
DE	600 N (C)
EA	0
DA	565.77 N (T)
DB	200 N (C)

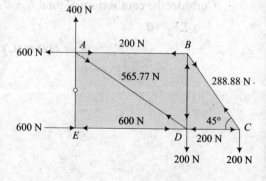

Example 8.12: Determine the support reactions and nature and magnitude of forces shown in the truss. [U.P.T.U I Sem., 2001-02]

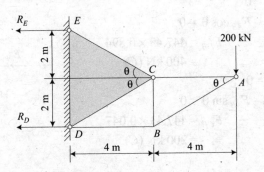

Solution:

(i) Check the stability of cantilever truss,

No. of members (excluding wall), $m = 6$

No. of joints = 5

For stability of cantilever truss,

$$m = 2j - 4 = 2 \times 5 - 4 = 6$$

$\therefore$ The truss is stable.

(ii) *Geometrical configuration*

$$AB = \sqrt{AC^2 + BC^2} = \sqrt{4^2 + 2^2} = 4.472 \text{ m}$$

$$\sin\theta = \frac{BC}{AB} = \frac{2}{4.472} = 0.447$$

$$\cos\theta = \frac{AC}{AB} = \frac{2}{4.472} = 0.892$$

(iii) *Pin A*

Consider the equilibrium of pin A and draw its FBD,

$$\Sigma F_X = 0$$

$$F_{BA} \cos\theta = F_{CA}$$

$$\Sigma F_y = 0$$

$$F_{BA} \sin\theta = 200$$

$$F_{BA} = \frac{200}{0.447} = 447.43 \text{ kN (C)}$$

$$F_{CA} = F_{BA} \cos\theta = 447.43 \times 0.84 = 400 \text{ kN } (T)$$

FBD of pin A

(iv) *Pin B*

Consider the equilibrium of pin B and draw its FBD,

$$\Sigma F_X = 0$$
$$F_{DB} - F_{AB}\cos\theta = 0$$
$$F_{DB} = 447.43 \times 0.894$$
$$= 400 \text{ kN } (C)$$

$$\Sigma F_Y = 0$$
$$F_{CB} - F_{AB}\sin\theta = 0$$
$$F_{CB} = 447.43 \times 0.047$$
$$= 200 \text{ kN } (C)$$

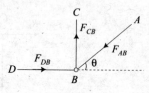

FBD of pin B

(v) *Pin C*

Consider the equilibrium of pin C and draw its FBD,

$$\Sigma F_X = 0$$
$$F_{AC} - F_{DC}\cos\theta - F_{EC}\cos\theta = 0$$

$$F_{DC} + F_{EC} = \frac{400}{0.894} = 447.427 \text{ kN} \qquad \ldots(1)$$

$$\Sigma F_Y = 0$$
$$F_{EC}\sin\theta - F_{DC}\sin\theta - F_{BC} = 0$$

$$F_{EC} - F_{DC} = \frac{F_{BC}}{\sin\theta} = \frac{200}{0.447} = 447.427 \text{ kN} \qquad \ldots(2)$$

FBD of pin C

From equations (1) and (2)

$$F_{EC} = 447.427 \text{ kN } (T)$$
$$F_{DC} = 0$$

(vi) *Joint D*

Consider the equilibrium of joint D and draw its FBD,

$$\Sigma F_X = 0$$

$$\therefore \qquad R_D = F_{BD} = 400 \text{ kN}$$
$$\therefore \qquad R_E = -R_D = -400 \text{ kN}$$

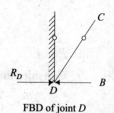

FBD of joint D

(vii) *Draw force table and diagram*

Member	Force
AC	400 kN (T)
AB	447.43 kN (C)
BC	200 kN (T)
BD	400 kN (C)
CD	0
CE	447.43 kN (T)

Example 8.13: A cantilever truss is loaded and supported as shown. Find the value of load P which would produce an axial force of magnitude 3 kN in the member AC using methods of section. [U.P.T.U. II Sem., 2002-03]

Solution: Let section K-K cut the members AC, AD and BD. Consider right hand part of cut.

Take moments about point C through which unknown forces F_{AD} and F_{BD} are passing.

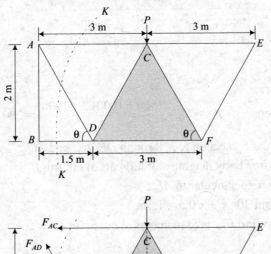

$$\Sigma M_D = 0$$
$$P \times 1.5 + F_{BD} \times 0 + F_{AD} \times 0 = F_{AC} \times 2$$
$$1.5\,P = 2\,F_{AC}$$
$$F_{AC} = 3 \text{ kN (given)}$$

$\therefore$
$$P = \frac{3 \times 2}{1.5} = 4 \text{ kN}$$

Example 8.14: Find the forces in the members DE, DC and AC respectively in the roof truss shown.

Solution:

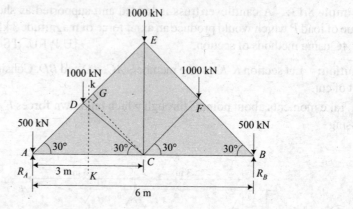

(i) *Reactions*

By symmetry, $R_A = R_B = \dfrac{(500 + 1000)\,2 + 1000}{2} = 200$ kN.

(ii) Cut the truss by section *K-K* through *DE*, *DC* and *AC*.

(iii) Draw the free body diagrams of left part of cut truss.

(iv) Draw *CG* perpendicular to *AE*.

$CG = AC \sin 30° = 3 \times 0.5 = 1.5$ m

(v) In order to find F_{DE} take members about point *C*.

$$\Sigma M_C = 0$$

$\therefore \qquad F_{DE} \times CG + 1000 \times 1.5 + 500 \times 3 = 2000 \times 3$

$\qquad\qquad F_{DE} \times 1.5 + 1000 \times 1.5 + 500 \times 3 = 2000 \times 3$

$\therefore \qquad\qquad F_{DE} = -2000$ kN (C)

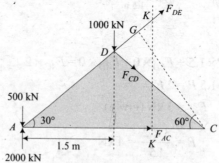

FBD of cut section *K-K*

(vi) In order to find F_{CD}, take moments about point *A*.

$$F_{CD} \times AG + 1000 \times 1.5 = 0$$

$$F_{CD} \times 3 \cos 60° + 1000 \times 1.5 = 0$$

$\therefore \qquad\qquad F_{CD} = -1000$ kN (C)

(vii) In order to find F_{AC} take moments about points D,

$$F_{AC} \times 1.5 \tan 30° + 500 \times 1.5 = 2000 \times 1.5$$

∴ $F_{AC} = 2598$ kN (T)

Example 8.15: It is desired to know the force in the member AB of the plane truss shown. The supporting forces have been determined and shown in the diagram.

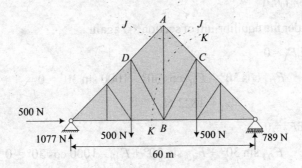

Solution:

  (i) Draw FBD of section J-J.

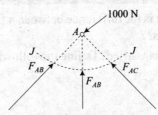

FBD of cut section J-J

There are three unknown forces F_{DA}, F_{AB} and F_{AC}, but only two equilibrium equations are available.

$$\Sigma F_X = 0 \text{ and } \Sigma F_Y = 0$$

Therefore, cut section K-K.

  (ii) Draw FBD of section K-K. Take moments about point B.

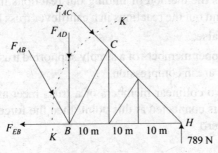

FBD of section K-K

$$\Sigma M_B = 0$$

$$(-500)(10) + (789)(30) - F_{AC}(\sin 30°)(30) = 0$$

The force F_{AC} has been transmitted to joint H.

$$\therefore \qquad F_{AC} = 1245 \text{ N}.$$

(iii) *Section (J-J)*

Consider the equilibrium of section J-J again

$$\Sigma F_X = 0$$

$$\therefore \qquad F_{DA} \cos 30° - F_{AC} \cos 30° - 1000 \sin 30° = 0$$

$$F_{DA} = 1822 \text{ N}$$

$$\Sigma F_Y = 0$$

$$F_{DA} \sin 30° + F_{AC} \sin 30° + F_{AB} - 1000 \cos 30° = 0$$

$$\therefore \qquad F_{AD} = -667 \text{ N (Tension member)}$$

QUESTION BANK NO. 8

1. Define a truss. What is the difference between a frame and truss? Where do you find trusses in use?

2. How the trusses are classified? Define the following trusses:
   (i) Perfect truss
   (ii) Imperfect truss
   (iii) Deficient truss
   (iv) Redundant truss

3. What is a simple truss? What assumptions are made in the analysis of a simple truss?

4. What are various methods of analysis of trusses? What is basically found when analysis of truss is done? What is the advantage of method of section over the method of joint?

5. What is the difference between a simply supported truss and a cantilever truss? Discuss the method of finding out reactions in both the cases. Is it essential to find out the reactions in a cantilever truss before analysing it?

6. State true or false:
   (i) All the upper members of a simply supported truss carrying only vertical loads are in compression.
   (ii) When two collinear members of a truss meet at a point, and a third member is connected at this point, then the force in the third member will be zero.

(iii) When two collinear members of a truss meet at a point and no external force or reaction act at this point, then the force in both the members will be zero.

(iv) Equation of equilibrium can be used to determine forces in the members of statically determinate truss.

PART III: STRESS AND STRAIN ANALYSIS (STRENGTH OF MATERIALS)

Simple and Compound Stresses and Strains

9.1 INTRODUCTION

The knowledge of strength of materials is very essential for the design of machines and structures. It has been observed that when a force is applied to a material, it first deforms and then failure takes place. Different types of metallic and non-metallic materials are available for use. The ability of a material to resist failure under the action of external forces or loads is called the strength of material.

9.1.1 Strength or Internal Resistance of Material

If a bar AB is pulled by an external force P' at both ends, the bar will be elongated (though negligible and may not be visible by naked eye) and will be under threat of breakage say at C. For equilibrium of portion of bar AC, there should be a force

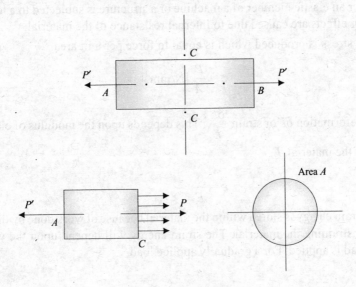

Fig. 9.1 Internal resistance

equal and opposite to force P' applied at section C. This force P is developed due to internal cohesion and attraction among molecules of the material. This is called internal resistance of the material and is measured as force per unit area.

$$P' = P = \sigma A$$

where
A = Area of cross-section of bar

σ = Internal resistance called stress.

If P developed is less than applied force P' failure will take place. The internal resistance (or strength) σ of the material depends upon the properties of the material and the total resistance, P depends upon σ and area of cross-section, A.

9.1.2 Elasticity

If external force, P' is removed, the internal resistance σ disappears and the bar springs back to its original length. This property of a material to spring back to its original size is called elasticity.

If there is no residual deformation (size and shape) left in the body after removal of external force, the material is called perfectly elastic. If there is some residual deformation left in the material after removal of force, it is called partially elastic. In most of the cases, elastic materials are used in the construction of machines and structures.

The strength and elasticity are exclusive properties of a material and depend upon the arrangement and distribution of molecules in the crystal lattice of the material.

9.2 LOADING OF AN ELASTIC MATERIAL

Whenever an elastic member of a machine or a structure is subjected to a load, the following effects are caused due to internal resistance of the material.

1. A stress is produced which is equal to force per unit area.

$$\sigma = \frac{P}{A} \ [\text{N/mm}^2]$$

2. Deformation δL or strain $\dfrac{\delta L}{L}$. This depends upon the modulus of elasticity of the material, E.

$$\varepsilon = \frac{\sigma}{E}$$

3. Strain energy is stored within the material because of work done by the force in straining the material. The strain energy will depend upon the way the load is applied. For a gradually applied load,

Strain energy = Work done

$$U = \frac{1}{2}\text{stress} \times \text{strain}$$

$$= \frac{\sigma\varepsilon}{2}.$$

9.2.1 Stress

The external force, P can be measured in N, kN, MN or GN.

where, $\text{kN} = 10^3 \text{ N (KILO)}$

$\text{MN} = 10^6 \text{ N (MEGA)}$

$\text{GN} = 10^9 \text{ N (GIGA)}$

Similarly, depending upon magnitude of the force, the area of cross-section of the material may be measured as mm^2 or cm^2 or m^2.

Therefore, stress is measured as N/mm^2 or MN/m^2, etc.

where, $1 \text{ N/mm}^2 = 10^6 \text{ N/m}^2 = 1 \text{ MN/m}^2 = 1 \text{ MPa}$.

and $1 \text{ N/m}^2 = 1 \text{ Pascal} = 1 \text{ Pa}$.

Types of Stress

The stress can be classified as simple stress or compound stress, normal (direct) stress and shear (tangential) stress and direct stress as tensile stress or compressive stress.

1. **Simple stress:** If the external load is applied on a body in one direction only, the stress developed is called simple stress.
2. **Compound stress:** If the external loads are applied in more than one direction, the stresses developed are called compound stresses.
3. **Normal stress (σ):** When a force acts longitudinally along the axis of the body, it produces a normal or direct stress which acts perpendicularly on the surface on which force is acting.
4. **Shear stress (τ):** When a force acts tangentially on the surface of the body, it produces shear stress parallel to the surface on which the force acts.
5. **Tensile stress:** If the length of the body increases due to the action of normal force or longitudinal force, the stress produced is called tensile stress and force producing it is called tensile force or a pull.

$$\sigma_t = \frac{P}{A}.$$

6. **Compressive stress:** If the length of the body decreases due to the action of the normal force or longitudinal force, the stress produced is called compressive stress and the force producing it is called compressive force or a push.

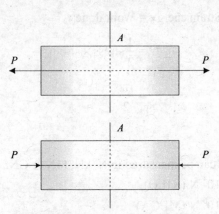

Fig. 9.2 Tensile and compressive stresses

9.2.2 Strain

There is a change in the length or dimension of the body subjected to external force or load. The ratio of change in length (δL) to the original length (L) of the material is called strain.

$$\varepsilon = \frac{\delta L}{L}$$

Tensile strain is the ratio of increase of length to its original length when subjected to tensile loading.

Compressive strain is the ratio of decrease in length to the original length of the material when subjected to compressive loading.

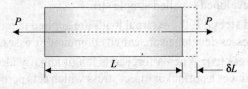

Fig. 9.3 Strain

9.2.3 Modulus of Elasticity

Hooke's law states that when a material is loaded within its elastic limit, the stress is proportional to strain. If a material is loaded within elastic limit, the body will return to its original size when load is removed.

$$\frac{\text{Stress}}{\text{Strain}} = \text{a constant} = E \qquad \text{(say)}$$

$$\therefore \qquad E = \frac{\sigma}{\varepsilon}$$

E is called modulus of elasticity or Young's modulus and has the same unit as stress. The value of E is more for ductile materials than for brittle materials. It is same for tensile and compressive loading.

Deformation of a body under external load

Stress, $\sigma = \dfrac{P}{A}$

Strain, $\varepsilon = \dfrac{\delta L}{L}$

Young's modulus,

$$E = \frac{\sigma}{\varepsilon}$$

$\therefore \qquad \varepsilon = \dfrac{\sigma}{E} = \dfrac{P}{AE}$

Now, $\qquad \varepsilon = \dfrac{\sigma L}{L}$

Deformation,

$$\delta L = \varepsilon L = \frac{\sigma L}{E} = \frac{PL}{AE}$$

The values of E for some important materials are given in Table 9.1.

Table 9.1 Young's modulus for engineering materials

Material	Young's modulus (E) (N/mm^2)
1. Mild steel	2.1×10^5
2. Cast iron	1.3×10^5
3. Aluminium	0.7×10^5
4. Copper	1.0×10^5
5. Timber	0.1×10^5

Young's modulus is a measure of stiffness of material or ductility of the material.

9.2.4 Strain Energy

Strain energy for a gradually applied load can be found out from the product of average force and deformation or area under the load-extension graph (shaded area).

$$U = \frac{P}{2}.\delta L$$

where, $\qquad P = \sigma.A$

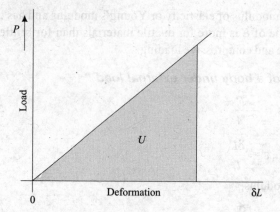

Fig. 9.4 Strain energy

and
$$\delta L = \varepsilon L = \frac{\sigma L}{E}$$

∴
$$U = \frac{1}{2}.\sigma A.\frac{\sigma L}{E} = \frac{1}{2}\frac{\sigma^2}{E}.AL$$

$$= \frac{1}{2}\frac{\sigma^2}{E} \times \text{Volume}.$$

The strain energy per unit volume is called resilience.

$$\text{Resilience} = \frac{1}{2}\frac{\sigma^2}{E}$$

$$= \frac{1}{2}.\sigma.\frac{\sigma}{E} = \frac{1}{2}\sigma.\varepsilon$$

∴
$$\text{Resilience} = \frac{\text{stress} \times \text{strain}}{2}$$

Resilience is a measure of toughness of a material.

Example 9.1: Find the diameter of a steel wire to raise a load of 10 kN, if the minimum stress is not to exceed 100 MN/m².

Solution:

$$\text{Load}, P = 10 \text{ kN} = 10 \times 10^3 \text{ N}.$$

Allowable stress, $\sigma = 100 \text{ MN/m}^2 = 100 \text{ N/mm}^2$

$$\text{Area}, A = \frac{P}{\sigma} = \frac{10 \times 10^3}{100} = 100 \text{ mm}^2$$

Now, $\qquad A = \dfrac{\pi}{4}\, d^2.$

Diameter of wire,

$$d = \sqrt{\dfrac{4A}{\pi}} = \sqrt{\dfrac{4 \times 100}{\pi}} = \mathbf{11.28\ mm} \quad \mathbf{Ans.}$$

Example 9.2: A steel wire 2 m long and 3 mm in diameter is extended by 0.75 mm due to a weight suspended from the wire. If the same weight is suspended from a brass wire, 2.5 m long and 2 mm in diameter, it is elongated by 5 mm. Determine the modulus of elasticity of brass, if that of steel is 2×10^5 N/mm^2.

Solution:

(i) *For steel wire*

$$L = 2\ m = 2000\ mm$$
$$d = 3\ mm$$
$$\delta L = 0.75\ mm$$
$$E = 2 \times 10^5\ N/mm^2$$

Area of wire,

$$A = \dfrac{\pi}{4} d^2 = \dfrac{\pi}{4}(3)^2$$
$$= 7.07\ mm^2$$

Strain, $\qquad \varepsilon = \dfrac{\delta L}{L} = \dfrac{0.75}{2000}$
$$= 0.000375$$

Stress, $\qquad \sigma = \varepsilon E = 0.000375 \times 2 \times 0^5$
$$= 75\ N/mm^2$$

Weight, $\qquad P = \sigma A = 75 \times 7.07$
$$= 530.\,25\ N.$$

(ii) *For brass wire*

$$L = 2.5\ m = 2500\ mm$$
$$d = 2\ mm$$
$$\delta L = 5\ mm$$
$$P = 530.25\ N$$

Strain,

$$\varepsilon = \dfrac{\delta L}{L} = \dfrac{5}{2500}$$
$$= 2 \times 10^{-3}$$

Stress, $\qquad \sigma = \dfrac{P}{A} = \dfrac{530.25}{\dfrac{\pi}{4}(2)^2}$

$\qquad\qquad\qquad = 168.78 \text{ N/mm}^2$

Modulus of elasticity,

$$E = \frac{\sigma}{\varepsilon} = \frac{168.78}{2 \times 10^{-3}}$$

$$= \mathbf{0.8439 \times 10^5 \, N/mm^2} \quad \mathbf{Ans.}$$

Example 9.3: A circular rod of diameter 20 mm and 500 mm long is subjected to a tensile load of 50 kN. The modulus of elasticity for steel may be taken as 200 kN/mm². Find the stress, strain and elongation of bar due to applied load.

Solutions:

$$d = 20 \text{ mm}$$
$$L = 500 \text{ mm}$$
$$P = 50 \text{ kN} = 50 \times 10^3 \text{ N.}$$
$$E = 200 \text{ kN/mm}^2 = 2 \times 10^5 \text{ N/mm}^2$$

Area of rod,

$$A = \frac{\pi}{4}(d)^2 = \frac{\pi}{4}(20)^2$$

$$= 314.16 \text{ mm}^2.$$

Stress, $\qquad \sigma = \dfrac{P}{A} = \dfrac{50 \times 10^3}{314.16}$

$$= \mathbf{159.15 \, N/mm^2} \quad \mathbf{Ans.}$$

Strain, $\qquad \varepsilon = \dfrac{\sigma}{E} = \dfrac{159.15}{2 \times 10^5}$

$$= \mathbf{0.00079577} \quad \mathbf{Ans.}$$

Elongation, $\qquad \delta L = \varepsilon L = 0.00079577 \times 500$

$$= \mathbf{0.39785 \, mm} \quad \mathbf{Ans.}$$

Example 9.4: Calculate the strain energy stored in a bar 250 cm long, 5 cm wide and 4 cm thick, when it is subjected to a tensile load of 6×10^4 N. Take $E = 2 \times 10^5 \text{ N/mm}^2$.

Solution:

$$L = 250 \text{ cm} = 2500 \text{ mm}$$
$$A = 5 \text{ cm} \times 4 \text{ cm} = 50 \text{ mm} \times 40 \text{ mm}$$
$$= 2000 \text{ mm}^2$$

$$P = 6 \times 10^4 \, N$$
$$E = 2 \times 10^5 \, N/mm^2$$

The extension,

$$\delta L = \frac{PL}{AE} = \frac{6 \times 10^4 \times 2500}{2000 \times 2 \times 10^5}$$

$$= 3.75 \, mm$$

The strain energy,

$$U = \frac{P\delta L}{2} = \frac{6 \times 10^4 \times 3.75}{2}$$

$$= 112500 \, N\text{-}m$$

$$= 112.5 \, kN\text{-}m$$

$$= \mathbf{112.5 \, kJ} \quad \mathbf{Ans.}$$

Example 9.5: The piston of a steam engine is 300 mm in diameter and the piston rod is of 50 mm diameter. The steam pressure is 1 N/mm². Find the stress in the piston rod and elongation in a length of 800 mm. Take $E = 200$ GPa.

Solution:

$$D = 300 \, mm$$

$$d = 50 \, mm$$

$$p = 1 \, N/mm^2$$

$$L = 800 \, mm$$

$$E = 200 \, GPa = 20 \times 10^9 \, Pa$$

$$= 2 \times 10^5 \, N/mm^2 \,.$$

The effective area of piston for tensile loading of rod,

$$A = \frac{\pi}{4}(D^2 - d^2) = \frac{\pi}{4}(300^2 - 50^2)$$

$$= 68722.34 \, mm^2.$$

Force, $P = pA = 1 \times 68722.34$

$$= 68722.34 \, N$$

Area of piston rod,

$$A_r = \frac{\pi}{4}d^2 = \frac{\pi}{4}(50)^2$$

$$= 1963.5 \, mm^2$$

Stress in piston rod,

$$\sigma = \frac{P}{A_r} = \frac{68722.34}{1963.5}$$

$$= 35 \text{ N/mm}^2 \quad \textbf{Ans.}$$

Elongation of piston rod,

$$\delta L = \frac{PL}{A_r E} = \frac{68722.34 \times 800}{1963.5 \times 2 \times 10^5}$$

$$= 0.14 \text{ mm} \quad \textbf{Ans.}$$

Example 9.6: Three bars of equal length and having cross-sectional areas in the ratio 1:2:4 are all subjected to equal load. Compare their strain energy.

[U.P.T.U. I Sem. 2003-04]

Solution:

The strain energy,

$$U = \frac{P^2 L}{2 AE}$$

$$A_1 = p$$
$$A_2 = 2A_1$$
$$A_3 = 4A_1$$

$\therefore$

$$U_1 = \frac{P^2 L}{2 A_1 E}$$

$$U_2 = \frac{P^2 L}{2 A_2 E} = \frac{P^2 L}{4 A_1 E} = \frac{U_1}{2}$$

$$U_3 = \frac{P^2 L}{2 A_3 E} = \frac{P^2 L}{8 A_1 E} = \frac{U_1}{4}$$

$\therefore$

$$U_1 : U_2 : U_3 = 1 : \frac{1}{2} : \frac{1}{4} \quad \textbf{Ans.}$$

Example 9.7: A metallic rectangular rod 1.5 m long and 40 mm wide and 25 mm thick is subjected to an axial tensile load of 120 kN. The elongation of the rod is measured as 0.9 mm. Calculate the stress, strain and modulus of elasticity.

[U.P.T.U. II Sem. 2003-04]

Solution:

$$L = 1.5 \text{ m} = 1500 \text{ mm}$$

$$A = 40 \text{ mm} \times 25 \text{ mm} = 1000 \text{ mm}^2$$

$$P = 120 \text{ kN} = 120 \times 10^3 \text{ N}$$

$$\delta L = 0.9 \text{ mm}$$

Stress, $\qquad \sigma = \dfrac{P}{A} = \dfrac{120 \times 10^3}{1000}$

$$= \mathbf{120 \ N/mm^2} \quad \textbf{Ans.}$$

Strain, $\qquad \varepsilon = \dfrac{\delta L}{L} = \dfrac{0.9}{1500}$

$$= 6 \times 10^4$$

Modulus of elasticity,

$$E = \frac{\sigma}{\varepsilon} = \frac{120}{6 \times 10^{-4}}$$

$$= \mathbf{2 \times 10^5 \, N/mm^2} \quad \textbf{Ans.}$$

Example 9.8: A 1 m long steel rod of rectangular section 80 mm × 40 mm is subjected to an axial tensile load of 200 kN. Find the strain energy and maximum stress produced in it when the load is applied gradually. Take $E = 2 \times 10^5 \text{ N/mm}^2$.

[U.P.T.U. II Sem. 2004-05]

Solution:

$$L = 1 \text{ m} = 1000 \text{ mm}$$
$$A = 80 \text{ mm} \times 40 \text{ mm} = 3200 \text{ mm}^2$$
$$P = 200 \text{ kN} = 200 \times 10^3 \text{ N}$$
$$E = 2 \times 10^5 \text{ N/mm}^2$$

Max. stress, $\qquad \sigma = \dfrac{P}{A} = \dfrac{200 \times 10^3}{3200} = 62.5 \text{ N/mm}^2$

Strain energy,

$$U = \frac{P^2 L}{2AE}$$

$$= \frac{\left(200 \times 10^3\right)^2 \times 1000}{2 \times 3200 \times 2 \times 10^5}$$

$$= 31250 \text{ N-mm}$$

$$= 31.25 \text{ N-m}$$

$$= \mathbf{31.25 \ J.} \quad \textbf{Ans.}$$

9.3 SHEAR STRESS

Stress produced by a force tangential to the surface of a body is called shear stress. The shear stress acts parallel to the surface on which the force acts. The corresponding strain is called shear strain.

A cube *ABCDEFG* of slide *L* is acted upon by a force *P* along the face *ABEF* tangentially. There is change in the shape of cube to *A'B'CDE'F'G*.

Shear stress is the force divided by surface area *ABEF*.

$$\tau = \frac{P}{L^2}.$$

Shear strain,

$$\gamma = \frac{BB'}{AB} = \frac{BB'}{L}$$

The ratio of shear stress to shear strain is called modulus of rigidity.

$$G = \frac{\tau}{\gamma}$$

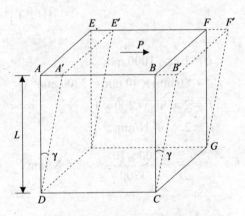

Fig. 9.5 Shear stress and strain

9.3.1 Complimentary Shear Stress

The block *ABCD* of unit thickness is subjected to a shear stress τ (clockwise) on faces *AB* and *CD*. Shear force on faces *AB* and *CD*.

$$= \tau \times AB \ \text{ or } \ \tau \times CD.$$

These parallel and equal forces form a couple.

The moment of couple

$$= \text{Force} \times \text{perpendicular distance}$$

$$= (\tau \times AB) \times AD \ \text{ or } \ (\tau \times CD) \times AD$$

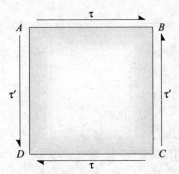

Fig. 9.6 Complimentary shear stress

In order to ensure equilibrium condition, a restoring couple equal and opposite to the above couple is generated.

Shear force acting on face AD or BC

$$= \tau' \times AD \text{ or } \tau' \times BC \qquad \text{(anticlockwise)}$$

These parallel forces will form the restoring couple.

The moment of the restoring couple

$$= (\tau' \times AD) \times AB \text{ or } (\tau' \times BC) \times AD$$

The moments of the two couples must be equal.

$$\therefore \qquad (\tau \times AB) \times AD = (\tau' \times AD) \times AD$$

or $$\tau = \tau'.$$

The stress τ' is called complimentary shear stress. Therefore, a set of shear stress across a plane is always accompanied by a set of balancing shear stress of the same intensity across the plane and normal to it.

9.3.2 Shear Strain Energy

Let the face $ABCD$ of a cube is distorted to $A'B'CD$ by a shear force P.

Shear strain, $$\gamma = \frac{AA'}{AD} = \frac{AA'}{L}$$

$$AA' = \gamma L$$

Shear stress, $$\tau = \frac{P}{\text{Area of top face}} = \frac{P}{L^2}$$

$$\therefore \qquad P = \tau \times L^2$$

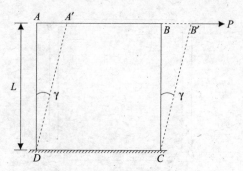

Fig. 9.7 Shear strain energy

Modulus of rigidity of the material of the cube,

$$G = \frac{\text{shear stress}}{\text{shear strain}} = \frac{\tau}{\gamma}$$

The strain energy stored in the cube due to shear stress is equal to work done by shear force, P. As the force is applied gradually, the average force $= \frac{P+o}{2} = \frac{P}{2}$.

Strain energy,
$$U_s = \frac{P}{2} \times AA' = \left(\frac{\tau \times L^2}{2}\right) \times (\gamma \times L)$$

$$= \left(\frac{\tau}{2} L^2\right) \times \left(\frac{\tau}{G} . L\right)$$

$$= \frac{\tau^2}{2G} L^3$$

$$= \frac{\tau^2}{2G} \times \text{Volume}$$

Modulus of resilience is strain energy per unit volume

$$= \frac{\tau^2}{2G}$$

9.4 COMPARISON OF NORMAL AND SHEAR STRESS

The main characteristics of normal and shear stress are compared in Table 9.2.

Table 9.2 Comparison of normal and shear stress

Normal stress	Shear stress
1. Normal stress or direct stress is produced by longitudinal force.	1. Shear stress is produced by tangential force.
2. 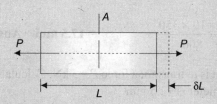	2. 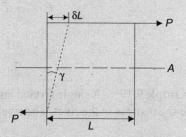
The normal stress is produced on a surface (A) perpendicular to the direction of force $\sigma = \dfrac{P}{A}$.	The shear stress is produced on a surface (A) parallel to the direction of force. $\tau = \dfrac{P}{A}$.
3. The normal strain is measured by the ratio of change of length and original length $\varepsilon = \dfrac{\delta L}{L}$.	3. The shear strain is measured by the ratio of deformation in the direction of force and original length. $\gamma = \dfrac{\delta L}{L}$.
4. The elastic constant is called Young's Modulus of Elasticity, $E = \dfrac{\sigma}{\varepsilon}$.	4. The elastic constant is called Modulus of Rigidity $G = \dfrac{\tau}{\gamma}$.
5. The strain energy due to tension or compression, $U = \dfrac{\sigma^2}{2E} \times$ Volume.	5. The shear strain energy, $U_S = \dfrac{\tau^2}{2G} \times$ Volume.
6. Resilience $= \dfrac{U}{V} = \dfrac{\sigma^2}{2E}$.	6. Resilience $= \dfrac{U_S}{V} = \dfrac{\tau^2}{2G}$.
7. *Examples* (i) Beams under pure bending (ii) Bolts of flange of a pressure vessel.	7. *Examples* (i) Shafts under pure torsion (ii) Rivets of a truss.

Example 9.9: A steel punch can sustain a compressive stress of 800 N/mm^2. Find the least diameter of hole which can be punched through a steel plate 10 mm thick if its ultimate shear stress is 350 N/mm^2.

Solution:

Diameter of punch = diameter of hole = d.

The maximum load which can be borne by the punch,

$$P = \frac{\pi}{4} d^2 \sigma_c.$$

The area of the hole which can be sheared by the punch $= \pi dt$.

∴   Force required $= \pi dt \tau$.

Equating the two forces,

$$P = \frac{\pi}{4} d^2 \sigma_c = \Pi dt \tau$$

∴ $$d = \frac{4t\tau}{\sigma_c} = \frac{4 \times 10 \times 350}{800} \qquad = \textbf{17.5 mm} \quad \textbf{Ans.}$$

Example 9.10:   A single riveted lap joint is to transmit a force $P = 5$ kN. Calculate the shear stress in the rivet. The rivet diameter is 12.5 mm.

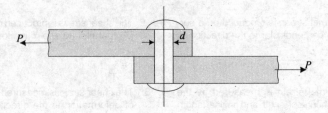

Solution:

$$\text{Load } P = 5 \text{ kN} = 5 \times 10^3 \text{ N}.$$

$$\text{Area of rivet, } A = \frac{\pi}{4} d^2$$

$$= \frac{\pi}{4} (12.5)^2 = 122.7 \text{ mm}^2$$

∴ $$\text{Shear stress, } \tau = \frac{P}{A} = \frac{5 \times 10^3}{122.7} = \textbf{40.75 N/mm}^2 \quad \textbf{Ans.}$$

9.5   STRESS-STRAIN CURVE

Stress-strain curves are obtained experimentally by applying load gradually on a test specimen. Such curves are plotted from the data collected of stress and strain during testing on Universal Testing Machine. The test can be tensile test or compression test. The following material properties are obtained from stress-strain curve.

1. Type of material ductile or brittle. Structural steel can withstand sudden loads and are called ductile materials. Cast iron and non-ferrous metals cannot withstand sudden loads and break easily. These are called brittle materials. It can be known from the shape of stress-strain diagram whether the material tested was ductile or brittle.

2. The ultimate stress
3. The yield stress
4. % age elongation of specimen
5. % age reduction in area
6. Type of fracture
7. Area under the curve.

9.5.1 Stress-Strain Curve for Structural Steel (Ductile Material)

A typical stress-strain for structural steel is shown in Fig. 9.8. The salient points are:

o : Starting point
a : Elastic limit
b : Yield point
c : Lower yield point
d : Ultimate strength
e : Breaking point
I : Elastic region
II : Plastic region

III = I + II : Area under the curve.

The curve can be divided into the following parts:

o – a: The strain is proportional to stress. The material obeys Hooke's Law. Point '*a*' is called elastic limit of proportionality. The slope of line *o – a* is called modulus of elasticity. This is a measure of stiffness of material. Young's

modulus, $E = \dfrac{\sigma}{\varepsilon}$ (N/mm^2).

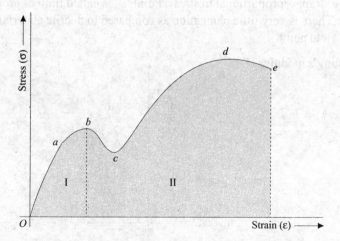

Fig. 9.8   Stress-strain diagram for ductile materials

a – b: The curve is not a straight line and material undergoes more rapid deformation. Hooke's Law is not obeyed although the material may still be elastic. The point '*b*' is called the yield point. The deformation is not fully recovered when the load is removed.

b – c: There is yielding of material and drop of load. The point '*c*' is called lower yield point.

c – d: The stress-strain curve continues to rise by increasing the load. The strain in portion *c-d* is nearly 100 times that in the portion *o-a*. There is a local neck formation in the material. The point '*d*' is called ultimate stress point and the ultimate strength of material is calculated at this point.

$$\text{Ultimate stress} = \frac{\text{Ultimate load}}{\text{Initial cross-sectional area}}.$$

d – e: The load falls off from maximum point and fracture takes place (at neck) at point '*e*' which is called breaking point.

9.5.2 Stress-Strain Curve for Cast Iron (Brittle Material)

A typical stress-strain curve for brittle materials is shown in Fig 9.9. The salient points on the curve are:

   a : Limit of proportionality

   b : Breaking point

   c : Proof point

   I . Elastic region

  II : Plastic region

 III = I + II: Area under the curve.

   The curve can be divided into the following parts.

o – a: The strain is proportional to stress. Point '*a*' is called limit of proportionality. There is very little elongation as compared to ductile material. There is no yield point.

Young's modulus, $E = \dfrac{\sigma}{\varepsilon}$.

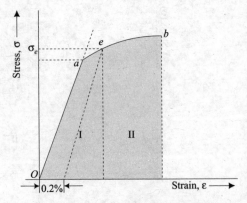

Fig. 9.9 Stress-strain curve for brittle materials

a – b: The curve is not a straight line. The material fractures at point '*b*' and is called breaking point.

Point *e*: In the design of machines and structures, the material of a member is not allowed to be stressed beyond yield point in order to avoid permanent set due to plastic deformation. In case of ductile materials, the yield point is well defined. But in case of brittle materials (cast iron, high carbon steel, non-ferrous metals), there is no defined yield point on the stress-strain diagram. In such a case, a proof stress (σ_e) is used to indicate the onset of plastic strain. A stress which produces a permanent strain of 0.2% is called the proof stress and is used for the design of machines and structures.

Strain energy stored in a material where it is stressed to proof point is called proof resilience (*Ue*)

$$Ue = \frac{1}{2} \frac{\sigma_e}{E} \times \text{volume}$$

$$\text{Modulus of resilience} = \frac{Ue}{V} = \frac{1}{2} \frac{\sigma_e}{E}.$$

9.6 DUCTILE AND BRITTLE MATERIALS

The difference in the behaviour of ductile and brittle materials can be studied with the help of their stress-strain curves. Comparison of ductile and brittle material is given in Table 9.3.

Table 9.3 Comparison of ductile and brittle materials

Ductile Material	Brittle Material
1. It can absorb vibration and withstand sudden and dynamic loads.	1. It fails to withstand vibration, sudden and dynamic loads.
2. The stress-strain curve is shown in Fig. 9.8.	2. The stress-strain curve is shown in Fig. 9.9.
3. Slope of line *o–a* is large and therefore value of modulus of elasticity (*E*) is more.	3. Slope of line *o–a* is small and therefore value of modulus of elasticity (*E*) is low.
4. There is a well-defined yield point.	4. There is no well-defined yield point.
5. There is a local neck formation in the specimen.	5. There is no neck formation in the specimen.
6. The shape of the fracture is *cup and cone* type.	6. The shape of fracture is along a plane at 45° to the axis of the specimen.
7. The plastic region (II) is well extended as compared to elastic region (I).	7. The plastic region (II) is not much extended as compared to elastic region (I).
8. The area under the curve (III) is very large and has high resilience and modulus of resilience to adsorb vibration and impact loads.	8. The area under the curve (III) is small and has low resilience and modulus of resilience. Therefore, brittle material is unsuitable for impact loads.

Example 9.11: A load of 50 kN is suspended from a steel pipe of 50 mm external diameter. If the ultimate tensile strength is 500 N/mm^2 and factor of safety is 4, determine:

   (i) internal diameter of pipe

   (ii) elongation of pipe over a length of 200 mm if stressed to its maximum permissible value.

      Take $E = 2 \times 10^5$ N/mm^2.

Solution:

   Load, $\qquad\qquad\qquad P = 50 \text{ kN} = 50 \times 10^3$ N.

   External diameter of pipe,

$$D = 50 \text{ mm}$$

$$L = 200 \text{ mm}$$

$$\sigma_{max} = 500 \text{ N/mm}^2$$

$$f.o.s. = 4.$$

$$E = 2 \times 10^5 \text{ N/mm}^2.$$

  (i) Allowable stress, $\sigma_a = \dfrac{\sigma_{max}}{f.o.s.} = \dfrac{500}{4} = 125 \text{ N/mm}^2.$

  Sectional area of pipe, $A = \dfrac{\pi}{4}(D^2 - d^2)$

$$\sigma_a = \frac{P}{A}$$

$\therefore\qquad\qquad\qquad A = \dfrac{P}{\sigma_a} = \dfrac{50 \times 10^3}{125}$

Also $\qquad\qquad\quad A = \dfrac{\pi}{4}(50^2 - d^2) = \dfrac{50 \times 10^3}{125}$

$\therefore\qquad\qquad\qquad d = \mathbf{44.62 \text{ mm}}$ **Ans.**

$$A = \frac{\pi}{4}(50^2 - 44.62^2) = 399.8 \text{ mm}^2$$

  (ii) $\qquad$ Elongation, $\delta L = \dfrac{PL}{AE} = \dfrac{50 \times 10^3 \times 200}{399.8 \times 2 \times 10^5}$

$$= \mathbf{0.125 \text{ mm}} \quad \textbf{Ans.}$$

Example 9.12: A steel bar 20 mm in diameter, 200 mm long was tested to destruction. Following observations were recorded.

Load at elastic point = 60 kN

Load at yield point = 75 kN.

Max. load = 135 kN

Extension at elastic limit = 0.22 mm

Final length = 250 mm

Final diameter = 14 mm

Find: (i) Modulus of elasticity

    (ii) Yield stress

    (iii) Ultimate stress

    (iv) % age elongation, and

    (v) % age contraction of area.

Solution:

(i) Modulus of elasticity,

$$E = \frac{\sigma}{\varepsilon} = \frac{PL}{A\delta L} = \frac{60 \times 10^3 \times 200}{\frac{\pi}{4}(20)^2 \times 0.22}$$

$$= 1.736 \times 10^5 \text{ N/mm}^2 \text{ Ans.}$$

(ii) Yield stress,

$$\sigma_e = \frac{\text{Yield load}}{\text{Area}} = \frac{75 \times 10^3}{\frac{\pi}{4}(20)^2}$$

$$= 238.73 \text{ N/mm}^2 \text{ Ans.}$$

(iii) Ultimate stress,

$$\sigma_{max} = \frac{\text{Maximum load}}{\text{Original area}} = \frac{135 \times 10^3}{\frac{\pi}{4}(20)^2}$$

$$= 429.72 \text{ N/mm}^2 \text{ Ans.}$$

(iv) % age elongation $= \dfrac{\text{Final length - Original length}}{\text{Original length}}$

$$= \frac{250 - 200}{200} \times 100 = 25\% \text{ Ans.}$$

(v) % age contraction $= \dfrac{\text{Original area} - \text{Final area}}{\text{Original area}}$

$$= \dfrac{\dfrac{\pi}{4}[(20)^2 - (14)]^2}{\dfrac{\pi}{4}(20)^2} \times 100$$

$= \mathbf{51\%}$ **Ans.**

Example 9.13: During a tensile test on a mild steel specimen, 40 mm diameter and 200 mm long, the following data was obtained.

Extension at 40 kN load = 0.0304 mm

Yield load = 161 kN.

Length of specimen at fracture = 249 mm.

Determine modulus of elasticity, yield stress and % age elongation.

Solution:

(i) Modulus of elasticity,

$$E = \dfrac{\sigma}{\varepsilon} = \dfrac{(P/A)}{(\delta L/L)} = \dfrac{40/\dfrac{\pi}{4}(40)^2}{0.0304/200}$$

$= \mathbf{209.4\ kN/mm^2}$ **Ans.**

(ii) Yield stress,

$$\sigma_e = \dfrac{\text{Yield load}}{\text{Area}} = \dfrac{161 \times 10^3}{\dfrac{\pi}{4}(40)^2}$$

$= \mathbf{128\ N/mm^2}$ **Ans.**

(iii) % age elongation $= \dfrac{\text{Final length} - \text{Original length}}{\text{Original length}} \times 100$

$$= \left(\dfrac{249 - 200}{200}\right) \times 100 = \mathbf{24.5\%}$$ **Ans.**

Example 9.14: A bar 1.5 m long and 40 mm × 20 mm cross-section has elastic limit stress of 160 N/mm² and $E = 2 \times 10^5$ N/mm². Calculate the modulus of resilience and proof resilience.

Solution:

Volume of specimen,

$$V = 40 \times 20 \times 1500 = 12 \times 10^5\ \text{mm}^3$$
$$\sigma_e = 160\ \text{N/mm}^2$$
$$E = 2 \times 10^5\ \text{N/mm}^2.$$

(i) Proof resilience,

$$U_e = \frac{1}{2}\frac{\sigma_e^2}{E} \times V$$

$$= \frac{1}{2}\frac{(160)^2}{2 \times 10^5} \times 12 \times 10^5$$

$$= 7.68 \times 10^4 \text{ N.mm.}$$

(ii) Modulus of resilience,

$$= \frac{U_e}{V} = \frac{7.68 \times 10^4}{12 \times 10^5}$$

$$= \textbf{0.064 N/mm}^2 \quad \textbf{Ans.}$$

Example 9.15: The following results were obtained in a tensile test on a rigid steel specimen of original diameter 20 mm and gauge length 50 mm. At the limit of proportionality, the load was 100 kN and extension was 0.5 mm. The specimen yielded at a load of 115 kN and the maximum load withstood was 200 kN. After rupture, the total length between gauge marks was found 6.67 cm and diameter of neck was 1.72 cm. Calculate Young's modulus, stress at limit of proportionality, the yield stress, ultimate tensile stress, percentage elongation and contraction.

Solution:

(i) Young's modulus,

$$E = \frac{\text{Stress at limit of proportionality}}{\text{Strain at limit of proportionality}}$$

$$= \frac{\dfrac{100 \times 10^3}{\dfrac{\pi}{4}(20)^2}}{\dfrac{0.5}{50}} = 0.318 \times 10^5 \text{ N/mm}^2 \quad \textbf{Ans.}$$

(ii) Stress at limit of proportionality,

$$\sigma = \frac{100 \times 10^3}{\dfrac{\pi}{4}(20)^2} = 318 \text{ N/mm}^2 \quad \textbf{Ans.}$$

(iii) Yield stress,

$$\sigma_e = \frac{115 \times 10^3}{\dfrac{\pi}{4}(20)^2} = 366 \text{ N/mm}^2 \quad \textbf{Ans.}$$

(iv) Ultimate tensile stress,

$$\sigma_{max} = \frac{200 \times 10^3}{\frac{\pi}{4}(20)^2} = 636.6 \text{ N/mm}^2 \quad \textbf{Ans.}$$

(v) % age elongation $= \frac{66.7 - 50}{50} \times 100$

$$= \textbf{33.4\%} \quad \textbf{Ans.}$$

(vi) % age contraction $= \dfrac{\frac{\Pi}{4}\left[(20)^2 - (17.2)^2\right]}{\frac{\pi}{4}(20)^2} \times 100$

$$= \textbf{26\%} \quad \textbf{Ans.}$$

9.7 ELASTIC CONSTANTS

The elastic constants are important properties of the material which influence the strains under different types of loading.

1. Modulus of Elasticity

Hooke's law staets that when a material is loaded within its elastic limits, the stress is proportional to strain,

$$\frac{\text{stress}}{\text{strain}} = \text{constant} = E$$

$\therefore \qquad E = \dfrac{\sigma}{\varepsilon}.$

E is called modulus of elasticity or Young's modulus and determines the linear strain under linear loading. It is same for both tension and compression.

2. Modulus of Rigidity

The ratio of shear stress and shear strain is constant and is called modulus of rigidity.

$$G = \frac{\tau}{r}.$$

G is the material property which determines the strain due to shear stress.

3. Bulk Modulus

Whenever a member is subjected to a force or a system of forces, there is change in its dimensions in all the directions and, hence, there is change of volume.

The ratio of change in volume to original volume is called volumetric strain (ε_v) or dilatational strain.

$$\varepsilon_v = \frac{\delta V}{V}.$$

Bulk modulus is a material property which determines volumetric strain. The bulk modulus,

$$K = \frac{\text{Direct stress}}{\text{Volumetric strain}} = \frac{\sigma}{\varepsilon_v}.$$

4. Poisson's Ratio

The deformation of a bar per unit length in the direction of load or force is called linear or primary strain.

$$\text{Linear strain} = \frac{\delta L}{L}.$$

The linear strain is always accompanied by lateral or secondary strain. Lateral strain is the ratio of change of lateral dimension or diameter of a bar and the lateral dimension or diameter.

$$\text{Lateral strain} = \frac{\delta D}{D} \text{ or } \frac{\delta b}{b}$$

where, D is the diameter and b is the lateral dimension.

It is observed that within elastic limits, the lateral strain is proportional to linear strain.

$$\therefore \quad \frac{\text{Lateral strain}}{\text{Linear strain}} = \text{constant} = \frac{1}{m} = \nu$$

This constant, ν or $\dfrac{1}{m}$ is called Poisson's ratio.

$$\text{Lateral strain} = \nu\varepsilon \quad \text{or } \frac{\varepsilon}{m}.$$

9.7.1 Relationship Between Elastic Constants

(a) Relation between E, K and ν

A cube is subjected to volumetric stress σ acting simultaneously along X, Y and Z directions.

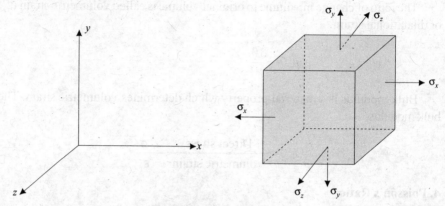

Fig. 9.10 Volumetric stress

$$\sigma_x = \sigma_y = \sigma_z = \sigma$$

Strain in x-direction,

$$\varepsilon_x = \frac{\sigma_x}{E} - v\frac{\sigma_y}{E} - v\frac{\sigma_z}{E} = \frac{\sigma}{E}(1-2v)$$

Strain in y-direction,

$$\varepsilon_y = \frac{\sigma_y}{E} - v\frac{\sigma_z}{E} - v\frac{\sigma_x}{E} = \frac{\sigma}{E}(1-2v)$$

Strain in z-direction,

$$\varepsilon_z = \frac{\sigma_z}{E} - v\frac{\sigma_x}{E} - v\frac{\sigma_y}{E} = \frac{\sigma}{E}(1-2v)$$

Volumetric strain,

$$\varepsilon_v = \varepsilon_x + \varepsilon_y + \varepsilon_z = \frac{3\sigma}{E}(1-2v)$$

Bulk Modulus,

$$K = \frac{\sigma}{\varepsilon_v} = \frac{\sigma}{\frac{3\sigma}{E}(1-2v)} = \frac{E}{3(1-2v)}$$

$\therefore$
$$E = 3K(1-2v)$$

(b) Relation between E, G and v

A cubic element is subjected to shearing force at the top face. The block experiences the following effects:

(i) Shearing stress, τ at face *AB* and *CD*.

(ii) Complimentary shear stress at faces *AD* and *BC*

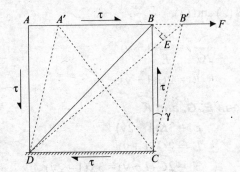

Fig. 9.11 Shearing stresses

(iii) Block is distorted as $A'B'CD$.

(iv) Diagonal DB elongates and diagonal AC is shortened. Therefore, longitudinal strain in diagonal DB,

$$= \frac{DB' - DB}{DB} = \frac{DB' - DE}{DB}$$

$$= \frac{EB'}{DB} \qquad \qquad \dots(\text{i})$$

BE is $\perp$ lar to DB'.

Let, $\qquad \angle DB'C = \angle DBC \qquad \qquad [\because BB' \text{ is small}]$
$\qquad \qquad = 45°.$

$\therefore \qquad EB' = BB' \cos 45° = \dfrac{BB'}{\sqrt{2}}$

$\therefore \qquad$ Longitudinal strain $= \dfrac{EB'}{BD} = \dfrac{BB'}{\sqrt{2}} \times \dfrac{1}{\sqrt{2}BC} = \dfrac{BB'}{2BC}$

$$= \frac{\tan r}{2} = \frac{\gamma}{2} \qquad \left[\because \gamma = \text{shear strain} = \frac{BB'}{BC} \right]$$

Now shear strain,

$$\gamma = \frac{\tau}{G}.$$

$\therefore$ Longitudinal strain of diagonal $DB = \dfrac{\tau}{2G}$ $\qquad \qquad \dots(2)$

Similarly, strain of diagonal DB = strain in DB due to tensile stress – strain in AC due to compresive stress.

$$= \frac{\tau}{E} - \left[-v \frac{\tau}{E} \right] = \frac{\tau}{E} (1 + v) \qquad \qquad \dots(3)$$

From equations (2) and (3),

$$\frac{\tau}{2G} = \frac{\tau}{E}(1+v)$$

or $$E = 2G(1+v).$$

(c) Relation between E, G, and K

$$E = 3K(1-2v)$$

and $$E = 2G(1+v)$$

$$\therefore \quad E = 2G(1+v) = 3K(1-2v)$$

$$\therefore \quad v = \frac{E}{2G} - 1$$

$$E = 3K\left[1 - 2\left(\frac{E}{2G} - 1\right)\right]$$

$$= 3K\left[1 - \left(\frac{E}{2G} - 2\right)\right]$$

or $$E = 3K\left[3 - \frac{E}{G}\right] = 9K - \frac{3KE}{G}$$

$$\therefore \quad E + \frac{3KE}{G} = 9K$$

$$E\left[\frac{G + 3K}{G}\right] = 9K$$

or $$E = \frac{9KG}{G + 3K}$$

The relations among elastic constants can be summarised on follows:

$$E = 3K(1 - 2v) = 3K\left(1 - \frac{2}{m}\right)$$

$$\therefore \quad K = \frac{mE}{3(m-2)}$$

$$E = 2G(1+v) = 2G\left(1 + \frac{1}{m}\right)$$

$$\therefore \quad G = \frac{mE}{2(m+1)}$$

Example 9.16: A circular rod of 100 mm diameter and 500 mm long is subjected to a tensile force of 1000 kN. Determine the modulus of rigidity, bulk modulus and change in volume, if Poisson's ratio is 0.3 and Young's modulus is 2×10^5 N/mm^2.

Solution:

$$P = 100 \text{ kN} = 1000 \times 10^3 \text{ N.}$$
$$D = 100 \text{ mm}$$
$$L = 500 \text{ mm}$$
$$E = 2 \times 10^5 \text{ N/mm}^2$$
$$v = 0.3 = \frac{1}{m} \qquad\qquad [m = 3.33.]$$

(i) Modulus of rigidity.

$$G = \frac{mE}{2(m+1)} = \frac{3.33 \times 2 \times 10^5}{2(3.33+1)}$$
$$= 0.78 \times 10^5 \text{ N/mm} \quad \textbf{Ans.}$$

(ii) Bulk modulus,

$$K = \frac{mE}{2(m-2)} = \frac{3.33 \times 2 \times 10^5}{2(3.33-2)}$$
$$= 1.67 \times 10^5 \text{ N/mm}^2 \quad \textbf{Ans.}$$

(iii) Volume,

$$V = \frac{\pi}{4} D^2 L = \frac{\pi}{4}(100)^2 \cdot 500$$
$$= 3926991 \text{ mm}^3.$$

$$\sigma = \frac{P}{A} = \frac{1000 \times 10^3}{\pi/4(100)^2} = 127.3 \text{ N/mm}^2.$$

$$K = \frac{\sigma}{\varepsilon_v} = \frac{\sigma V}{\delta V}$$

$$\therefore \quad \delta V = \frac{\sigma V}{K} = \frac{127.3 \times 3926991}{1.67 \times 10^5}$$

$$= 2994 \text{ mm}^2 \quad \textbf{Ans.}$$

Example 9.17: A bronge specimen has $E = 1.2 \times 10^5$ N/mm^2 and $G = 0.45 \times 10^5$ N/mm. Determine the Poisson's ratio of the material.

Solution:

$$G = \frac{mE}{2(m+1)}$$

$$\therefore \quad 2mG + 2G = mE$$

$$m = \frac{-2G}{(2G-E)} = \frac{-2 \times 0.45 \times 10^5}{(2 \times .45 \times 10^5 - 1.2 \times 10^5)}$$

$$= \frac{0.9}{0.3} = 3$$

$$\therefore \quad v = \frac{1}{m} = \frac{1}{3} = 0.33.$$

Example 9.18: Find the bulk modulus and lateral contraction of a round bar of 50 mm diameter and 2.5 m long when stretched to 2.5 mm. Take $E = 1.2 \times 10^5$ N/mm^2 and $v = 0.25$.

Solution:

Longitudinal strain,

$$\varepsilon_L = \frac{\delta L}{L} = \frac{2.5}{2500} = 0.001$$

Lateral strain,

$$\frac{\delta d}{d} = v\varepsilon_L = 0.25 \times 0.001$$

$$= 0.00025$$

$$\therefore \quad \delta d = 0.00025 \times 50$$

$$= .0125 \text{ mm} \quad \textbf{Ans.}$$

Bulk modulus,

$$K = \frac{mE}{3(m-2)} = \frac{4 \times 1.2 \times 10^5}{3(4-2)}$$

$$= 0.8 \times 10^5 \text{ N/mm}^2$$

Example 9.19: A steel bar 300 mm long, 50 mm wide and 12 mm thick is subjected to a tensile pull of 100 kN. If $E = 2 \times 10^5$ N/mm^2 and $v = 0.32$, determine the change in volume of the bar

Solution:

Volume,

$$V = Lbt = 300 \times 50 \times 12$$

$$= 180,000 \text{ mm}^3$$

Strain,
$$\varepsilon = \frac{\sigma}{E} = \frac{P/A}{E} = \frac{100 \times 10^3}{50 \times 12 \times 2 \times 10^5}$$
$$= 8.33 \times 10^{-4}$$

Volumetric strain,

$$\varepsilon_v = \frac{\delta V}{V} = \varepsilon \left[1 - \frac{2}{m} \right]$$

$$\delta_V = 180,000 \times 8.33 \times 10^{-4} (1 - 2 \times 0.32)$$
$$= 54 \text{ mm}^3.$$

Example 9.20: A 2 m long rectangular bar of 7.5 cm $\times$ 5 cm is subjected to an axial tensile load of 1000 kN. Bar gets elongated by 2 mm in length and decreases in width by 10×10^{-6} m. Determine the modulus of elasticity E and Poisson's ratio v of the material of the bar.

$$L = 2 \text{ m}$$
$$b = 7.5 \text{ cm} = 7.5 \times 10^{-2} \text{ m}$$
$$t = 5 \text{ cm} = 5 \times 10^{-2} \text{ m}$$
$$P = 1000 \text{ kN}$$
$$\delta L = 2 \text{ mm} = 2 \times 10^{-3} \text{ m}$$
$$\delta b = 10 \times 10^{-6} \text{ m}.$$

The longitudinal stress,

$$\sigma = \frac{P}{A} = \frac{P}{b \times t} = \frac{1000}{7.5 \times 10^{-2} \times 5 \times 10^{-2}}$$
$$= 26.67 \times 10^4 \text{ kN/m}^2$$

$$\varepsilon = \frac{\delta L}{L} = \frac{2 \times 10^{-3}}{2} = 10^{-3}$$

$$E = \frac{\sigma}{\varepsilon} = \frac{26.67 \times 10^4}{10^{-3}} = 26.67 \times 10^7 \text{ kN/m}^2$$

Lateral strain,
$$= \frac{\delta b}{b} = \frac{10 \times 10^{-6}}{7.5 \times 10^{-2}} = 1.333 \times 10^{-4}$$

∴ Poisson's ratio,
$$v = \frac{1.333 \times 10^{-4}}{10^{-3}} = 0.1333$$

Example 9.21: A bar of 25 mm diameter is subjected to a pull of 60 kN. The measured extension over a gauge length of 250 mm is 0.15 mm and change in diameter is 0.004 mm. Calculate the modulus of elasticity, modulus of rigidity and Poisson's ratio. [U.P.T.U. II Sem., 2005-06]

Solution:

$$d = 25 \text{ mm}$$
$$P = 60 \text{ kN}$$
$$L = 250 \text{ mm}$$
$$\delta L = 0.15 \text{ mm}$$
$$\delta d = 0.004 \text{ mm}$$

Stress,
$$\sigma = \frac{P}{A} = \frac{60 \times 10^3}{\frac{\pi}{4}(25)^2}$$

$$= 122.23 \text{ N/mm}^2.$$

Strain,
$$\varepsilon = \frac{\delta L}{L} = \frac{0.15}{250} = 6 \times 10^{-4}$$

Modulus of elasticity,

$$E = \frac{\sigma}{\varepsilon} = \frac{122.23}{6 \times 10^{-4}} = 203716.67 \text{ N/mm}^2$$

$$= \mathbf{2.037 \times 10^5 \, N/mm^2} \quad \textbf{Ans.}$$

Lateral strain,
$$= \frac{\delta d}{d} = \frac{0.004}{25}$$

$$= 1.6 \times 10^{-4}$$

Poisson's ratio,
$$\nu = \frac{\text{Lateral strain}}{\text{Linear strain}} = \frac{1.6 \times 10^{-4}}{6 \times 10^{-4}}$$

$$= 0.2667$$
$$m = 3.75$$

Modulus of rigidity,
$$G = \frac{mE}{2(m+1)}$$

$$= \frac{3.75 \times 2.037 \times 10^5}{2(3.75 + 1)}$$

$$= \mathbf{0.80 \times 10^5 \, N/mm^2} \quad \textbf{Ans.}$$

Example 9.22: A metallic rectangular rod s1.5 m long and 40 mm wide and 25 mm thick is subjected to an axial tensile load of 120 kN. The elongation of the rod is measured as 0.9 mm. Calculate the stress, strain and modulus of elasticity.

[U.P.T.U. II Sem., 2003-04]

Solution:
$$L = 1.5 \text{ m} = 1500 \text{ mm}$$
$$b = 40 \text{ mm}$$
$$t = 25 \text{ mm}$$
$$P = 120 \text{ kN} = 120 \times 10^3 \text{ N}$$
$$\delta L = 0.9 \text{ mm}.$$

Stress,
$$\sigma = \frac{P}{A} = \frac{120 \times 10^3}{40 \times 25} = 120 \text{ N/mm}^2$$

Strain,
$$\varepsilon = \frac{\delta L}{L} = \frac{0.9}{1500} = 6 \times 10^{-4}$$

Modulus of elasticity, $E = \dfrac{\sigma}{\varepsilon} = \dfrac{120}{6 \times 10^{-4}} = 2 \times 10^5 \text{ N/mm}^2$ **Ans.**

9.8 MEMBERS OF VARYING CROSS-SECTION

Members of varying cross-section are widely used in the construction of machines and structures. These members can be of the following shape:

1. Stepped bars.
2. Tapered bars.
3. Composite bars.

9.8.1 Stepped Bars

In stepped bars, the forces may be acting at some interior section of a body along its length, in addition to forces acting at the ends. The bar is divided in different lengths (sections) and forces for each section/length are worked out on the principle of equilibrium of each section. Free body diagram is drawn for each length and deformation is determined. The deformation for the entire length is worked out by summing the sectional deformations.

$$\delta L = \frac{PL}{AE} = \frac{1}{E}\left[\frac{P_1 L_1}{A_1} + \frac{P_2 L_2}{A_2} + \cdots \right]$$

The deformation of individual sections can be worked out as follows.

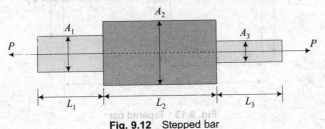

Fig. 9.12 Stepped bar

$$\delta L_1 = \frac{PL_1}{A_1 E}$$

$$\delta L_2 = \frac{PL_2}{A_2 E}$$

$$\delta L_3 = \frac{PL_3}{A_3 E}$$

Total elongation of the bar,

$$\delta L = \delta L_1 + \delta L_2 + \delta L_3$$

$$= \frac{P}{E}\left[\frac{L_1}{A_1} + \frac{L_2}{A_2} + \frac{L_3}{A_3}\right]$$

If the stepped bar is made from sections of different materials

$$\delta L = P\left[\frac{L_1}{A_1 E_1} + \frac{L_2}{A_2 E_2} + \frac{L_3}{A_3 E_3}\right]$$

where, E_1, E_2 and E_3 are modulus of elasticity of materials of different steps.

9.8.2 Tapered Bars

A tapered circular bar of big-end diameter D_1 and small-end diameter D_2 and length L is loaded with axial load P. It is required to find the extension of the bar. Consider a small strip of length dX at a distance of X from big-end.

Diameter of bar at this section, $D_x = D_1 - (D_1 - D_2)\dfrac{X}{L}$

$$= D_1 - mX$$

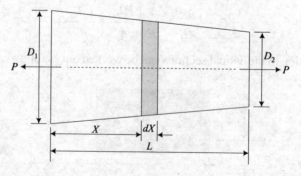

Fig. 9.13 Tapered bar

where, $$m = \frac{D_1 - D_2}{L}$$

Cross-sectional area of strip,

$$A_x = \frac{\pi}{4}(D_1 - mX)^2$$

Stress in the strip,

$$\sigma_x = \frac{P}{A_x} = \frac{P}{\frac{\pi}{4}(D_1 - mX)^2} = \frac{4P}{(D_1 - mX)^2}$$

Axial strain in the strip,

$$\varepsilon_x = \frac{\sigma_x}{E} = \frac{4P}{\pi(D_1 - mX)^2 E}$$

Extension of strip,

$$\delta L_x = \varepsilon_x\, d_x = \frac{4P d_x}{\pi(D_1 - mX)^2 E}$$

Total extension of bar

$$\delta L = \int_O^L \delta L_x = \int_O^L \frac{4P d_x}{\pi(D_1 - mX)^2 E}$$

$$= \frac{4P}{\pi E} \int_O^L (D_1 - mX)^{-2} dx$$

$$= \frac{4P}{\pi E}\left[\frac{(D_1 - mX)^{-1}}{(-1) \times (-m)}\right]_o^L$$

$$= \frac{4P}{\pi E}\left[\frac{1}{(D_1 - mX)}\right]_o^L$$

$$= \frac{4P}{\pi E\left(\dfrac{D_1 - D_2}{L}\right)}\left[\frac{1}{\left\{D_1 - \dfrac{(D_1 - D_2)}{L}\right\}L} - \frac{1}{D_1}\right]$$

$$\delta L = \frac{4P}{\pi E \left(\dfrac{D_1 - D_2}{L} \right)} \times \frac{(D_1 - D_2)}{D_1 D_2}$$

$$\delta L = \frac{4PL}{\pi E D_1 D_2}$$

Note: This is the relation for a circular taper bar. For other shapes of bar, proceed from first principles.

9.8.3   Composite Bars

Composite columns and ties are used by joining together two or more elements of different materials. These may be subjected to tensile or compressive loads axially. As the members are joined rigidly at both ends, the following principles apply.

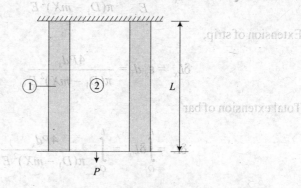

Fig. 9.14   Composite bar

1. The deformation (extension or contraction) of each member is same.

$$\delta L_1 = \delta L_2 = \dots$$

2. All the elements of composite bar should have same strain as they have same length.

$$\frac{\delta L_1}{L} = \frac{\delta L_2}{L} = \dots$$

$$\varepsilon_1 = \varepsilon_2 = \dots$$

3. Total axial load must be the sum of loads carried by different members of the composite bar.

$$P = P_1 + P_2 + \dots$$

$$= \sigma_1 A_1 + \sigma_2 A_2 + \dots$$

$$\varepsilon_1 = \varepsilon_2 = \dots$$

$$\therefore \qquad \frac{\sigma_1}{E_1} = \frac{\sigma_2}{E_2} = \dots$$

$$\therefore \qquad \frac{\sigma_1}{\sigma_2} = \frac{E_1}{E_2} = m_r$$

m_r is called the modular ratio.

Example 9.23:   A round bar is subjected to an axial load of 100 kN. Calculate the diameter 'D' if the stress there is 100 MN/m². Find also the total elongation ($E = 200$ GPa).

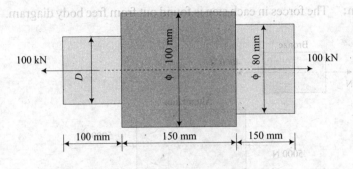

Solution:

Stress,
$$\sigma = \frac{P}{A}$$

$$\therefore \qquad A = \frac{P}{\sigma} = \frac{100 \times 10^3}{100} = 1000 \text{ mm}^2$$

$$D = \sqrt{\frac{4A}{\pi}} = \sqrt{\frac{4 \times 1000}{\pi}} = \textbf{35.68 mm} \quad \textbf{Ans.}$$

$$\delta L = \delta L_1 + \delta L_2 + \delta L_3$$

$$= \frac{P}{E}\left[\frac{L_1}{A_1} + \frac{L_2}{A_2} + \frac{L_3}{A_3}\right]$$

$$= \frac{100 \times 10^3}{2 \times 10^5} \times \frac{4}{\pi}\left[\frac{100}{(35.68)^2} + \frac{150}{(100)^2} + \frac{150}{(80)^2}\right]$$

$$= \frac{100 \times 10^3}{2 \times 10^5} \times \frac{4}{\pi}\,[0.07855 + 0.015 + 0.0234]$$

$$= \textbf{0.0745 mm} \quad \textbf{Ans.}$$

Example 9.24: A composite bar consists of aluminium section rigidly fastened between a bronze section and steel section as shown. Axial loads are applied at points indicated. Determine the stress in each section.

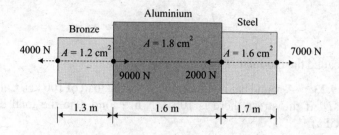

Solution: The forces in each step is found out from free body diagram.

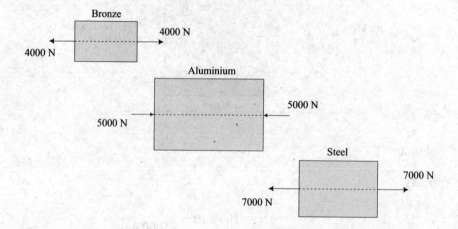

The bronze specimen is under a tensile load of 4000 N as applied at the end.

The steel specimen is under a tensile load of 7000 N as applied at the end.

Therefore, the central aluminium step will be under a compressive load of 5000 N (i.e., 9000 N – 4000 N) for equilibrium stress in bronze bar,

$$\sigma_b = \frac{P}{A} = \frac{4000}{1.2 \times 10^2} = 33.33 \text{ N/mm}^2 \text{(tensile)}$$

Stress in aluminium bar,

$$\sigma_a = \frac{5000}{1.8 \times 10^2} = 27.78 \text{ N/mm}^2 \text{ (compressive)}$$

Stress in steel bar,

$$\sigma_s = \frac{7000}{1.6 \times 10^2} = 43.75 \text{ N/mm}^2 \text{ (tensile)}$$

Example 9.25:   A member *ABCD* is subjected to point loads P_1, P_2, P_3 and P_4 as shown. Calculate the force P_2 necessary for equilibrium, if $P_1 = 4.5$ kN, $P_3 = 45$ kN and $P_4 = 13$ kN. Determine the total elongation of the member, assuming modulus of elasticity to be 2×10^5 N/mm^2.

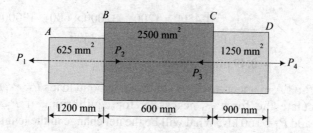

Solution:

For equilibrium of bar,

$$\sum_i (P_i)_{\text{left}} = \sum_i (P_i)_{\text{right}}$$

$\therefore$

$$P_1 + P_3 = P_2 + P_4$$

$\therefore$

$$P_2 = P_1 + P_3 - P_4 = 4.5 + 45 - 13$$

$$= \textbf{36.5 kN}\quad \textbf{Ans.}$$

The forces in each step are found out from free body diagram of individual step.

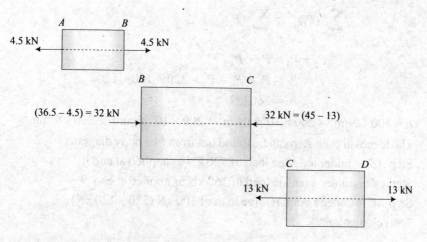

Step *AB* is under a tensile load of 4.5 kN as applied at end A.

Step *CD* is under a tensile load of 13 kN as applied at end *D*.

Step *BC* is under a compressive load of 32 kN (36.5 kN – 4.5 kN)

Total elongation of bar,

$$\sigma L = \sigma L_1 + \sigma L_2 + \sigma L_3 = \frac{1}{E}\left[\frac{P_1 L_1}{A_1} + \frac{P_2 L_2}{A_2} + \frac{P_3 L_3}{A_3}\right]$$

$$= \frac{1}{2 \times 10^5}\left[\frac{4500 \times 1200}{625} - \frac{32000 \times 600}{2500} + \frac{13000 \times 900}{1250}\right]$$

$$= 0.49 \text{ mm} \quad \textbf{Ans.}$$

Example 9.26: A member $ABCD$ is subjected to point loads P_1, P_2, P_3 and P_4 as shown. Calculate the force P_3 necessary for equilibrium, if $P_1 = 120$ kN, $P_2 = 220$ kN and $P_4 = 160$ kN. What will be the net change in the length of the bar, if $E = 200$ GN/m^2.

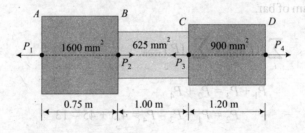

Solution:

For equilibrium of bar,

$$\sum_i (P_i)_{\text{left}} = \sum_i (P_i)_{\text{right}}$$

$$P_1 + P_3 = P_2 + P_4$$

$$\therefore \qquad P_3 = P_2 + P_4 - P_1 = 220 + 160 - 120$$

$$= 260 \text{ kN}$$

$E = 200$ GN/m$^2 = 200 \times 10^3$ MN/m$^2 = 200 \times 10^3$ N/mm$^2 = 200$ kN/mm^2

The forces in each step can be found out from free body diagram.

Step AB is under a tensile load of 120 kN as applied at end A.

Step CD is under a tensile load of 160 kN as applied at end D.

Step BC is under a compressive load of 100 kN (220 – 120 kN)

Total elongation,

$$\delta L = \delta L_1 + \delta L_2 + \delta L_3 = \frac{1}{E}\left[\frac{P_1 L_1}{A_1} + \frac{P_2 L_2}{A_2} + \frac{P_3 L_3}{A_3}\right]$$

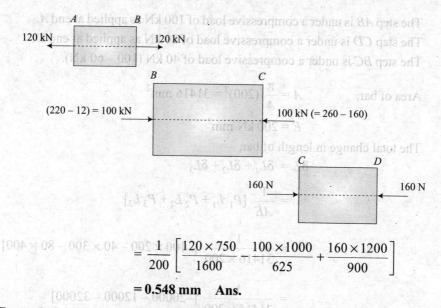

$$= \frac{1}{200}\left[\frac{120 \times 750}{1600} - \frac{100 \times 1000}{625} + \frac{160 \times 1200}{900}\right]$$

$$= 0.548 \text{ mm} \quad \textbf{Ans.}$$

Example 9.27: A steel bar is subjected to loads as shown. If Young's moduleus for the bar material is 200 kN/mm², determine the change in length of bar. The bar is 200 mm in diameter. [U.P.T.U. I Sem., 2005-06]

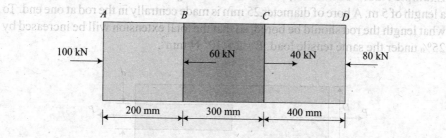

Solution: The forces in each step can be calculated from free body diagram.

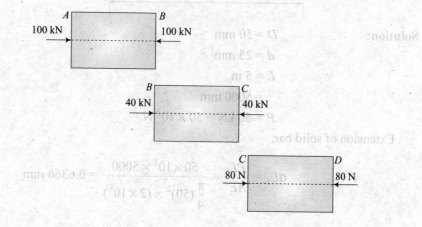

The step *AB* is under a compressive load of 100 kN as applied at end *A*.

The step *CD* is under a compressive load of 80 kN as applied at end *D*.

The step *BC* is under a compressive load of 40 kN (100 – 60 kN).

Area of bar, $\quad\quad\quad A = \dfrac{\pi}{4}\,(200)^2 = 31416\text{ mm}^2$

$$E = 200\text{ kN/mm}^2$$

The total change in length of bar,

$$\delta L = \delta L_1 + \delta L_2 + \delta L_3$$

$$= \frac{1}{AE}\,[P_1\,L_1 + P_2\,L_2 + P_3\,L_3]$$

$$= \frac{1}{31416 \times 200}\,[-100 \times 200 - 40 \times 300 - 80 \times 400]$$

$$= \frac{1}{31416 \times 200}\,[-20000 - 12000 - 32000]$$

$$= \textbf{0.01 mm (decrease)}\quad \textbf{Ans.}$$

Example 9.28: A tensile load of 50 kN is acting on a rod of diameter 50 mm and a length of 5 m. A bore of diameter 25 mm is made centrally in the rod at one end. To what length the rod should be bored, so that the total extension will be increased by 25% under the same tensile load. $E = 2 \times 10^5$ N/mm^2.

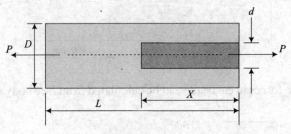

Solution: $\quad\quad\quad\quad D = 50\text{ mm}$

$$d = 25\text{ mm}$$

$$L = 5\text{ m}$$

$$= 5000\text{ mm}$$

$$P = 50\text{ kN} = 50 \times 10^3\text{ N}$$

Extension of solid bar,

$$\sigma L_1 = \frac{PL}{AE} = \frac{50 \times 10^3 \times 5000}{\dfrac{\pi}{4}(50)^2 \times (2 \times 10^5)} = 0.6366\text{ mm}$$

Extension of bar required after boring,

$$\delta L_2 = 0.6366 \times 1.25 = 0.7957 \text{ mm.}$$

Let the length of bore = X.

∴ Total extension of bar = Extension of solid portion + Extension of hollow portion.

$$\delta L_2 = \frac{P}{E}\left[\frac{L_1}{A_1} + \frac{L_2}{A_2}\right]$$

$$= \frac{P}{E}\left[\frac{X}{\frac{\pi}{4}(D^2 - d^2)} + \frac{(L - X)}{\frac{\pi}{4}D^2}\right]$$

$$= \frac{50 \times 10^3}{2 \times 10^5}\left[\frac{X}{\frac{\pi}{4}(50^2 - 25^2)} + \frac{(5000 - X)}{\frac{\pi}{4}50^2}\right] = 0.7957$$

∴ $X = 3748 \text{ mm} = \textbf{3.748 m}$ **Ans.**

Example 9.29: A steel rod 40 mm in diameter and 2 m long is subjected to a pull of 80 kN. To what length the rod should be bored centrally so that the total extension may increase by 20% under the same pull, the bore being 20 mm in diameter $E = 2 \times 10^5$ N/mm².

Solution:

$$D = 40 \text{ mm}$$
$$d = 20 \text{ mm}$$
$$L = 2 \text{ m} = 2000 \text{ mm}$$
$$P = 80 \text{ kN} = 80 \times 10^3 \text{ N}$$
$$E = 2 \times 10^5 \text{ N/mm}^2$$

Extension of rod before boring,

$$\delta L_1 = \frac{PL}{AE} = \frac{(80 \times 10^3) \times 2000}{\left(\frac{\pi}{4} 40^2\right) \times (2 \times 10^5)} = 0.6366 \text{ mm}$$

Let the length of bore = X mm.

Extension of rod required after boring = 1.2×0.6366

$$\delta L_2 = 0.764 \text{ mm}.$$

Extension of hollowed rod

= Extension of solid portion + Extension of hollow portion

$$\delta L_2 = \frac{P}{E}\left[\frac{L_1}{A_1} + \frac{L_2}{A_2}\right]$$

$$= \frac{80 \times 10^3}{2 \times 10^5}\left[\frac{X}{\frac{\pi}{4}(40^2 - 20^2)} + \frac{(2000 - X)}{\frac{\pi}{4}(40^2)}\right] = 0.764$$

$\therefore$ $X = \mathbf{1198.6 \text{ mm}}$ **Ans.**

Example 9.30: A vertical rod of 4 m length is rigidly fixed at upper end and carries an axial load of 50 kN force. Calculate the total extension of the bar if it tapers uniformly from a diameter of 50 mm at top to 30 mm at bottom. $E = 210 \text{ GN/m}^2$.

Solution:

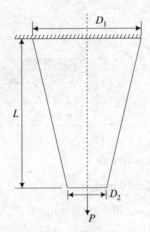

$$P = 50 \text{ kN} = 50 \times 10^3 \text{ N}$$

$$L = 4 \text{ m} = 4000 \text{ mm}$$

$$E = 210 \text{ GN/m}^2 = 2.1 \times 10^5 \text{ N/mm}^2$$

$$D_1 = 50 \text{ mm}$$

$$D_2 = 30 \text{ mm}$$

For a circular taper bar,

$$\delta L = \frac{4PL}{\pi E D_1 D_2} = \frac{4 \times (50 \times 10^3) \times 4000}{\pi \times (2.1 \times 10^5) \times 50 \times 30}$$

$$= \textbf{0.8 mm} \quad \textbf{Ans.}$$

Example 9.31: Two solid circular cross-section conical bars, one of titanium and the other of steel, are joined together as shown. The system is subjected to concentric axial tensile force of 500 kN together with an axisymmetric ring load applied at the junction of the bars having a horizontal resultant of 1000 kN. Determine the change in length of the system. For titanium, $E = 110$ GPa and for

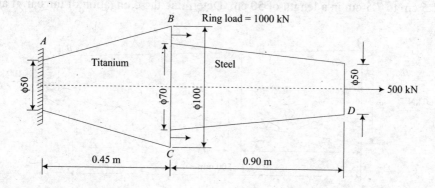

steel, $E = 200$ GPa.

Solution:

For titanium cone AB,

$$P = 1000 + 500 = 1500 \text{ kN} = 1500 \times 10^3 \text{ N}$$

$$L = 0.45 \text{ m} = 450 \text{ mm}$$

$$D_1 = 100 \text{ mm}$$

$$D_2 = 50 \text{ mm}$$

$$E = 110 \text{ GPa} = 1.1 \times 10^5 \text{ N/mm}^2$$

$$\delta L_1 = \frac{4PL}{\pi E D_1 D_2} = \frac{4 \times (1500 \times 10^3) \times 450}{\pi \times (1.1 \times 10^5) \times 100 \times 50} = 1.5626 \text{ mm}$$

For steel cone CD,

$$P = 500 \text{ kN} = 500 \times 10^3 \text{ N}$$

$$L = 0.90 \text{ mm} = 900 \text{ mm}$$

$$D_1 = 70 \text{ mm}$$

$$D_2 = 50 \text{ mm}$$

$$E = 2 \times 10^5 \text{ N/mm}^2$$

$$\delta L_2 = \frac{4PL}{\pi E D_1 D_2} = \frac{4 \times (50 \times 10^3) \times 900}{\pi \times (2 \times 10^5) \times 70 \times 50} = 0.8185 \text{ mm}$$

Total increase of length of the system,

$$\delta L = \delta L_1 + \delta L_2 = 1.5626 + 0.8185 \text{ mm}$$

$$= \mathbf{2.38 \text{ mm}} \quad \mathbf{Ans.}$$

Example 9.32: A steel bar *AB* of uniform thickness 2 cm, tapers uniformly from 1.5 cm to 7.5 cm in a length of 50 cm. Determine the elongation of the bar, if an

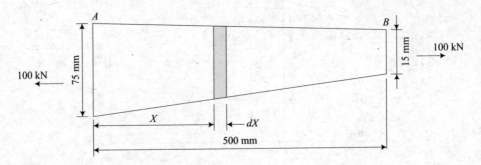

axial tensile force of 100 kN is applied on it. $E = 2 \times 10^5$ N/mm.

Solution:

Width of bar at a distance *X* from *A*,

$$b_X = b_A - \left(\frac{b_A - b_B}{L}\right)X = 75 - \left(\frac{75-15}{500}\right)X$$

$$= 75 - 0.12X$$

Area, $\quad A_X = 20\, b_X = 1500 - 2.4X$

Stress, $\quad \sigma_X = \dfrac{P}{A_X} = \dfrac{100 \times 10^3}{(1500 - 2.4X)} = \dfrac{41667}{(625 - X)}$

Strain, $\quad \varepsilon_X = \dfrac{\sigma_X}{E} = \dfrac{41667}{2 \times 10^5 (625 - X)} = \dfrac{1}{4.8(625 - X)}$

$$\delta L_X = \varepsilon_X d_X = \frac{dX}{4.8(625 - X)}$$

Total elongation of bar.

$$\delta L = \int_0^{500} \frac{dX}{4.8(625-X)} = \frac{1}{4.8(-1)} \left| \ln(625-X) \right|_0^{500}$$

$$= -\frac{1}{4.8}(\ln 125 - \ln 625)$$

$$= \frac{1}{4.8}(\ln 625 - \ln 125) = \frac{1}{4.8} \ln \frac{625}{125}$$

$$= 0.335 \text{ mm} \quad \textbf{Ans.}$$

Example. 9.33:   A beam weighing 500 N is held in horizontal position by three wires. The outer wires are of brass of 1.2 mm diameter and attached to each end of the beam. The central wire is of steel of 0.6 mm diameter and attached to the middle of the beam. The beam is rigid and the wires are of the same length and unstrained before the beam is attached. Determine the stress induced in each wire. The Young's modulus of brass is 80 GN/m² and that of steel is 200 GN/m².

Solution:

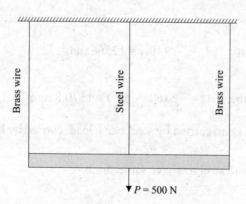

The total load,

$$P = 2P_b + P_S$$

where P_b = load on brass wire

P_s = load on steel wire

∴ $P = 2\sigma_b A_b + \sigma_s A_s$

Now, the wires have same strains.

$$\varepsilon_s = \varepsilon_b \text{ and } \delta L_s = \delta L_b$$

∴ $\dfrac{\sigma_s}{E_s} = \dfrac{\sigma_b}{E_b}$

or,
$$\sigma_s = \frac{E_S}{E_b} \cdot \sigma_b = \frac{200}{80} \sigma_b = 2.5\sigma_b$$

$$500 = 2\sigma_b\left(\frac{\pi}{4} \times 1.2^2\right) + 2.5\,\sigma_b\left(\frac{\pi}{4} \times 0.6^2\right) = 2.9673\sigma_b$$

$\therefore$
$$\sigma_b \frac{500}{2.9673} = 168.5 \text{ N/mm}^2 \quad \text{Ans.}$$

$$\sigma_s = 2.5\sigma_b = 2.5 \times 168.5$$
$$= 421.25 \text{ N/mm}^2 \quad \text{Ans.}$$

Example 9.34: A brass tube of 60 mm outside diameter completely encloses a steel bar of 40 mm diameter. The composite system measures 300 mm in length and carries an axial thrust which induces a stress of 50 N/mm^2 in brass tube. Determine:
(a) Stress developed in steel bar.
(b) Magnitude of compressive force and change in length of composite bar.
   $(E_s = 210 \text{ GPa}, E_b = 105 \text{ GPa})$

Solution:

Area of steel bar, $\quad A_s = \dfrac{\pi}{4}\,(40^2) = 12566 \text{ mm}^2.$

Area of brass tube, $\quad A_b = \dfrac{\pi}{4}\,(60^2 - 40^2) = 1570.8 \text{ mm}^2.$

(a) Total load = Load carried by steel bar + load carried by brass tube.
$$P = P_S + P_b$$
The brass tube and steel rod have same strain
$$\varepsilon_b = \varepsilon_s$$

$\therefore$
$$\frac{\sigma_b}{E_b} = \frac{\sigma_s}{E_s}$$

or
$$\sigma_s = \frac{E_s}{E_b} \cdot \sigma_b = \frac{210}{105} \times 50 = 100 \text{ N/mm}^2 \quad \text{Ans.}$$

(b) Again, $\quad\quad P = P_b + P_s$
$$= \sigma_b A_b + \sigma_s A_s$$
$$= 50 \times 1570.8 + 100 \times 12566$$
$$= 204200 \text{ N} = 204.2 \text{ kN}.$$

Extension of tube = Extension of rod
$$\delta L_b = \delta L_s$$

$$= \frac{\sigma_b L_b}{E_b} = \frac{P_b}{A_b} \cdot \frac{L_b}{E_b}$$

$$= \frac{50 \times 300}{1.05 \times 10^5}$$

$$= 0.14285 \text{ mm} \quad \textbf{Ans.}$$

9.9 EXTENSION OF BAR UNDER OWN WEIGHT

9.9.1 Bar of Uniform Section

A bar of cross-sectional area A and length L hanging vertical will extend under its own weight.

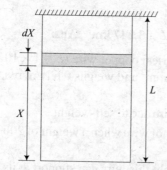

Fig. 9.15 Uniform bar under own weight

The weight of bar of length X will be:
$$P = \rho g A X$$
This load will extend from dX by say δX.

∴
$$\delta X = \frac{P dX}{AE} = \frac{\rho g A X}{AE} dx = \frac{\rho g}{E} \times dx.$$

Total change in length of bar under its own weight can be found out by integrating the above equation.

∴
$$\delta L = \int_{O}^{L} \frac{\rho g}{E} \times dX = \frac{\rho g}{E} \left| \frac{y^2}{2} \right|_{0}^{L}$$

$$= \frac{\rho g}{E} \cdot \frac{L^2}{2}.$$

If $\qquad W = \rho g A L = $ Total weight of bar

$\therefore \qquad \delta L = \dfrac{W}{AL} \cdot \dfrac{L^2}{2E} = \dfrac{WL}{2AE}$.

Therefore, extension of bar under its own weight is equal to extension of bar if effective load is half of its weight applied at the bar end.

Example 9.35: Determine the maximum length of mild steel bar of uniform cross-section that can be suspended vertically if the maximum stress is not to exceed 250 MN/m^2. Given, $g = 9.81 \text{ m/sec}^2$ and density of steel, $\rho = 7.85 \times 10^3 \text{ kg/m}^3$.

Solution: The maximum tensile force at the suspension point due to weight of rod,

$$P = \rho g A L = (7.85 \times 10^3 \times 9.81) A L (N)$$

$$\sigma_{max} = \dfrac{P}{A} = \dfrac{(7.85 \times 10^3 \times 9.81) A L}{A}$$

$$= 77 \times 10^3 \, L \text{ N/m}^2$$

$\therefore \qquad 77 \times 10^3 \, L = 250 \times 10^6$

$$L = \textbf{3.247 km} \quad \textbf{Ans.}$$

Example 9.36: An electrical copper wire ($E = 1 \times 10^5 \text{ N mm}^2$), 40 m long has cross-sectional area of 80 mm^2 and weighs 0.6 N/m run. If the wire is suspended vertically, calculate:

    (a) the elongation of wire due to self-weight.

    (b) the total elongation of wire when a weight of 200 N is attached to its lower end, and

    (e) the maximum weight this wire can support at its lower end if the limiting value of stress is 65 N/mm^2.

Solutions:

    (a) Elongation of wire due to self-weight,

$$\delta L = \dfrac{WL}{2AE} = \dfrac{(wL)L}{2AE} = \dfrac{(0.6 \times 40) \times 40 \times 10^3}{2 \times 80 \times (1 \times 10^5)}$$

$$= \textbf{0.06 mm} \quad \textbf{Ans.}$$

    (b) Elongation of wire due to external load,

$$\delta L_1 = \dfrac{PL}{AE} = \dfrac{200 \times (40 \times 10^3)}{80 \times (1 \times 10^5)}$$

$$= 1.00 \text{ mm.}$$

Total extension of wire $= \delta L + \delta L_1 = 0.01 + 1.0$

$$= 1.00 \text{ mm} \quad \textbf{Ans.}$$

(c) Stress due to self-weight,

$$\sigma = \frac{W/2}{A} = \frac{wL^2}{2A} = \frac{0.6 \times 40 \times (40 \times 10)}{2 \times 80}$$

$$= 0.15 \text{ N/mm}^2$$

The balance stress, $\sigma_1 = 650 - 0.15 = 64.85$ N/mm^2

Maximum weight which can be supported by wire

$$P = \sigma_1 A = 64.85 \times 80 = \textbf{5188 N} \quad \textbf{Ans.}$$

9.10 STRAIN ENERGY

When an external force acts on an elastic material it deforms and internal resistance is developed in the material due to molecular cohesion. The work done by the interval resistance is stored as strain energy within elastic limits. This strain energy is called the resilience.

The maximum strain energy absorbed in this material when loaded upto elastic limit is called *proof resilience*.

Resilience, $\qquad U = \dfrac{\sigma^2}{2E} \times$ volume

Proof resilience, $\qquad U_e = \dfrac{\sigma_e^{\ 2}}{2E} \times$ volume.

where, σ = stress developed when the material is loaded within elastic limit.

σ_e = Proof stress developed when the material is loaded upto elastic limit.

The proof resilience (maximum strain energy at elastic limit) per unit volume of material is called *modulus of resilience*.

Modulus of resilience, $\qquad \dfrac{U_e}{V} = \dfrac{\sigma_e^{\ 2}}{2E}$.

9.10.1 Strain Energy due to Gradually Applied Load

During gradually applied load the load starts from zero and reaches upto a value P linearly. The work done in straining the material will be given by the shaded area under the load-extension curve.

The strain energy,

$$U = \frac{P}{2} \delta L$$

But, $\qquad P = \sigma A$

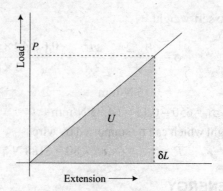

Fig. 9.16 Strain energy due to gradually applied load

and
$$\delta L = \varepsilon L = \frac{\sigma L}{E}$$

$\therefore$
$$U = \frac{1}{2} \sigma A \times \frac{\sigma L}{E}$$

$$= \frac{1}{2} \frac{\sigma^2}{E} AL = \frac{1}{2} \frac{\sigma^2}{E} \times \text{volume}$$

Modulus of resilience $= \dfrac{U}{V} = \dfrac{1}{2} \dfrac{\sigma^2}{E} = \dfrac{1}{2} \sigma . \dfrac{\sigma}{E} = \dfrac{\sigma . \varepsilon}{2}$

$$= \frac{\text{Stress} \times \text{Strain}}{2}$$

9.10.2 Suddenly Applied Load

The load is applied suddenly and it remains constant during extension of the material. The area of load – extension curve will be a rectangle.

Strain energy,
$$U = P.\delta L$$

$$\sigma = \frac{2P}{A} \text{ for suddenly applied load.}$$

The strain energy,

$$U = P. \left(\frac{PL}{AE} \right)$$

$$= \left(\frac{P}{A} \right)^2 . \frac{AL}{E} = \frac{\sigma^2}{E} . \text{volume}$$

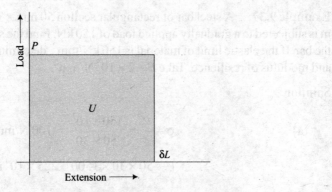

Fig. 9.17 Strain energy due to suddenly applied load

Modulus of resilience, $= \dfrac{U}{V} = \dfrac{\sigma^2}{E}$.

9.10.3 Impact Loading

When a weight P is dropped on a member from a height h, the falling weight causes impact loading. The strain energy is equal to the loss of potential energy of the weight.

$$U = P(h + \delta L)$$

where, $h =$ Free height of fall of weight

$\delta L =$ Extension due to loading

$\therefore$ $U = Ph + P.\delta L$

$$= Ph + \sigma^2 \frac{AL}{E}$$

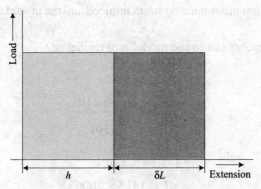

Fig. 9.18 Strain energy due to impact loading

Example 9.37: A steel bar of rectangular section 50 mm × 30 mm and length 1.5 m is subjected to a gradually applied load of 150 kN. Find the strain energy stored in the bar. If the elastic limit of material is 150 kN/mm², determine the proof resilience and modulus of resilience. Take $E = 2 \times 10^5$ N/mm².

Solution:

(a)
$$\sigma = \frac{P}{A} = \frac{150 \times 10^3}{50 \times 30} = 100 \text{ N/mm}^2$$

$$V = 50 \times 30 \times 1500 = 225 \times 10^4 \text{ mm}^2$$

∴
$$U = \frac{\sigma^2}{2E} V = \frac{(100)^2}{2 \times (20 \times 10^5)} \times (225 \times 10^4)$$

$$= 56250 \text{ N-mm.}$$

(b) Proof resilience
$$= U_e = \frac{\sigma_e^2}{2E} \cdot V$$

$$= \frac{(150)^2}{2 \times (2 \times 10^5)} \times (225 \times 10^4)$$

$$= 144000 \text{ N-mm.}$$

(c) Modulus of resilience $= \dfrac{U_e}{V} = \dfrac{144000}{225 \times 10^4} = 0.064$ N/mm²

Example 9.38: A steel rod of 30 mm diameter and 2 m long has a pull of 50 kN suddenly applied to it. Calculate:

(a) Maximum instantaneous stress induced and the instantaneous elongation, and

(b) Strain energy stored and modulus of resilience.

Solution:

(a)
$$\sigma = \frac{2P}{A} = \frac{2 \times 50 \times 10^3}{\frac{\pi}{4}(30)^2}$$

$$= 141.54 \text{ N/mm}^2$$

$$\sigma_L = \frac{\sigma L}{E} = \frac{141.5 \times 2000}{(2 \times 10^5)} = 1.415 \text{ mm}$$

(b) Volume, $V = \dfrac{\pi}{4} D^2 \; L = \dfrac{\pi}{4}(30)^2 \times 2000$

$$= 14.13 \times 10^5 \, \text{mm}^3$$

$$U = \dfrac{\sigma^2}{2E} \cdot V = \dfrac{(141.54)^2}{2 \times (2 \times 10^5)} \times 14.13 \times 10^5$$

$$= 70768.6 \, \text{N-mm}$$

Modulus of resilience,

$$\dfrac{U}{V} = \dfrac{70768.6}{14.13 \times 10^5} = \textbf{0.05008 N/mm}^2$$

9.11 COMPOUND STRESS AND STRAIN

The tensile and compressive stresses act on planes normal to the direction of external tensile and compressive forces. The shear stresses are induced on planes parallel to the direction of external shear forces. On planes inclined to applied external forces, both normal and shear stresses are induced. The combination of normal and shear stress is called compound stress. In most of the situations, the machine and structure members are subjected to two or more stresses simultaneously. These members have to be designed and analysed for compound stresses and strains.

9.11.1 State of Plane Stress

A cubic element *ABCD* within a strained material, when isolated, is found to be subjected to normal stress σ_x, σ_y and σ_z and shear stresses parallel to planes. Such a system is called three-dimensional stress system.

However, if on planes parallel to plane *ABCD*, there is no stress ($\sigma_z = 0$), the cubic element is said to be under 2-D stress system or plane stress condition.

The following stresses will be acting on the isolated element:

1. The parallel planes *AD* and *BC* are under tensile stress, σ_x and shear stress, τ_{xy}.

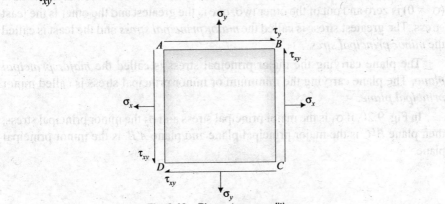

Fig. 9.19 Plane stress condition

2. The parallel planes AB and CD are under tensile stress, σ_y and shear stress (or complimentary shear stress), τ_{xy}.

9.11.2 Principal Planes

At any point within a stressed body, no matter how complex the state of stress may be, there always exist three mutually perpendicular planes on each of which the resultant stress is a normal stress (shear stress being zero). These mutually perpendicular planes are called principal planes.

In Fig. 9.20, the isolated element $ABCD$ has been rotated anti-clockwise through an angle θ to new position $A'B'CD'$. The stresses acting in the new position are normal stresses σ_1 and σ_2 and the shear stresses on all faces are absent. Then these mutually perpendicular planes, $B'C$ and $A'B'$ are called *principal planes*.

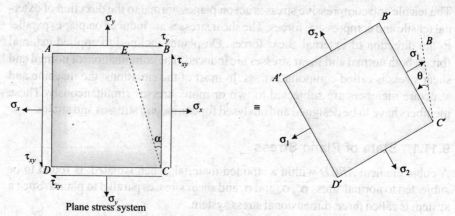

Plane stress system

Fig. 9.20 Principal planes

9.11.3 Principal Stresses and Strains

The resultant normal stresses (σ_1 and σ_2) acting on principal planes are called *principal stresses*. In case of two-dimentional system, one of the principal stresses ($\sigma_3 = 0$) is zero and out of the other two, one is the greatest and the other is the least stress. The greatest stress is called the *major principal stress* and the least is called the *minor principal stress*.

The plane carrying the major principal stress is called the *major principal plane*. The plane carrying the minimum or minor principal stress is called minor *principal plane*.

In Fig. 9.20, if σ_1 is the major principal stress and σ_2 the minor principal stress, then plane $B'C$ is the major principal plane and plane $A'B'$ is the minor principal plane.

9.12 DETERMINATION OF PRINCIPAL STRESSES AND STRAINS

There are two methds for determination of principal stresses and strains.

1. Analytical method by the use of mathematical expression.
2. Graphical method by the use of Mohr's circle.

(a) The principal stresses can be calculated from the following general expresson.

The principal stress,

$$\sigma_p = \frac{1}{2}(\sigma_x + \sigma_y) \pm \frac{1}{2}\sqrt{(\sigma_x - \sigma_y)^2 + 4\tau_{xy}^2}$$

Major principal stress,

$$\sigma_1 = \frac{1}{2}(\sigma_x + \sigma_y) + \frac{1}{2}\sqrt{(\sigma_x - \sigma_y)^2 + 4\tau_{xy}^2}$$

Minor principal stress

$$\sigma_2 = \frac{1}{2}(\sigma_x + \sigma_y) - \frac{1}{2}\sqrt{(\sigma_x - \sigma_y)^2 + 4\tau_{xy}^2}$$

where, σ_x = Normal stress along x-axis.

σ_y = Normal stress along y-axis.

τ_{xy} = Shear stress acting parallel to planes.

Sign Convention

1. Tensile stress is positive
2. Compressive stress is negative
3. Clockwise shear stress is positive
4. Anticlockwise shear stress is negative.

(b) Maximum shear stress can be calculated as:

$$\tau_{max} = \frac{1}{2}\sqrt{(\sigma_x - \sigma_y)^2 + 4\tau_{xy}^2}$$

$$= \frac{1}{2}(\sigma_1 - \sigma_2).$$

(c) Principal planes

If θ_1 is the angle between plane of major principal stress σ_1 with the plane of σ_x and θ_2 is the angle of plane of minor principal stress σ_2 with the plane of σ_x, then

$$\tan 2\theta = \frac{2\tau_{xy}}{(\sigma_x - \sigma_y)}$$

$$\tan \theta_1 = \frac{\sigma_1 - \sigma_x}{\tau_{xy}}$$

$$\tan \theta_2 = \frac{\sigma_2 - \sigma_x}{\tau_{xy}}$$

Maximum shear stress occurs on planes at 45° to the principal planes.

(a) **Principal strains.**

Principal strains are the greatest and least direct strains in a material subjected to complex stresses. These strains are produced in the direction of the principal stresses.

The major principal strain,

$$\varepsilon_1 = \frac{\sigma_1}{E} - \frac{\sigma_2}{mE} = \frac{\sigma_1}{E} - \frac{v\sigma_2}{E}$$

The minor principal strain,

$$\varepsilon_2 = \frac{\sigma_2}{E} - \frac{v\sigma_1}{E}.$$

σ_1 and σ_2 are the major and minor principal stresses and v or $\frac{1}{m}$ is the Poisson's ratio.

Example 9.39: In an elastic material, the direct stresses of 120 MN/m² and 90 MN/m² are applied at a certain point on planes at right angles to each other in tension and compression respectively. Estimate the shear stress to which material could be subjected, if the maximum principal stress is 150 MN/m².

Also find the magnitude of other principal stress and its inclination to 120 MM/m². [U.P.T.U. II Sem., 2000-01]

Solution:

$$\sigma_x = 120 \text{ MN/m}^2 \text{ (T)}$$
$$\sigma_y = -90 \text{ MN/m}^2 \text{ (C)}$$
$$\sigma_1 = 150 \text{ MN/m}^2$$

Now,

$$\sigma_1 = \frac{\sigma_x + \sigma_y}{2} + \sqrt{\left(\frac{\sigma_x - \sigma_y}{2}\right)^2 + \tau_{xy}^2}$$

$$150 = \frac{120 - 90}{2} + \sqrt{\left(\frac{120 + 90}{2}\right)^2 + \tau_{xy}^2}$$

∴

$$\tau_{xy} = 84.85 \text{ MN/m}^2.$$

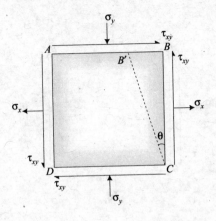

$$\therefore \quad \sigma_2 = \frac{\sigma_x + \sigma_y}{2} - \sqrt{\left(\frac{\sigma_x - \sigma_y}{2}\right)^2 + \tau_{xy}^2}$$

$$= \frac{120 - 90}{2} - \sqrt{\left(\frac{120 + 90}{2}\right)^2 + (84.85)^2}$$

$$= -120 \text{ N/mm}^2$$

$$= 120 \text{ N/mm}^2 \text{ (C)}.$$

Angle of minor principal plane,

$$\tan \theta_2 = \frac{\sigma_2 - \sigma_x}{\tau_{xy}} = \frac{-120 - 120}{84.85}$$

$$\theta_2 = -70.5° \text{ or } 19.5°$$

Angle of plane of σ_2 with plane of σ_x i.e., angle of σ_2 with σ_x.

Example 9.40: A load-carrying member is subjected to the following stress conditions:

Tensile stress, $\sigma_x = 400$ MPa

Tensile stress, $\sigma_y = 300$ MPa

Shear stress, $\tau_{xy} = 200$ MPa clockwise

Obtain :

(i) Principal stresses and their planes

(ii) The maximum shear stress and its plane.   [U.P.T.U. II Sem., 2001-02]

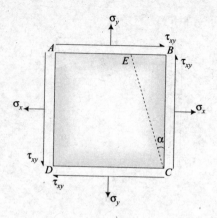

Solution:

(1) Principal stresses

$$\sigma_1 = \frac{\sigma_x + \sigma_y}{2} + \sqrt{\left(\frac{\sigma_x - \sigma_y}{2}\right)^2 + \tau_{xy}^2}$$

$$= \frac{400 - 300}{2} + \sqrt{\left(\frac{400 + 300}{2}\right)^2 + 200^2}$$

$$= 50 + 403$$

$$= 453 \text{ MN/ m}^2 \text{ (T)}$$

$$\sigma_2 = 50 - 403$$

$$= 353 \text{ MN/m}^2 \text{ (C)}.$$

$$\tan \theta_1 = \frac{\sigma_1 - \sigma_x}{\tau_{xy}} = \frac{453 - 400}{200}$$

$\therefore$ $\theta_1 = 14.84°$ Angle of plane of σ_1 with plane of σ_x (i.e. *BC*)

$$\tan \theta_2 = \frac{\sigma_2 - \sigma_x}{\tau_{xy}} = \frac{353 - 400}{200}$$

$\therefore$ $\theta_2 = -13.2°$ or $90 + 14.84° = 104.84°$. Angle of plane of σ_2 will plane of σ_x.

$$\tau_{max} = \sqrt{\frac{(\sigma_x - \sigma_y)^2}{2} + 4\tau_{xy}^2} = 403 \text{ MN/m}^2$$

Maximum shear stress marks an angle of 45° with principal plane.

Example 9.41: Find the principal stresses for the state of stress given below:

[U.P.T.U. II Sem., 2001-02]

Solution:

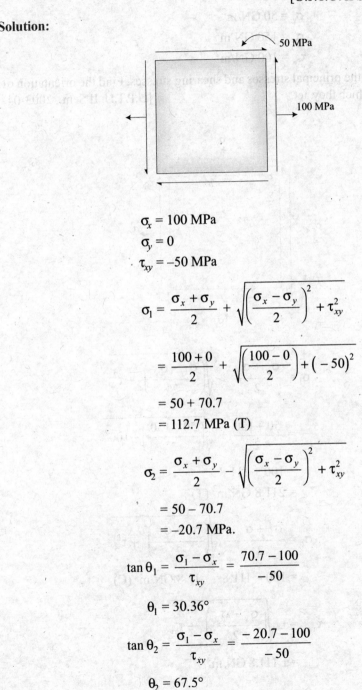

$$\sigma_x = 100 \text{ MPa}$$

$$\sigma_y = 0$$

$$\tau_{xy} = -50 \text{ MPa}$$

$$\sigma_1 = \frac{\sigma_x + \sigma_y}{2} + \sqrt{\left(\frac{\sigma_x - \sigma_y}{2}\right)^2 + \tau_{xy}^2}$$

$$= \frac{100 + 0}{2} + \sqrt{\left(\frac{100 - 0}{2}\right) + (-50)^2}$$

$$= 50 + 70.7$$

$$= 112.7 \text{ MPa (T)}$$

$$\sigma_2 = \frac{\sigma_x + \sigma_y}{2} - \sqrt{\left(\frac{\sigma_x - \sigma_y}{2}\right)^2 + \tau_{xy}^2}$$

$$= 50 - 70.7$$

$$= -20.7 \text{ MPa}.$$

$$\tan \theta_1 = \frac{\sigma_1 - \sigma_x}{\tau_{xy}} = \frac{70.7 - 100}{-50}$$

$$\theta_1 = 30.36°$$

$$\tan \theta_2 = \frac{\sigma_1 - \sigma_x}{\tau_{xy}} = \frac{-20.7 - 100}{-50}$$

$$\theta_2 = 67.5°$$

Example 9.42: The state of stress at a point in a loaded component is found to be as given below:

$$\sigma_x = 50 \text{ GN/m}^2$$
$$\sigma_y = 150 \text{ GN/m}^2$$
$$\tau_{xy} = 100 \text{ GN/m}^2$$

Determine the principal stresses and shearing stresses. Find the orientation of the planes on which they act. [U.P.T.U. II Sem., 2003-04]

Solution:

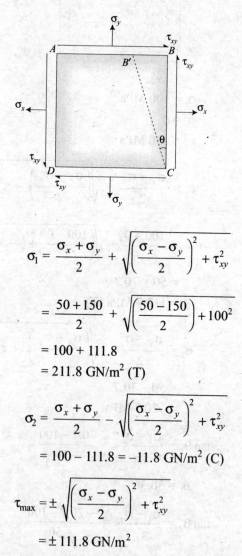

$$\sigma_1 = \frac{\sigma_x + \sigma_y}{2} + \sqrt{\left(\frac{\sigma_x - \sigma_y}{2}\right)^2 + \tau_{xy}^2}$$

$$= \frac{50 + 150}{2} + \sqrt{\left(\frac{50 - 150}{2}\right) + 100^2}$$

$$= 100 + 111.8$$

$$= 211.8 \text{ GN/m}^2 \text{ (T)}$$

$$\sigma_2 = \frac{\sigma_x + \sigma_y}{2} - \sqrt{\left(\frac{\sigma_x - \sigma_y}{2}\right)^2 + \tau_{xy}^2}$$

$$= 100 - 111.8 = -11.8 \text{ GN/m}^2 \text{ (C)}$$

$$\tau_{max} = \pm \sqrt{\left(\frac{\sigma_x - \sigma_y}{2}\right)^2 + \tau_{xy}^2}$$

$$= \pm 111.8 \text{ GN/m}^2$$

$$\tan \theta_1 = \frac{\sigma_1 - \sigma_x}{\tau_{xy}} = \frac{211.8 - 50}{100}$$

$$\theta_1 = 58.28°$$

$$\tan \theta_2 = \frac{\sigma_2 - \sigma_x}{\tau_{xy}} = \frac{-11.8 - 50}{100}$$

$$\theta_2 = -31.7° \text{ or } 58.28 + 90 = 148.28°$$

τ_{max} marks an angle of 45° with the principal stress or 45° + 58.28 = 103.28°.

Example 9.43: A rectangular element is subjected to a plane stress system as shown. Determine the principal stresses. [U.P.T.U. II Sem., 2003-04]

Solution:

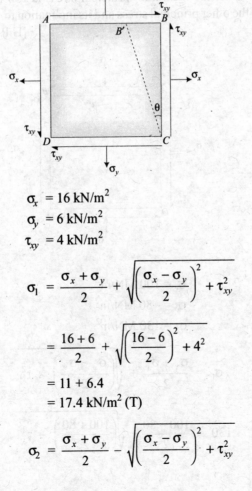

$$\sigma_x = 16 \text{ kN/m}^2$$
$$\sigma_y = 6 \text{ kN/m}^2$$
$$\tau_{xy} = 4 \text{ kN/m}^2$$

$$\sigma_1 = \frac{\sigma_x + \sigma_y}{2} + \sqrt{\left(\frac{\sigma_x - \sigma_y}{2}\right)^2 + \tau_{xy}^2}$$

$$= \frac{16 + 6}{2} + \sqrt{\left(\frac{16 - 6}{2}\right)^2 + 4^2}$$

$$= 11 + 6.4$$

$$= 17.4 \text{ kN/m}^2 \text{ (T)}$$

$$\sigma_2 = \frac{\sigma_x + \sigma_y}{2} - \sqrt{\left(\frac{\sigma_x - \sigma_y}{2}\right)^2 + \tau_{xy}^2}$$

$$= 11 - 6.4$$
$$= 4.6 \text{ kN/m}^2 \text{(T)}$$

$$\tan \theta_1 = \frac{\sigma_1 - \sigma_x}{\tau_{xy}} = \frac{17.4 - 16}{4} = 0.35°$$

$$\theta = 19.3°$$

$$\tan \theta_2 = \frac{\sigma_2 - \sigma_x}{\tau_{xy}} = \frac{4.6 - 16}{4} = -2.85$$

$$\theta_2 = -70.67°.$$

Example 9.44: In an elastic material the direct stresses of 100 MN/m^2 and 80 MN/m^2 are applied at certain point on planes at right angle to each other in tension and compression respectively. Estimate the shear stress to which material can be subjected, if the maximum principal stress is 130 MN/m^2. Also find the magnitude of the other principal stress and its inclination to 100 MN/m^2 stress.

[U.P.T.U. I Sam., 2005-06]

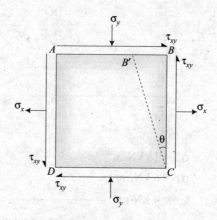

Solution:

$$\sigma_x = 100 \text{ MN/m}^2$$
$$\sigma_y = -80 \text{ MN/m}^2$$
$$\sigma_1 = 130 \text{ MN/m}^2.$$

$$\sigma_1 = \frac{\sigma_x + \sigma_y}{2} + \sqrt{\left(\frac{\sigma_x - \sigma_y}{2}\right)^2 + \tau_{xy}^2}$$

$$130 = \frac{100 - 80}{2} + \sqrt{\left(\frac{100 + 80}{2}\right)^2 + \tau_{xy}^2}$$

$$90^2 + \tau_{xy}^2 = 120^2$$

$$\tau_{xy} = \sqrt{120^2 - 90^2} = \textbf{79.4 MN/m}^2 \quad \textbf{Ans.}$$

$$\sigma_2 = \frac{\sigma_x + \sigma_y}{2} - \sqrt{\left(\frac{\sigma_x - \sigma_y}{2}\right)^2 + \tau_{xy}^2}$$

$$= \frac{100 - 80}{2} - \sqrt{\left(\frac{100 + 80}{2}\right)^2 + (79.4)^2}$$

$$= 20 - 120 = -100 \text{ MN/m}^2 \text{ (compressive)}$$

$$\tan \theta_2 = \frac{\sigma_2 - \sigma_x}{\tau_{xy}} = \frac{-100 - 100}{79.4} = -2.51889$$

$$\theta_2 = -68.3° \text{ or } 21.65°.$$

Example 9.45: At a point in a beam, the bending stress is 50 MPa tensile and shear stress is 20 MPa. Determine:

(i) The principal stresses and maximum shear stress.

(ii) The tensile stress which acting alone would produce the same maximum shear stress as in (i) above.

(iii) The shear stress which acting alone would produce the same principal stress as in (i) above.

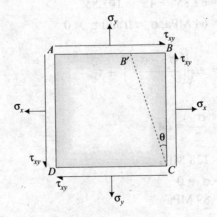

Solution:

(i) $\qquad\qquad \sigma_x = \sigma_b = \text{MPa}$

$\qquad\qquad\quad \sigma_y = 0$

$$\tau_{xy} = 20 \text{ MPa}$$

$$\sigma_1 = \frac{\sigma_x + \sigma_y}{2} + \sqrt{\left(\frac{\sigma_x - \sigma_y}{2}\right)^2 + \tau_{xy}^2}$$

$$= \frac{50 - 0}{2} + \sqrt{\left(\frac{50 - 0}{2}\right)^2 + 20^2}$$

$$= 25 + 64 = 89 \text{ MPa}.$$

$$\sigma_2 = \frac{\sigma_x + \sigma_y}{2} - \sqrt{\left(\frac{\sigma_x - \sigma_y}{2}\right)^2 + \tau_{xy}^2}$$

$$= 25 - 64 = -39 \text{ MPa}.$$

$$\tan \theta_1 = \frac{\sigma_1 - \sigma_x}{\tau_{xy}} = \frac{89 - 56}{20} = 1.95$$

$\theta_1 = 62.85°$ i.e., angle of plane of σ_1 will plane of σ_x.

$\theta_2 = \theta_1 + 90° = 62.85 + 90° = 152.85$, i.e., angle of plane of σ_2 with plane of σ_x.

$$\tau_{max} = \frac{\sigma_1 - \sigma_2}{2} = \frac{89 + 39}{2} = 64 \text{ MPa}$$

τ_{max} acts on a plane at 45° to the principal plane.

$$\theta_s = 62.85° + 45° = 107.85°.$$

(ii) If $\quad \tau_{max} = 64 \text{ MPa}, \sigma_y = 0$ and $\tau_{xy} = 0$

$$\tau_{max} = \pm\sqrt{\left(\frac{\sigma_x - \sigma_y}{2}\right)^2 + \tau_{xy}^2}$$

$$= \frac{\sigma_x}{2} = 64$$

$\therefore \qquad \sigma_x = 128 \text{ MPa}$

(iii) If $\quad \sigma_x = \sigma_y = 0$

$\qquad\qquad \sigma_1 = 89 \text{ MPa}$

$$\sigma_1 = \frac{\sigma_x + \sigma_y}{2} + \sqrt{\left(\frac{\sigma_x - \sigma_y}{2}\right)^2 + \tau_{xy}^2}$$

$$= \tau_{xy}$$

$\therefore \qquad \tau_{xy} = 89 \text{ MPa}$

Example 9.46: The state of stress at a given point in a material component is shown. The loading on the component is increased by a factor of k. Determine the maximum value of k if the material can withstand maximum normal and shear stress of 300 MPa and 200 MPa respectively.

Solution: The new values of stresses are

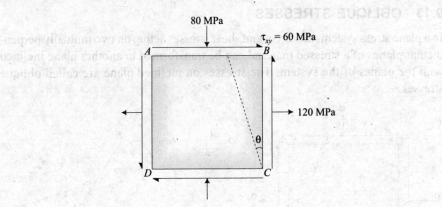

$$\sigma_x = 120 \, k \text{ MPa}$$
$$\sigma_y = -80 \, k \text{ MPa}$$
$$\tau_{xy} = 60 \, k \text{ MPa}.$$

The maximum normal stress,

$$\sigma_1 = \frac{\sigma_x + \sigma_y}{2} + \sqrt{\left(\frac{\sigma_x - \sigma_y}{2}\right)^2 + \tau_{xy}^2}$$

$$\sigma_1 = \frac{120k - 80k}{2} + \sqrt{\left(\frac{12k + 80k}{2}\right)^2 + (60k)^2}$$

$$= 20\,k + 116.6\,k$$

$$= 136.62\,k = 300 \text{ MPa}$$

$$\therefore \qquad k = \frac{300}{136.62} = 2.196$$

Maximum shear stress,

$$\tau_{max} = \sqrt{\left(\frac{\sigma_x - \sigma_y}{2}\right)^2 + \tau_{xy}^2}$$

$$= \sqrt{\left(\frac{12k + 80k}{2}\right)^2 + (60k)^2}$$

$$\tau_{max} = 116.6 \ k = 200 \text{ MPa}$$

$$\therefore \qquad k = \frac{200}{116.6} = 1.715$$

Maximum permissible value of k is **1.715 Ans.**

9.13 OBLIQUE STRESSES

In a plane stress system, the direct and shear stresses acting on two mutually perpendicular planes of a stressed material can be transformed to another plane inclined with the planes of the system. The stresses on inclined plane are called oblique stresses.

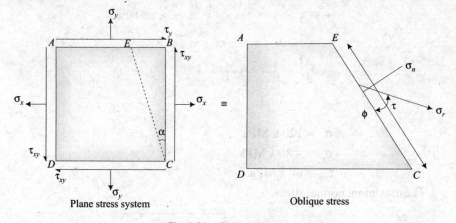

Plane stress system   Oblique stress

Fig. 9.21 Oblique stresses

Oblique stresses consisting of a normal stress σ_n and shear stress τ are acting on oblique plane CE which is at an angle α with plane BC in the anti-clockwise direction.

$$\sigma_n = \frac{\sigma_x + \sigma_y}{2} + \frac{\sigma_x - \sigma_y}{2} \cos 2\alpha + \tau_{xy} \sin 2\alpha$$

$$\tau = \frac{\sigma_x - \sigma_y}{2} \sin 2\alpha - \tau_{xy} \cos 2\alpha.$$

The resultant stress,

$$\sigma_r = \sqrt{\sigma_n^2 + \tau^2}$$

The inclination of resultant stress with the oblique plane can be calculated as:

$$\tan \phi = \frac{\sigma_n}{\tau}.$$

The oblique stresses can also be found out graphically by the use of Mohr's circle.

Note: 1. If σ_y is compressive stress,

$$\sigma_n = \frac{\sigma_x - \sigma_y}{2} + \frac{\sigma_x + \sigma_y}{2}\cos2\alpha + \tau_{xy}\sin2\alpha$$

$$\tau = \frac{\sigma_x + \sigma_y}{2}\sin 2\alpha - \tau_{xy}\cos 2\alpha.$$

2. If shear stress is absent,

$$\tau_{xy} = 0$$

$$\sigma_n = \frac{\sigma_x + \sigma_y}{2} + \frac{\sigma_x - \sigma_y}{2}\cos2\alpha$$

$$\tau = \frac{\sigma_x + \sigma_y}{2}\sin 2\alpha.$$

Example 9.47: The principal stresses at a point in a plane are 200 N/mm² (tensile) and 100 N/mm² (compressive). Determine the magnitude and direction of the resultant stress on a plane making an angle of 60° with the axis of major principal stress.

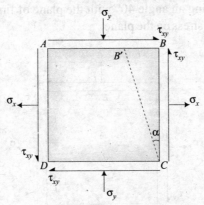

Solution:

$$\sigma_x = 200 \text{ N/mm}^2$$

$$\sigma_y = -100 \text{ N/mm}^2$$

$$\alpha = 90° - 60° = 30°$$

$$\sigma_n = \frac{\sigma_x + \sigma_y}{2} + \frac{\sigma_x - \sigma_y}{2}\cos2\alpha + \tau_{xy}\sin2\alpha$$

$$= \frac{200 - 100}{2} + \frac{200 + 100}{2}\cos 60° - 0$$

$$= 125 \text{ N/mm}^2.$$

$$\tau = \frac{\sigma_x - \sigma_y}{2} \sin 2\alpha - \tau_{xy} \cos 2\alpha$$

$$= \frac{200 + 100}{2} \cos 60°$$

$$= 129.9 \text{ N/mm}^2$$

The resultant stress,

$$\sigma_r = \sqrt{\sigma_n^2 + \tau^2}$$

$$= \sqrt{(125)^2 + (129.9)^2} = 180.27 \text{ N/mm}^2.$$

$$\tan \phi = \frac{\sigma_n}{\tau} = \frac{125}{129.9}$$

$\phi = 43.9°$ i.e., the inclination of σ_r with oblique plane.

Example 9.48: At a point in a bracket, the stresses on two mutually perpendicular planes are 350 N/mm² and 150 N/mm² both tensile in nature. The shear stress across these planes is 90 N/mm². Find the magnitude and direction of the resultant stress on a plane making an angle 40° with the plane of first stress. Also find the normal and tangential stress on the plane.

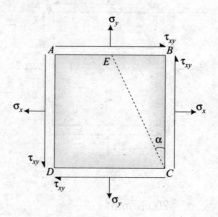

Solution:

$$\sigma_x = 350 \text{ N/mm}^2$$
$$\sigma_y = 150 \text{ N/mm}^2$$
$$\tau_{xy} = 90 \text{ N/mm}^2$$
$$\alpha = 40°.$$

Normal stress,

$$\sigma_n = \frac{\sigma_x + \sigma_y}{2} + \frac{\sigma_x - \sigma_y}{2} \cos 2\alpha + \tau_{xy} \sin 2\alpha$$

$$= \frac{350+150}{2} + \frac{350-150}{2} \cos 80° + 90 \sin 80°$$

$$= 250 + 17.365 + 88.632$$

$$= 366 \text{ N/mm}^2.$$

Tangential stress,

$$\tau = \frac{\sigma_x - \sigma_y}{2} \sin 2\alpha - \tau_{xy} \cos 2\alpha$$

$$= \frac{350-150}{2} \sin 80° - 90 \cos 80°$$

$$= 82.85 \text{ N/mm}^2$$

Resultant stress,

$$\sigma_r = \sqrt{\sigma_n^2 + \tau^2} = \sqrt{366^2 + 82.85^2}$$

$$= 375.28 \text{ N/mm}^2$$

If ϕ is the angle of resultant with the oblique plane,

$$\tan \phi = \frac{\sigma_n}{\tau} = \frac{366}{82.85}$$

$$\phi = 77°.$$

Example 9.49: The principal stresses at a point in a body are 30 and 50 MN/m², both tensile. Determine by calculations or by Mohr's circle, the following on a plane inclined at 40° to the plane on which the major principal stress acts:

(i) Normal and tangential compartments of stress

(ii) Magnitude of resultant stress

(iii) Angle of obliquity. [U.P.T.U. II Sem., 2002-03]

Solution:

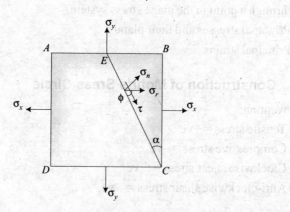

$$\sigma_x = 30 \text{ MN/m}^2$$
$$\sigma_y = 50 \text{ MN/m}^2$$
$$\tau_{xy} = 0$$
$$\alpha = 40°$$

$$\sigma_n = \frac{\sigma_x + \sigma_y}{2} + \frac{\sigma_x - \sigma_y}{2} \cos 2\alpha + \tau_{xy} \sin 2\alpha$$

$$= \frac{30 + 50}{2} + \frac{30 - 50}{2} \cos 80° - 0$$

$$= 40 - 3.47$$

$$= 36.527 \text{ MN/m}^2.$$

$$\tau = \frac{\sigma_x - \sigma_y}{2} \sin 2\alpha - \tau_{xy} \cos 2\alpha$$

$$= \frac{30 - 50}{2} \sin 80° - 0$$

$$= -9.848 \text{ MN/m}^2$$

$$\tan \phi = \frac{\sigma_n}{\tau} = \frac{36.527}{-9.848}$$

$$\phi = -74.9° \text{ or } 105.1°.$$

9.14 MOHR'S CIRCLE

A graphical method was developed by Otto Mohr in 1882 to determine the following:

1. Oblique stresses, i.e., normal and shear stresses on any inclined plane through a point in the plane stress system.
2. Principal stresses and their planes.
3. Principal strains.

9.14.1 Construction of Mohr's Sress Circle

Sign convention:

Tensile stress = +ve

Compressive stress = –ve

Clockwise shear stress = + ve

Anti-clockwise shear stress = – ve.

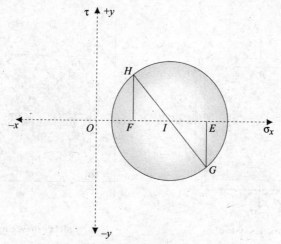

Fig. 9.22 Mohr's stress circle

1. Select a suitable scale on a graph paper.
2. Plot normal stresses on x-axis and shear stress on y-axis.
3. Cut $OE = \sigma_x$ and $OF = \sigma_y$.
4. Draw EG and FH perpendicutar to x-axis and cut $EG = -\tau_{xy}$ and $FH = +\tau_{xy}$.
5. Join GH intersecting x-axis in I.
6. With I as centre and HG as diameter, draw a circle. This circle is called

Mohr's stress circle. The coordinates of I are $\dfrac{\sigma_x + \sigma_y}{2}$ and diameter of

circle is $\sqrt{\left(\dfrac{\sigma_x - \sigma_y}{2}\right)^2 + \tau_{xy}^2}$

9.14.2 Use of Mohr's Circle for Finding Principal Stresses

(a) Convention
1. All normal stresses $(\sigma_x, \sigma_y, \sigma_1, \sigma_2)$ are shown on x-axis
2. All shear stresses (τ_{xy}, τ_{max}) are shown on y-axis.
3. All angles of planes are taken in anticlockwise direction from plane BC of stress σ_x.
4. Tensile stresses are +ve. Compressive stresses are −ve.
5. Shear stresses in clockwise direction are +ve and in anticlockwise direction are −ve.

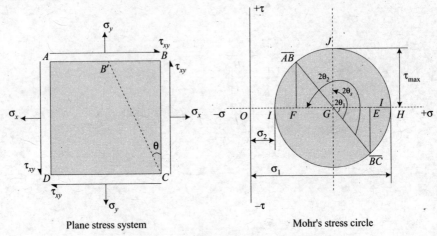

Plane stress system   Mohr's stress circle

Fig. 9.23 Principal stresses by Mohr's circle

(b) Construction

Draw or cut to scale:

1. $OE = \sigma_x$ along x-axis
2. $OF = \sigma_y$ along x-axis
3. $F\overline{AB} = +\tau_{xy}$ and perpendicular to x-axis
4. $E\overline{BC} = -\tau_{xy}$ and perpendicular to x-axis
5. Join $\overline{AB}$ and $\overline{BC}$ intersecting x-axis in G. G is the centre of Mohr's stress
   circle and has coordinates $\left(\dfrac{\sigma_x + \sigma_y}{2}, 0\right)$
6. Draw a circle with G as centre and $\overline{AB} - \overline{BC}$ as diameter. The radius of
   Mohr's circle is $\dfrac{\sigma_1 - \sigma_2}{2}$ or τ_{max}. The circle is the Mohr's stress circle.

(c) Measurements

Measure the following and multiply by scale selected.

1. $OH = \sigma_1$, major principal stress
2. $OI = \sigma_2$, minor principal stress.
3. $GJ = \tau_{max}$, maximum shear stress.
4. Measure $\angle \overline{BC}\ GH$ as $2\theta_1$.

   θ_1 is the angle of major principal plane $(G - \overline{BC})$ with plane BC.
5. Measure $\angle \overline{BC} - G - I$ as $2\theta_2$.

   θ_2 is the angle of minor principal plane $(\overline{AB} - G)$ with plane BC.
6. Measure $\angle \overline{BC} - GJ$ as $2\theta_s$.

   θ_s is the angle of plane of maximum shear stress with plane BC.

9.14.3 Mohr's Stress Circle for Different Loadings

(a) Uniaxial stress system

Uniaxial stress system is shown in Fig. 9.24.

$$\sigma_y = 0 \text{ and } \tau_{xy} = 0$$

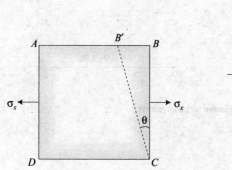 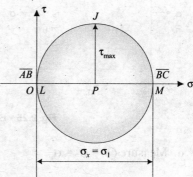

Fig. 9.24 Uniaxial stress system

1. Draw $OM = \sigma_x$ and $OL = \sigma_y = 0$.
2. Take $LP = PM$.
3. With P as centre and LM as diameter draw Mohr's stress circle.
4. $\overline{BC}$ coincides with M and $\overline{AB}$ coincides with L as $\tau_{xy} = 0$.
5. $\sigma_1 = \sigma_x = OM$
6. $\sigma_2 = \sigma_y = OL = 0$.

7. $\tau_{max} = JP = \dfrac{OM}{2} = \dfrac{\sigma_x}{2}$.

8. $\theta_1 = \dfrac{1}{2} \angle MP \ \overline{BC} = 0°$

9. $\theta_2 = \dfrac{1}{2} \angle MOL = \dfrac{180°}{2} = 90°$.

10. $\theta_5 = \dfrac{1}{2} \angle MPJ = \dfrac{90°}{2} = 45°$.

(b) Biaxial stress system

Biaxial stress system is shown in Fig. 9.25. $\tau_{xy} = 0$.
1. Draw $OM = \sigma_x$ and $OL = \sigma_y$
2. Take P as bisector of LM.
3. With P as centre and LM as diameter, draw Mohr's stress circle.
4. As $\tau_{xy} = 0$, $\overline{BC}$ coincides with M and $\overline{AB}$ coincides with L.
5. Measure $OM = \sigma_1 = \sigma_x$.

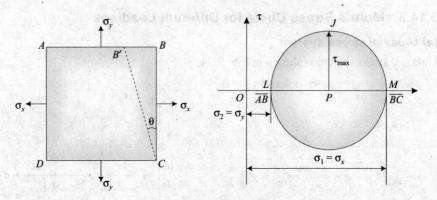

Fig. 9.25 Biaxial stress system

6. Measure $OL = \sigma_2 = \sigma_y$

7. Measure $JP = \tau_{max} = \dfrac{\sigma_x - \sigma_y}{2} = \dfrac{\sigma_1 - \sigma_2}{2}.$

8. $\theta_1 = \dfrac{1}{2} \angle \overline{BC}\text{-}P\text{-}M = 0.$

9. $\theta_2 = \dfrac{1}{2} \angle \overline{BC}\text{-}P\text{-}L = \dfrac{180°}{2} = 90°$

10. $\theta_s = \dfrac{1}{2} \angle \overline{BC}\text{-}P\text{-}J = \dfrac{90°}{2} = 45°.$

(c) Biaxial stress system with equal tension and compression

The stress system is shown in Fig. 9.26. $\sigma_x = \sigma_y$ and $\tau_{xy} = 0$.

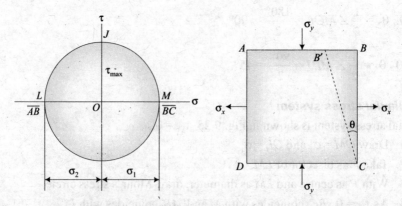

Fig. 9.26 Biaxial stress system with equal tension and compression

1. Draw $OM = \sigma_x$ centre and $OL = \sigma_y$.
2. With O as centre and LM as diameter, draw Mohr's stress circle.
3. $\overline{BC}$ coincides with M and $\overline{AB}$ coincides with L as $\tau_{xy} = 0$.
4. Measure $OM = \sigma_x = \sigma_1$ (tensile)

$$OL = \sigma_y = \sigma_2 \text{ (compression)}$$
$$OJ = \tau_{max} = \sigma_1 = \sigma_2$$

5. $\theta_1 = \dfrac{1}{2} \angle MO\,\overline{BC} = 0°.$

$$\theta_2 = \frac{1}{2} \angle MOL = \frac{180°}{2} = 90°$$

$$\theta_s = \frac{1}{2} \angle MOJ = \frac{90°}{2} = 45°.$$

(d) Pure shear

The pure shear stress system is shown in Fig. 9.27. $\sigma_x = 0$ and $\sigma_y = 0$.

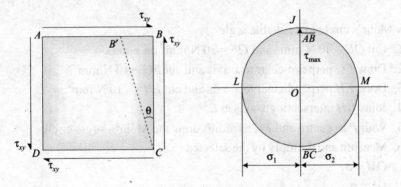

Fig. 9.27   Pure shear

1. Draw $OJ = \tau_{xy}$.
2. With O as centre and OJ as radius draw Mohr's stress circle cutting x-axis in L and M.
3. Measure $OL = \sigma_2 = \tau_{xy}$ $OJ = \tau_{max} = \tau_{xy}$
$$OM = \sigma_1 = \tau_{xy}$$

4. Measure $\theta_1 = \dfrac{1}{2} \angle \overline{BC}\, OM = \dfrac{90°}{2} = 45°.$

$$\theta_2 = \frac{1}{2} \angle \overline{BC}\, OL = \frac{270°}{2} = 135°$$

$$\theta_s = \frac{1}{2} \angle \overline{BC}\, OJ = \frac{180°}{2} = 90°.$$

Example 9.50: The stresses on two mutually perpendicular planes are 40 N/mm^2 (tensile) and 20 N/mm^2 (tensile). The shear stress across these planes is 10 N/mm^2. Find out principal stresses and their orientation using Mohr's circle. Also find the magnitude and direction of maximum shear stress.

Solution:

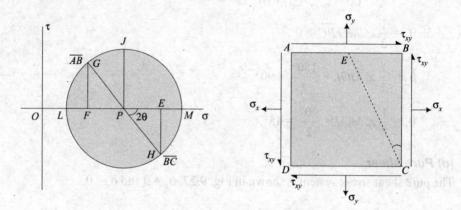

Draw Mohr's circle to a suitable scale

1. Cut $OE = 40$ N/mm^2 and $OF = 20$ N/mm^2 on x-axis.
2. Draw FG perpendicular to x-axis and cut $FG = 10$ N/mm^2.
3. Draw EH perpendicular to x-axis and cut $EN = -10$ N/mm^2.
4. Joins GH intersecting x-axis in P.
5. With P as centre and PG as radius draw the Mohr's stress circle.
6. Measure and multiply by the selected scale.

$$OM = \sigma_1$$

$$OL = \sigma_2$$

$$PJ = \tau_{max}$$

$$\theta_1 = \frac{1}{2}\angle EPH$$

$$\theta_2 = \frac{1}{2}\angle HPF \text{ or } \theta_1 + 90°$$

$$\theta_s = \frac{1}{2}\angle HPJ \text{ or } \theta_1 + 45°$$

9.14.4 Use of Mohr's Circle for Oblique Stresses

Mohr's circle is a powerful tool for finding oblique stress. Cube *ABCD* represents a plane stress system. It is required to find normal, shear and resultant stress on an

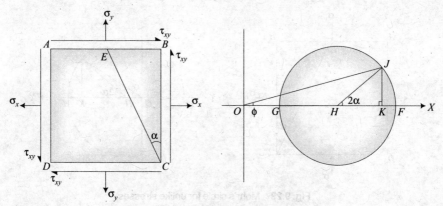

Fig. 9.28 Mohr's circle for oblique stress

obliqe plane CE making an angle α with BC, i.e., plane of major stress σ_x in the anti-clockwise direction.

(a) Construction
1. Draw OX horizontally.
2. Cut $OF = \sigma_x$ and $OG = \sigma_y$ to a selected scale.
3. Bisect GF so that $GH = HF$.
4. With H as centre, draw a circle with HF as radius.
5. Draw HJ so that $\angle FHJ = 2\alpha$.
6. Draw JK perpendicular to HF.
7. Join OJ.

(b) Measurements
Measure the following and multiply by scale.
1. $OK = \sigma_n$
2. $JK = \tau$
3. $OJ = \sigma_r$
4. $\angle KOJ = \phi$, i.e., angle of obliquity.

Note: If σ_x is tensile and σ_y is a compressive stress, the Mohr's circle can be drawn as shown in Fig. 9.29.
1. Draw XOX horizontally.
2. Cut $OF = \sigma_x$ and $OG = \sigma_y$ on left hand side to scale.
3. Bisect GF in H.
4. With H as centre and HF as radius, draw a circle.
5. Draw the line HJ so that $\angle FHJ = 2\alpha$.
6. Draw line JK perpendicular on XOX.
7. Join OJ.

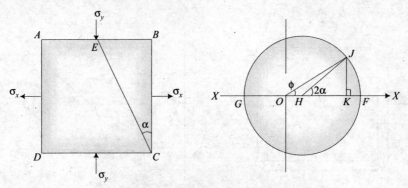

Fig. 9.29   Mohr's circle for unlike stresses

Measure the following and multiply by scale.

1. $OK = \sigma_n$
2. $JK = \tau$
3. $OJ = \sigma_r$
4. $\angle KOJ = \phi$, i.e., angle of obliquity.

Example 9.51:   A uniform steel bar of 2 cm × 2 cm area of cross-section is subjected to an axial pull of 4000 kg. Calculate the intensity of normal stress, shear stress on and resultant stress a plane, normal to which is inclined at 30° to the axis of the bar. Solve the problem graphically by drawing Mohr's circle.

[U.P.T.U. I Sem., 2002-03]

Solution:

$$\sigma_x = \frac{P}{A} = \frac{4000 \times 9.81}{20 \times 20} = 98 \text{ N/mm}^2$$

$$\alpha = 30°$$

$$\sigma_n = \sigma_x \cos^2\alpha = 98 \cos^2 30° = 73.5 \text{ N/mm}^2$$

$$\tau = \sigma_x \cos\alpha \sin\alpha = 98 \cos 30° \sin 30° = 42.435 \text{ N/mm}^2$$

$$\sigma_r = \sqrt{\sigma_n^2 + \tau^2} = \sqrt{73.5^2 + 42.435^2}$$

$$= 84.87 \text{ N/mm}^2.$$

Graphical Solution

1. Draw a horizontal line *X–O–X*.
2. Cut $OF = \sigma_x = 98$ N/mm^2 to a suitable scale.
3. Bisect the line *OF* in *H*
4. Draw a circle with *H* as centre and *HF* as radius.
5. Draw a line *HJ* so that $\angle FHJ = 2\alpha = 60°$.

6. Draw *JK* perpendicular to *XOX*.

7. Join *OJ*.

Measure the following values and multiply by selected scales:

$$OK = \sigma_n = 73.5 \text{ N/mm}^2$$
$$JK = \tau = 42.5 \text{ N/mm}^2$$
$$OJ = \sigma_r = 85 \text{ N/mm}^2.$$

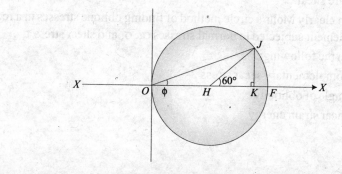

QUESTION BANK NO. 9

1. Discuss the effect of loading of an elastic material.

2. How do you differentiate between normal and shear stress?

3. Draw stress-strain diagram for structural steel and cast iron and briefly explain the various salient points.

4. With the help of suitable stress-strain diagram, explain the difference in the behaviour of ductile and brittle materials.

5. Explain the following elastic constants and establish relation between them:
   (i)   Modulus of Elasticity
   (ii) Modulus of Rigidity
   (iii) Bulk Modulus
   (iv) Poisson's Ratio.

6. (a) Define resilience, proof resilience and modulus of resilience.

   (b) What is the value of strain energy for gradually applied load, sudden load and impact load?

7. Describe briefly:
   (i)   State of plane stress
   (ii) Principal planes
   (iii) Principal stresses and strains.

8. Describe the construction of Mohr's stress circle for a plane stress system. Explain how is it used to find:

    (i) Principal stresses and their planes.

    (ii) Maximum shear stress and its plane.

9. Draw Mohr's stress circle for the following loadings:

    (i) Uniaxial stress

    (ii) Biaxial stress

    (iii) Equal tension and compression

    (iv) Pure shear.

10. Explain clearly Mohr's circle method of finding oblique stresses in a rectangular element subjected to normal stresses σ_x, σ_y and shear stress τ_{xy}.

11. Explain the following:

    (i) Complementary shear stress

    (ii) Angle of obliquity

    (iii) Shear strain energy.

Bending and Torsion

10

10.1 INTRODUCTION

Beam is a structural member to support external loads at right angle to its axis. The load applied on a beam gets transferred to its supports. Any section of the beam experiences the following:

1. Shear force that tries to shear off the section. This force is resisted by the beam and shear stress is induced.

2. Bending moment which tries to bend it. Consequently, longitudinal or bending stresses are induced in the cross-section of the beam.

The bending moment and shear force vary from point to point along the length of the beam. However, in a large number of practical cases, the bending moment is maximum when the shear force is zero or changes the sign and passes through zero value. Therefore, it is desired to design the beam for maximum bending moment.

If a length of a beam is subjected to a constant bending moment and no shear force, it is said to be in *pure bending* or simple bending. The bending stress for that part of the beam remains constant.

10.2 THEORY OF SIMPLE BENDING

When a beam is subjected to simple bending moment M, the beam will bend with some radius of curvature R. The curvature can be taken as circular.

The maximum compression develops at the top-most layer of beam and maximum tension develops at the bottom-most layer in a hogging beam.

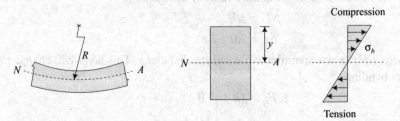

Fig. 10.1 Stress diagram in simple bending

The neutral axis passing through the centroid of the beam section is neither in tension nor compression and value of stress is zero. The magnitude of bending stress, σ_b and bending strain, ε_b at any depth of beam are proportional to its distance from neutral layer of the beam. The neutral layer is called neutral axis. The stress variation along the beam depth is linear and can be represented by triangles.

10.2.1 Assumptions in the Theory of Simple Bending

The following assumptions are made in the theory of simple bending:

1. The material of the beam is homogeneous and isotropic. The material properties are same in all directions.
2. The beam is initially straight and of uniform section throughout the length.
3. The radius of curvature of bending is very large in comparison to cross-sectional dimensions.
4. The stresses developed are within elastic limit of the material.
5. The material is elastic and obeys Hooke's law.
6. The value of Young's modulus of elasticity, E is same in tension and compression.
7. The loads are perpendicular to longitudinal axis of the beam.
8. Only longitudinal stresses are present in the beam and there is no lateral stresses.
9. The transverse section of the beam remains plane before and after bending.
10. Each layer of beam is free to expand (elongate) or contract (shorten) independently of the layer below or above it. There is no friction or shear force between the layers.

10.2.2 Bending Stress

The neutral axis of a beam is bent under the action of bending moment or a couple M, and forms an arc of the circle of radius R.

A beam section is shown before and after bending in Fig. 10.2. Consider a small length δx of a beam subjected to bending moment M.

Original length of neutral axis $NL = \delta x$

Changed length of neutral axis $N'L' = R\theta$

As there is no change in the length of neutral axis,

$$NL = N'L'$$

$$\therefore \qquad \delta x = R\theta = EF$$

where, EF is a layer parallel to NA and at a distance y. This layer will change to $E'F'$ after bending,

$$\therefore \qquad E'F' = (R + y)\,\theta$$

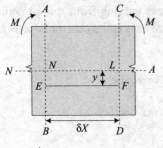

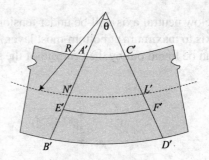

Beam before bending   Beam after bending

Fig. 10.2 Bending stress

Change in the length of layer EF,

$$\delta EF = E'F' - EF$$
$$= (R + y)\theta - R\theta = y\theta.$$

$$\text{Strain} = \frac{\text{Change in length}}{\text{Original length}}$$

$$\varepsilon = \frac{Y\theta}{R\theta} = \frac{Y}{R}$$

$$\text{Modulus of elasticity} = \frac{\text{stress}}{\text{strain}}$$

$\therefore$

$$E = \frac{\sigma_b}{\varepsilon} = \frac{\sigma_b}{Y/R}$$

$\therefore$ Bending stress,

$$\sigma_b = \frac{YE}{R}$$

Or,

$$\frac{\sigma_b}{Y} = \frac{E}{R}$$

Since, $\dfrac{E}{R}$ is a constant for a beam, bending stress, σ_b is proportional to its distance from neutral axis and varies linearly.

10.3 BENDING EQUATION

For pure sagging moment on a beam, all layers above neutral axis undergo compressive stress varying from zero at neutral axis and maximum at top layer. Layers below neutral axis will be under tension with stress varying from zero at neutral axis to maximum at bottom-most layer. The neutral axis is the centroidal axis and can be found out by the centroid of the section.

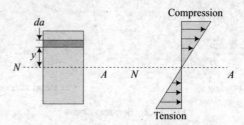

Fig. 10.3 Stress distribution

The section of a rectangular beam is shown in Fig. 10.3.

The force acting on an elementary area, *da*

$$= \sigma_b \times da$$

But
$$\sigma_b = \frac{E}{R} \times y$$

∴
$$\text{Force} = \left(\frac{E}{R} \times y \right) da$$

The bending moment will be resisted by an equal moment to ensure equilibrium. This is called moment of resistance and is calculated as product of force and its distance from neutral axis.

Moment of resistance of force acting on elementary area about *N-A*.

$$\delta MR = \left(\frac{E}{R} \times y \right) da \times y.$$

The total moment of resistance of beam section can be found out by integration. Total moment of resistance of the beam section,

$$MR = \int \delta MR = \int \left(\frac{E}{R} \times y \right).da \times y$$

$$= \frac{E}{R} \int y^2 da.$$

Now,
$$MR = M$$

where,
$$M = \text{bending moment on the beam section}$$

∴
$$M = MR = \frac{E}{R} \int y^2 da$$

But moment of inertia (M.O.I.) of beam section about N-A,

$$I = \int y^2 da$$

$\therefore$

$$M = \frac{E}{R} I$$

But

$$\frac{\sigma_b}{y} = \frac{E}{R}$$

$\therefore$

$$M = \frac{\sigma_b}{y} . I$$

$\therefore$

$$\frac{M}{I} = \frac{E}{R} = \frac{\sigma_b}{y}$$

This is called *bending equation* based on theory of bending.

Example 10.1: A steel bar 10 cm wide and 8 mm thick is subjected to bending moment. The radius of neutral surface is 100 cm. Determine the maximum and minimum bending stresses in the beam. [U.P.T.U. II Sem., 2001-02]

Solution:

$b = 10$ cm $= 100$ mm

$d = 8$ mm

$R = 100$ cm $= 1000$ mm

Assume, $E = 210$ kN/mm^2.

From bending equation,

$$\frac{\sigma_b}{y} = \frac{E}{R}$$

$\therefore$

$$\sigma_b = \frac{Ey}{R} = \frac{210 \times 10^3 \times 4}{1000} = 840 \text{ N/mm}^2$$

The maximum stress is 840 N/mm^2 at the outer fibres. Minimum tress is zero at neutral axis.

Example 10.2: Determine the least radius to which a steel bar 100 mm wide and 10 mm thick can be bent if the maximum allowable bending stress is 500 N/mm^2. Take $E = 2 \times 10^5$ N/mm^2.

Solution:
$$b = 100 \text{ mm}$$
$$d = 10 \text{ mm}$$
$$\sigma_b = 500 \text{ N/mm}^2$$
$$E = 2 \times 10^5 \text{ N/mm}^2$$

From bending equation,

$$\frac{\sigma_b}{y} = \frac{E}{R}$$

$$\therefore \quad R = \frac{Ey}{\sigma_b} = \frac{2 \times 10^5}{500} \times \left(\frac{10}{2}\right) = 2000 \text{ mm}.$$

Radius at the inner surface of bar $= R - y$
$$= 2000 - 5$$
$$= 1995 \text{ mm}$$

10.4 SECTION MODULUS

The moment of inertia of an area about any axis is the product of the area and square of the perpendicular distance of its centre of gravity from the axis. The moment of inertia is also called second moment of area.

Moment of inertia of area A about x-axis,

$$I_{xx} = A(\bar{y})^2$$

where, $\quad A = $ Area of section

$\bar{y} = $ Distance of $c.g.$ of area from x-axis.

Similay, moment of inertia of area A about y-axis

$$I_{yy} = A(\bar{X})^2$$

where, $\bar{X} = $ Distance of $c.g.$ of area from y-axis. Section modulus is the ratio of moment of inertia of a beam section about neutral axis to the distance of extreme fibre from neutral axis.

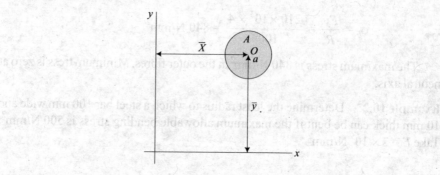

Fig. 10.4 Moment of inertia

Section modulus, $Z = \dfrac{I}{y_{\max}}$

where, $I = M.O.I.$ of section about neutral axis.

$y_{\max} =$ Distance of extreme fibre of beam from $N - A$.

From equation of bending,

$$\frac{M}{I} = \frac{\sigma_b}{y}$$

$\therefore$ $$\sigma_b = \frac{My}{I} = \frac{M}{I/y} = \frac{M}{Z}$$

The strength of the beam section depends upon the section modulus to resist bending moment. The section modulus of standard section is available in hand books. The section modulus of common section used for beams is calculated below.

10.4.1 Rectangular Section

$$Z = \frac{I}{y_{\max}}$$

But, $$I = \frac{bd^3}{12}$$

and $$y_{\max} = \frac{d}{2}$$

$\therefore$ $$Z = \frac{bd^3/12}{d/2} = \frac{bd^2}{6}.$$

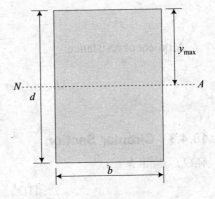

Fig. 10.5 Rectangular section

$$M = \sigma Z = \frac{\sigma}{6}\, bd^2.$$

Refer Fig. 10.5 for dimensions of section.

10.4.2 Hollow Rectangular Section

The dimensions of a hollow rectangular section are shown in Fig. 10.6.

$$I = \frac{BD^3}{12} - \frac{bd^3}{12}\,\frac{(BD^3 - bd^3)}{12}$$

$$y_{\max} = D/2$$

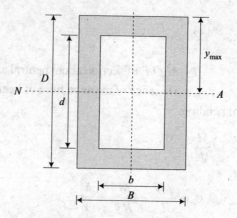

Fig. 10.6 Hollow rectangular section

$$\therefore \qquad Z = \frac{I}{y_{max}} = \frac{(BD^3 - bd^3) \times 2}{D \times 12}$$

$$= \frac{BD^3 - bd^3}{6D}$$

Moment of resistance

$$M = \sigma Z = \sigma \left[\frac{BD^3 - bd^3}{6D} \right]$$

10.4.3   Circular Section

M.O.I. about $N - A$,

$$I = \frac{\Pi D^4}{64}$$

$$y_{max} = D/2$$

$\therefore$ Section modulus,

$$Z = \frac{I}{y_{max}} = \frac{\Pi D^4 / 64}{D/2}$$

$$= \frac{\Pi D^3}{32}$$

Moment of resistance,

$$M = \sigma Z = = \sigma Z \frac{\Pi D^3}{32}$$

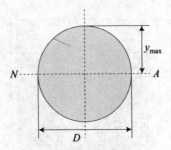

Fig. 10.7 Circular section

10.4.4 Hollow Circular Section

The dimensions are shown in Fig. 10.8.

M.O.I. of section about N–A,

$$I = \frac{\Pi}{64}(D^4 - d^4).$$

$$y_{max} = D/2$$

∴ Section modulus,

$$Z = \frac{I}{y_{max}} = \frac{\Pi}{64}(D^4 - d^4) \times \frac{2}{D}$$

$$= \frac{\Pi}{32}\left(\frac{D^4 - d^4}{D}\right)$$

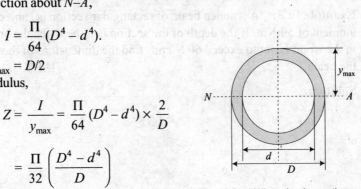

Fig. 10.8 Hollow circular section

Moment of resistance,

$$M = \sigma Z = \frac{\Pi\sigma}{32}\left(\frac{D^4 - d^4}{D}\right)$$

10.4.5 I-Section

The dimensions of a symmetrical *I*-section are shown in Fig. 10.9.

M.O.I. about $N - A$ = Area of rectangle $(B \times D)$ – Area of rectangle $[(B - b) \times d]$

$$I = \frac{BD^3}{12} - \frac{(B - b)d^3}{12}$$

$$y_{max} = D/2$$

∴

$$Z = \frac{I}{y_{max}} = \frac{\dfrac{BD^3}{12} - \dfrac{(B-b)d^3}{12}}{D/2}$$

$$= \frac{BD^3 - (B - b)d^3}{6D}$$

Moment of resistance, $M = \sigma Z$.

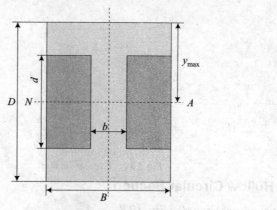

Fig. 10.9 I-section

Example 10.3: A wooden beam of rectangular section is subjected to a bending moment of 5 kNm. If the depth of the section is to be twice the breadth and stress in the wood is not to exceed 60 N/cm², find the dimensions of the cross-section of the beam. [U.P.T.U. I Sem., 2005-06]

Solution:

$$M = 5 \text{ kNm} = 5 \times 10^3 \text{ Nm}$$
$$d = 2b$$
$$\sigma_b = 60 \text{ N/cm}^2 = 6 \times 10^4 \text{ N/m}^2$$
$$M = \sigma_b Z$$

$$\therefore \qquad Z = \frac{M}{\sigma_b} = \frac{bd^2}{6}$$

$$\therefore \qquad \frac{d^3}{3} = \frac{M}{\sigma_b}$$

$$\therefore \qquad d = \sqrt[3]{\frac{3M}{\sigma_b}} = \sqrt[3]{\frac{3 \times 5 \times 10^3}{60 \times 10^4}} = 1.357 \text{ m}$$

$$\therefore \qquad b = d/2 = 0.6786 \text{ m}$$

The size of beam = **0.6786 m × 1.357 m** **Ans.**

Example 10.4: A rectangular beam with depth 150 mm and width 100 mm is subjected to a maximum bending moment of 300 kN-m. Calculate the maximum stress in the beam. [U.P.T.U. II Sem., 2003-04]

Solution:

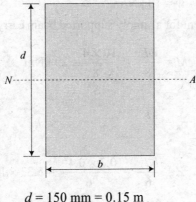

$$d = 150 \text{ mm} = 0.15 \text{ m}$$
$$b = 100 \text{ mm} = 0.10 \text{ m}$$
$$M = 300 \text{ kNm}$$

Bending stress,

$$\sigma_b = \frac{M}{Z} = \frac{300}{bd^2/6}$$

$$= \frac{300 \times 6}{0.1 \times 0.15^2}$$

$$= 8 \times 10^5 \text{ kN/m}^2$$
$$= 80 \text{ kN/cm}^2$$

Example 10.5: A rectangular beam of 200 mm in width and 400 mm in depth is simply supported over a span of 4 m and carried a distributed load of 10 kN/m. Determine the maximum bending stress in the beam.

Solution:

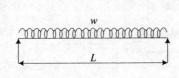

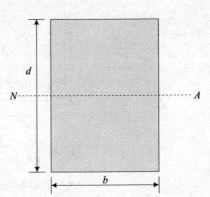

$$b = 200 \text{ mm} = 0.2 \text{ m}$$
$$d = 400 \text{ mm} = 0.4 \text{ m}$$
$$L = 4 \text{ m}$$
$$w = 10 \text{ kN/m}$$

Max. bending moment for a simply supported beam carrying *UDLw*,

$$M = \frac{wL^2}{8} = \frac{10 \times 4^2}{8} \quad 20 \text{ kNm}$$

Max. bending stress,

$$\sigma_b = \frac{M}{Z}$$

$$Z = \frac{bd^2}{6} = \frac{0.2 \times 0.4^2}{6} = 5.33 \times 10^{-3} \text{ m}^3$$

$$\therefore \qquad \sigma_b = \frac{20}{5.33 \times 10^{-3}} = 3750 \text{ kN/m}^2$$

Example 10.6: A rectangular beam of cross-section (300×200) mm^2 is simply supported over a span of 5 m. What uniformly distributed load per meter the beam may carry in two options:

(i) when height is 300 mm

(ii) when height is 200 mm

The bending stress is not to exceed 130 N/mm^2.

Solution:

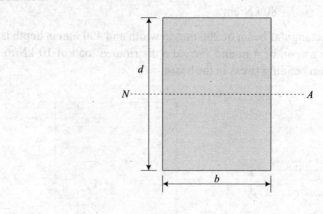

$$M = \frac{wL^2}{8} = \frac{25}{8} \ w \text{ Nm}$$

$$\sigma_b = 130 \text{ N/mm}^2 = 130 \times 10^6 \text{ N/m}^2$$

(i)
$$b = 200 \text{ mm} = 0.2 \text{ m}$$
$$d = 300 \text{ mm} = 0.3 \text{ m}$$
$$Z = \frac{bd^2}{6} = \frac{0.2 \times 0.3^2}{6} = 3 \times 10^{-3} \text{ m}^3$$

$$\sigma_b = \frac{M}{Z}$$

$$\therefore \qquad M = \sigma_b Z$$

$$= \frac{25}{8} w = (130 \times 10^6) \times (3 \times 10^{-3})$$

$$\therefore \qquad w = (130 \times 10^6) \times (3 \times 10^{-3}) \times \frac{8}{25} = 124.8 \times 10^3 \text{ kN/m}.$$

(ii)
$$b = 0.3 \text{ m}$$
$$d = 0.2 \text{ m}$$

$$Z = \frac{0.3 \times 0.2^2}{6} = 2 \times 10^{-3} \text{ m}^3$$

$$w = (130 \times 10^6) \times (2 \times 10^{-3}) \times \frac{8}{25} = 83.2 \times 10^3 \text{ kN/m}.$$

Example 10.7: A beam made of C.I. having a circular section of 50 mm external diameter and 25 mm internal diameter is supported at two points 4 m apart. The beam carries a concentrated load of 100 N at the centre. Find the maximum bending stress induced in the beam. [U.P.T.U. I Sem., 2001-02]

Solution:
Bending moment is maximum at the centre of beam span.

$$M = \frac{WL}{4} \quad \frac{100 \times 4000}{4} = 10^5 \text{ N-mm}$$

$$Z = \frac{\Pi}{32} \left(\frac{D^4 - d^4}{D} \right) = \frac{\Pi}{32} \left| \frac{50^4 - 25^4}{50} \right| = 0.011506 \times 10^6 \text{ mm}^3$$

$$\sigma_b = \frac{M}{Z} = \frac{10^5}{0.011506 \times 10^6}$$

$$= \textbf{8.69 N/mm}^2 \quad \textbf{Ans.}$$

Example 10.8: A hollow circular bar having outside diameter twice the inside diameter is used as a beam subjected to a bending moment of 50 kNm. Determine the inside diameter of the bar if the allowable bending stress is limited to 100 MN/m^2.

Solution:

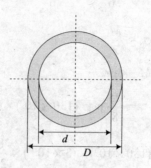

$$M = 50 \text{ kNm,}$$
$$\sigma_b = 100 \text{ MN/m}^2 = 100 \times 10^3 \text{ kN/m}^2$$
$$D = 2\ d.$$

$$Z = \frac{\Pi}{32} \left(\frac{D^4 - d^4}{D} \right)$$

$$= \frac{\Pi}{32} \left| \frac{(2d)^4 - d^4}{2d} \right|$$

$$= \frac{15 \Pi d^3}{64}.$$

$$\sigma_b = \frac{M}{Z}.$$

$$Z = \frac{M}{\sigma_b} = \frac{50}{100 \times 10^3}$$

$$\therefore \qquad d^3 = \frac{64}{15\Pi} \times \frac{50}{100 \times 10^3}$$

$$d = 0.08812 \text{ m} = \textbf{88.12 mm} \quad \textbf{Ans.}$$

Example 10.9: A rectangular beam 300 mm deep has $I = 8.5 \times 10^6 \text{ mm}^4$. The beam is simply supported and has a span of 5 m. Calculate *UDL* for a bending stress of 130 N/mm^2.

Solution:

$$I = 8.5 \times 10^6 \text{ mm}^4$$

$$y = \frac{300}{2} = 150 \text{ mm}$$

$$\sigma_b = 130 \text{ N/mm}^2$$

$$M = \frac{wL^2}{8} = \frac{\sigma_b I}{y}$$

$$\therefore \quad w = \frac{8\sigma_b I}{yL^2} = \frac{8 \times 130 \times 8.5 \times 10^6}{150 \times (5000)^2} = 2.357 \text{ N/mm}$$

$$= 2.36 \text{ N/mm} = \textbf{2.36 kN/m.} \quad \textbf{Ans.}$$

Example 10.10: Find the dimension of the strongest rectangular beam that can be cut out of a log of 250 mm diameter.

Solution:

Log diameter, $\qquad D = 250$ mm

$$D^2 = b^2 + d^2$$

$$\therefore \qquad d^2 = B^2 - b^2$$

Now

$$Z = \frac{bd^2}{6} = \frac{b}{6}(D^2 - b^2)$$

$$= \frac{1}{6}(bD^2 - b^3)$$

For strongest beam, Z should be maximum.

$$\therefore \qquad \frac{dZ}{db} = 0$$

$$\frac{d}{db}\left[\frac{1}{6}bD^2 - \frac{b^3}{6}\right] = 0.$$

$$\therefore \qquad \frac{D^2}{6} - \frac{3b^2}{6} = 0$$

$$b = \frac{D}{\sqrt{3}} = \frac{250}{\sqrt{3}} = 144 \text{ mm}$$

$$d = \sqrt{D^2 - b^2} = \sqrt{250^2 - 144^2} = 204 \text{ mm}$$

The size of the strongest beam = 204 mm × 144 mm.

Example 10.11: Three beams have the same length, the same allowable stress and the same bending moment. The cross-sections of the beam are a square, a rectangle with depth twice the width and a circle. Determine the ratios of the weights of circular and rectangle beams with respect to that of square beam.

[U.P.T.U. I Sem., 2000-01]

Solution: The dimensional sketches of the beam are shown.

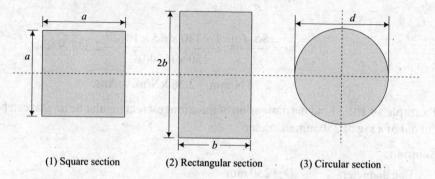

(1) Square section   (2) Rectangular section   (3) Circular section

The three beams have same bending moment and same allowable stress.

$$\sigma_b = \frac{M}{Z}$$

$$\therefore \quad Z_1 = Z_2 = Z_3$$

$$\frac{a \times a^2}{6} = \frac{b(2b)^2}{6} = \frac{\Pi}{32} d^3$$

$$\frac{a^3}{6} = \frac{2b^3}{3} = \frac{\pi}{32} d^3$$

$$4b^3 = a^3$$

$$b = 0.63\,a$$

$$\frac{\pi d^3}{32} = \frac{a^3}{6}$$

$$d = \sqrt[3]{\frac{32\,a^3}{\pi \times 6}} = 1.19a.$$

$$\frac{\text{Weight of rectangular beam}}{\text{Weight of square beam}} = \frac{\rho g A_2 L}{\rho g A_1 L} = \frac{A_2}{A_1} = \frac{2b^2}{a^2}$$

$$= \frac{2(0.63a)^2}{a^2} = 0.71$$

$$\frac{\text{Weight of circular beam}}{\text{Weight of square beam}} = \frac{\rho g A_3 L}{\rho g A_1 L} = \frac{A_3}{A_1} = \frac{\frac{\pi}{4}d^2}{a^2}$$

$$= \frac{\pi}{4}\frac{(1.19a)^2}{a^2} = 1.12$$

Example 10.12: For a given stress, compare the moment of resistance of a beam of square section when placed (i) with two sides horizontal and (ii) with its diagonal horizontal.

Solution: The moment of resistance is proportional to section modulus because $M = \sigma Z$.

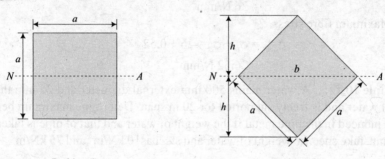

Position (i)   Position (ii)

Case 1 $Z_1 = \dfrac{bd^2}{6} = \dfrac{a \times a^2}{6} = \dfrac{a^3}{6}$.

Case 2 $Z_2 = 2 \times \dfrac{bh^3}{12} \times \dfrac{1}{h} = \dfrac{bh^2}{6}$

where, $b = \sqrt{a^2 + a^2} = \sqrt{2} \cdot a$

$$h = \frac{b}{2} = \frac{\sqrt{2}a}{2} = \frac{a}{\sqrt{2}}$$

$$Z_2 = \frac{\sqrt{2}\,a.}{6}\left(\frac{a}{\sqrt{2}}\right)^2 = \frac{a^3}{6\sqrt{2}}$$

$\therefore$ $\dfrac{M_1}{M_2} = \dfrac{Z_1}{Z_2} = \dfrac{(a^3/6)}{(a^3/6\sqrt{2})} = \sqrt{2} = $ **1.414 Ans.**

Example 10.13: Calculate the maximum compressive stress in a beam of a rectangular section 80×120 mm carrying an *UDL* of 40 kN/m over a span of 1 m. There is additional compressive force of 5 kN.

Solution:

Direct compressive stress,

$$\sigma_d = \frac{P_c}{A} = \frac{5000}{80 \times 120} = 0.52 \text{ N/mm}^2$$

Compressive stress due to bending,

$$\sigma_b = \frac{M}{Z} = \frac{(wL^2/8)}{\dfrac{bd^2}{6}} = \frac{\left(\dfrac{40 \times 10^3 \times 1}{8}\right) \times 10^3}{\left(\dfrac{80 \times 120^2}{6}\right)}$$

$$= 26 \text{ N/mm}^2$$

∴ Maximum fibre stress,

$$\sigma_{max} = \sigma_b + \sigma_d = 26 + 0.52$$
$$= 26.52 \text{ N/mm}^2.$$

Example 10.14: A water mains 500 mm external diameter and 25 mm thick is full of water and is freely supported for 20 m span. Determine maximum bending stress induced in the pipe metal if the weight of water and that of pipe is taken into account. Take specific weight of water and steel as 10 kN/m³ and 75 kN/m³.

Solution: The water mains acts as a simply supported beam carrying *UDL* consisting of weight of water and weight of steel pipe.

$D = 500 \text{ mm} = 0.5 \text{ m}$,

$d = 500 - 2 \times 25 = 450 \text{ mm} = 0.45 \text{ m}$.

Weight of steel pipe per meter length $= \dfrac{\Pi}{4}(D^2 - d^2)\gamma_s L$

$$= \frac{\Pi}{4}(0.5^2 . 0.45^2) \times 75 \times 1$$

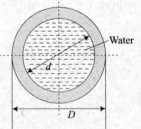
Water

$$= 2.8 \text{ kN/m}.$$

Weight of water per meter length of pipe $= \dfrac{\Pi}{4}d^2\gamma_w L$

$$= \frac{\Pi}{4} \times (0.45)^2 \times 10 \times 1 = 1.59 \text{ kN/m}$$

Total $UDL = 2.8 + 1.59 = 4.39 \text{ kN/m}$.

Bending moment, $M = \dfrac{wL^2}{8} = \dfrac{4.39 \times 20^2}{8} = 219.5 \text{ KNm}$

Moment of inertia, $I = \dfrac{\Pi}{64}(D^2 - d^4)$

$$= \dfrac{\Pi}{64}(0.5^4 - 0.45^4)$$

$$= 1.055 \times 10^{-3} \text{ m}^4$$

Bending stress, $\sigma_b = \dfrac{MY}{I} = \dfrac{219.5 \times 0.5/2}{1.055 \times 10^{-3}}$

$$= 52014 \text{ kN/m}^2$$

$$= \textbf{52.014 N/mm}^2 \quad \textbf{Ans.}$$

Example 10.15:   A cantilever with a constant breadth of 100 mm has a span of 2.5 m. It carries a uniformly distributed load of 20 kN/m. Determine the depth of the section at the middle of the length of the cantilever and also at the fixed end if the stress remains the same throughout and is equal to 120 MN/m².

Solution:

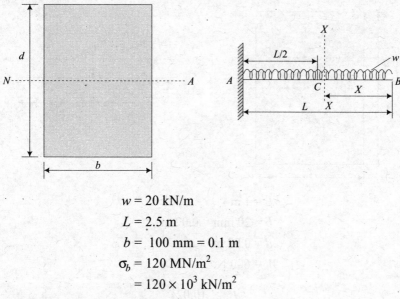

$$w = 20 \text{ kN/m}$$

$$L = 2.5 \text{ m}$$

$$b = 100 \text{ mm} = 0.1 \text{ m}$$

$$\sigma_b = 120 \text{ MN/m}^2$$

$$= 120 \times 10^3 \text{ kN/m}^2$$

Bending moment at section $X-X$,

$$M_x = \dfrac{wx^2}{2} = \dfrac{20}{2}x^2 = 10x^2 \text{ kN-m.}$$

$$Z = \dfrac{M}{\sigma_b}$$

$$\frac{bd^2}{6} = \frac{10x^2}{120 \times 10^3}$$

$$\therefore \qquad d = \sqrt{\frac{6 \times 10 \times x^2}{0.1 \times 120 \times 10^3}} = 0.0707x$$

Depth at fixed end,

$$x = 2.5 \text{ m}$$

$$\therefore \qquad d = 0.176776 \text{ m} = 176.777 \text{ mm}$$

Depth at midpoint, $\quad x = 1.25 \text{ m}$

$$\therefore \qquad d = 0.0707 \times 1.25 = 0.088 \text{ m} = \textbf{88 mm} \quad \textbf{Ans.}$$

Example 10.16: A simply supported beam 1 m long and 20 mm × 20 mm in cross-section fails when a central load of 600 N is applied to it. What tensity of UDL would cause failure of a cantilever beam 2 m long and 40 mm wide × 80 mm deep made of same material?

Solution: *Simply supported beam,*

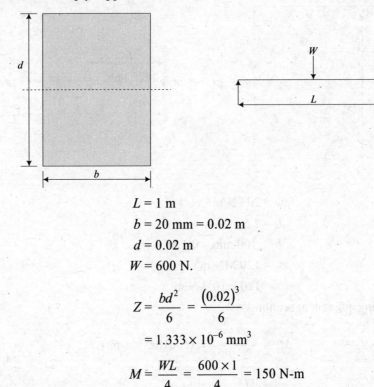

$$L = 1 \text{ m}$$

$$b = 20 \text{ mm} = 0.02 \text{ m}$$

$$d = 0.02 \text{ m}$$

$$W = 600 \text{ N.}$$

$$Z = \frac{bd^2}{6} = \frac{(0.02)^3}{6}$$

$$= 1.333 \times 10^{-6} \text{ mm}^3$$

$$M = \frac{WL}{4} = \frac{600 \times 1}{4} = 150 \text{ N-m}$$

$$\sigma_b = \frac{M}{Z} = \frac{150}{1.333 \times 10^{-6}} = 112.5 \times 10^6 \text{ N/m}^2$$

$$= 112.5 \text{ N/mm}^2 \quad \textbf{Ans.}$$

Example 10.17: Determine the longest span of a simply supported beam to carry a UDL of 6 kN/m without exceeding a bending stress of 120 MN/m². The depth and moment of inertia of the symmetrical *I*-section are 20 cm and 2640 cm⁴ respectively.

Solution:

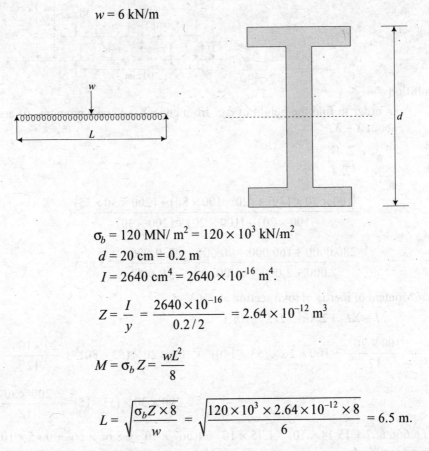

$w = 6$ kN/m

$$\sigma_b = 120 \text{ MN/ m}^2 = 120 \times 10^3 \text{ kN/m}^2$$
$$d = 20 \text{ cm} = 0.2 \text{ m}$$
$$I = 2640 \text{ cm}^4 = 2640 \times 10^{-16} \text{ m}^4.$$

$$Z = \frac{I}{y} = \frac{2640 \times 10^{-16}}{0.2/2} = 2.64 \times 10^{-12} \text{ m}^3$$

$$M = \sigma_b Z = \frac{wL^2}{8}$$

$$L = \sqrt{\frac{\sigma_b Z \times 8}{w}} = \sqrt{\frac{120 \times 10^3 \times 2.64 \times 10^{-12} \times 8}{6}} = 6.5 \text{ m.}$$

Example 10.18: The cross-section of a cast iron beam is shown and is simply supported at the ends and carries a uniformly distributed load of 20 kN/m. If the span of the beam is 3 m, determine the maximum tensile and compressive bending stress in the beam.

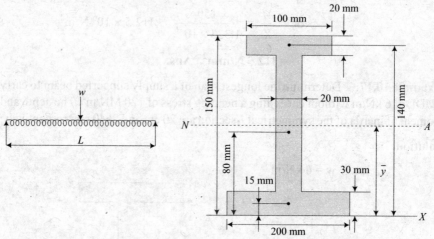

Solution:

1. In order to find the height of *c.g.* from edge $X - X$, take moments of area about $X - X$,

$$\bar{y} = \frac{\Sigma AY}{\Sigma A}$$

$$= \frac{(100 \times 20 \times 140) + (20 \times 100 \times 80) + (200 \times 30 \times 15)}{(100 \times 20) + (100 \times 20) + (200 \times 30)}$$

$$= \frac{280{,}0000 + 160{,}000 + 90{,}000}{2{,}000 + 2{,}000 + 6{,}000} = \frac{530{,}000}{10{,}000} = 53 \text{ mm}$$

Moment of inertia of total section about N–A

$$I = \Sigma I_G + \Sigma Ah^2 = \Sigma(I_G + Ah^2)$$

$$= \frac{100 \times 20^3}{12} + 100 \times 20 \times (53 - 140)^2 + 100 \times 20 \times (53 - 80)^2 + \frac{20 \times 100^3}{12}$$

$$+ 200 \times 30 \times (53 - 15)^2 + \frac{200 \times 30^3}{12}$$

$$= 66{,}666.667 + 15.14 \times 10^6 + 1.45 \times 10^6 + 1.667 \times 10^6 + 8.66 \times 10^6 + 0.45 \times 10^6$$

$$= 0.06667 \times 10^6 + 1.1667 \times 10^6 + 0.45 \times 10^6 + 15.14 \times 10^6 + 1.45 \times 10^6 + 8.66 \times 10^6$$

$$= 2.18367 \times 10^6 + 2525 \times 10^6 = 27.43 \times 10^6 \text{ mm}^4$$

$$M = \frac{wL^2}{8} = \frac{20 \times (3000)^2}{8} = 22.5 \times 10^6 \text{ kN-mm}$$

$$\sigma_c = \frac{MYc}{I} = \frac{22.5 \times 10^6 \times (150 - 53)}{27.43 \times 10^6} = 79.566 \text{ kN/mm}^2 \qquad \text{(at top)}$$

$$\sigma_t = \frac{MYt}{I} = \frac{22.5 \times 10^6 \times 53}{27.43 \times 10^6} = 43.474 \text{ kN/mm}^2 \qquad \text{(at bottom)}$$

10.5 TORSION OF SHAFTS OF CIRCULAR SECTION

Solid and hollow circular shafts are used in industry to transmit power from prime movers to various types of machines. Other shapes of shafts are seldom used. The shafts are subjected to the following loads.

1. Torsion or twisting moment or torque.
2. Bending moment due to transverse loads or weights.
3. Axial thrust due to longitudinal loads.
4. Combination of above three loads.

A shaft is under *pure torsion*, if it is not associated with bending moment or axial force. If a shaft is under pure torsion, its cross-section experiences the following effects,

1. Shear stress or torsional shear stress due to torque.
2. Twisting or torsional shear strain measured as angle of twist,
3. Strain energy.

Therefore, shafts are designed so that shear stress induced and angle of twist produced are within allowable limits.

10.6 POWER TRANSMISSION

The shaft rotated by a motor or a prime mover (engine, turbine, etc.) has rotational energy. This energy can be expressed by shaft work.

Shaft work,

$$W = \int_{1}^{2} T\, d\theta$$

where, T = Torque or turning moment applied to the shaft (Nm)

  $d\theta$ = Angular displacement of shaft

Shafts power,

$$P = \int_{1}^{2} T \frac{d\theta}{dt} = T\omega$$

where, ω = Angular velocity of shaft

or $P = 2\Pi NT$ (W)

where, N = shaft speed (rps)

$$P = \frac{2\Pi NT}{60 \times 1000} \text{ (kW)}$$

where N = shaft speed (rpm).

The rotating shaft possesses kinetic energy which can be expressed as:

$$KE = \frac{1}{2} I \omega^2$$

where,
I = Mass moment of inertia

$= mk^2$

m = mass of rotating shaft (kg)

k = radius of gyration (m)

The work done by a rotating shaft is equal to change of kinetic energy.

$$W_{1-2} = KE_2 - KE_1$$

1 stands for initial and 2 for final condition.

10.7 ASSUMPTIONS IN PURE TORSION OF CIRCULAR SHAFTS

Certain assumptions are made to consider pure torsion in circular shafts. These assumptions are made to develop the torsion equation.

1. Material of the shaft is homogeneous and isotropic. The material is assumed free of impurities and mechanical properties are same in all directions.
2. The shaft is perfectly elastic and obeys Hooke's law.
3. The stress does not exceed the limit of proportionality.
4. The value of Young's modulus of elasticity is same in tension and compression.
5. The shaft is subjected to pure twisting moment or torque acting in a plane perpendicular to the axis of the shaft.
6. The twist is uniform along the total length of the shaft.
7. The shaft remains circular in section before and after application of torque.
8. The plane cross-sections of shafts remain plane before and after application of torque.
9. All diameters of the cross-section of the shaft remain straight without change in their magnitude before and after twisting.
10. The relative rotation between any two cross-sections of the shaft is proportional to the distance between them.
11. There is no change in the length of the shaft between any two normal cross-sections after application of torque.

10.8 TORSIONAL SHEAR STRESS

When a shaft is transmitting power, it behaves as if a force F acts tangentially on the shaft of diameter D at one end and is fixed at the other end. The twisting moment can be represented by a couple $F \times l$ as shown.

The torque, $T = F \times l$.

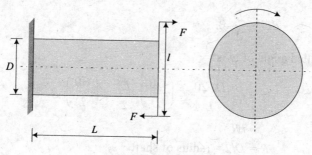

Fig. 10.10   Torque

Torque can be expressed in terms of shaft diameter, the force being assumed acting on the surface of the shaft. Then,

$$T = F \times D.$$

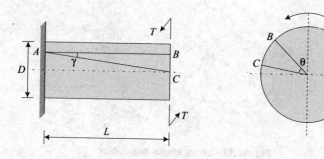

Fig. 10.11   Torsion of circular shaft

A solid circular shaft of diameter D and length L is subjected to a torque T at one end and fixed at the other end (by a couple equal to torque). A straight line AB parallel to shaft axis on the surface of the shaft will take the form of helix AC after twisting. Assume γ is the angle of shear strain on the surface.

In the front view,

$$BC = L\gamma$$

or $$\gamma = \frac{BC}{L}$$

But shear strain,

$$\gamma = \frac{\tau}{G}$$

where, τ = shear stress

G = Modulus of rigidity.

or $\qquad \tau = G\gamma = G \times \dfrac{BC}{L}$

In the end view of the shafts,

$$BC = \dfrac{D}{2}.\theta.$$

where, θ is called angle of twist.

$\therefore \qquad \tau = G.\dfrac{D/2\theta}{L} = \left(\dfrac{G\theta}{L}\right)\dfrac{D}{2} = \left(\dfrac{G\theta}{L}\right) R$

$\qquad\qquad\qquad = kR$

where, $\qquad R = D/2 =$ radius of shaft

$\qquad k = \dfrac{G\theta}{L} =$ constant for a given shaft.

Therefore, shear stress is proportional to the radius of the shaft and varies linearly with the radius.

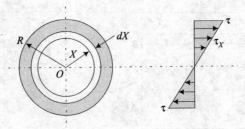

Fig. 10.12 Shear stress distribution

At any radius X,

$$\dfrac{\tau_x}{\tau} = \dfrac{X}{R}$$

$\therefore \qquad \tau_x = \dfrac{X}{R}.\tau$

But $\qquad \dfrac{\tau}{R} = k = \dfrac{G\theta}{L}$

Consider an elementary ring of the shaft at radius X and thickness dX.

The turning force on the ring,

$$FX = \tau_x . X \, dA = \tau_x \times 2\pi X dX$$

Moment of turning of this force about shaft axis

$$dT = 2\pi X dX \, \tau_x . X = 2\pi X^2 dX \, \tau_x$$

$$= 2\pi X^2 dX \frac{X}{R} \tau = 2\pi X^3 dX \frac{\tau}{R}$$

Total turning moment of resistance,

$$TR = \int dT = \int_O^R 2\pi X^3 \times \frac{\tau}{R} dX$$

$$= \frac{2\pi\tau}{R} \int_O^R X^3 dX = \frac{2\pi\tau}{R} \left| \frac{X^4}{4} \right|_o^R$$

$$= \frac{2\pi\tau}{R} \frac{R^4}{4} = \frac{\pi R^3}{2} \tau = \frac{\pi}{16} D^3 \tau$$

But applied torque, T = Total turning moment of resistance,

$$TR = T = \frac{\pi}{16} D^3 \tau.$$

$\therefore$

$$T = \frac{\tau}{(D/2)} \cdot \frac{\pi D^4}{32} = \frac{\tau}{D/2} \cdot J$$

where,

$$J = \frac{\pi D^4}{32}$$

J is the polar moment of inertia of the shaft cross-section

$\therefore$

$$\frac{T}{J} = \frac{\tau}{R} = \frac{G\theta}{L}.$$

This is called the *torsion equation* for the circular shaft.

10.9 TORSION EQUATION FOR HOLLOW CIRCULAR SHAFT

Figure 10.13 shows a hollow circular shaft with outside radius R and inside radius r, and subjected to a torque T. The turning moment of resistance on elementary ring,

$$dT = \tau_X . 2\pi X . dX . X$$

The total tunning moment resistance shaft cross-section

$$T = \int dT = \int_r^R 2\pi X^3 dX \frac{\tau}{R}$$

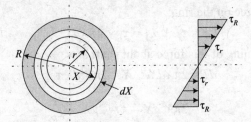

Fig. 10.13 Shear stress distribution for hollow circular shaft

$$T = \frac{2\pi\tau}{R} \int_r^R X^3 dX = \frac{2\pi\tau}{R} \left|\frac{X^4}{4}\right|_r^R$$

$$= \frac{\pi}{2} \cdot \frac{\tau}{R} \cdot (R^4 - r^4)$$

$$= \frac{\pi}{16} \tau \left[\frac{D^4 - d^4}{D}\right]$$

But
$$J = \frac{\pi}{32} (D^4 - d^4) = \frac{\pi}{2} (R^4 - r^4) \text{ (for a hollow shaft)}$$

∴
$$T = \frac{\tau}{R} \cdot J.$$

or
$$\frac{T}{J} = \frac{\tau}{R} = \frac{G\theta}{L}.$$

The torsion equation equally holds good for solid and hollow circular shafts.

10.10 POLAR SECTION MODULUS

Polar section modulus is the ratio of polar moment of inertia and outer radius of the shaft. It is analogous to section modulus in case of beam subjected to bending. The polar section modulus is a measure of strength and rigidity of a shaft against a twisting moment or torque. Polar moment of inertia is the moment of inertia about z-axis or longitudinal axis of the shaft.

$$J = I_{zz} = I_{xx} + I_{yy}$$

Polar section modulus,

$$Z_p = \frac{J}{R} = \frac{\frac{\pi}{32} \times D^4}{D/2} = \frac{\pi}{16} D^3 \quad \text{for solid circular shaft.}$$

$$Z_p = \frac{\pi/32(D^4 - d^4)}{D/2} = \frac{\pi}{16}\left(\frac{D^4 - d^4}{D}\right) \text{ for hollow circular shaft.}$$

10.10.1   Strength of Circular Shaft

Strength of a shaft may be defined as the maximum torque which can be applied to the shaft without exceeding allowable shear stress and angle of twist.

(a) Solid shaft

The torque transmitted by a shaft,

$$T = \tau Z_p = \frac{\pi}{16} D^3 \tau.$$

The maximum torque which can be transmitted,

$$T_{max} = \frac{\pi}{16} D^3 \tau_{max} \qquad \qquad ...(1)$$

where, τ_{max} = maximum allowable shear stress.

Also, $\dfrac{T}{J} = \dfrac{G\theta}{L}$

$\therefore$ $T = \dfrac{G\theta}{L} \cdot J = \dfrac{G\theta}{L} \cdot \dfrac{\pi}{32} D^4$

Similar, $T_{max} = \dfrac{\pi}{32} \dfrac{GD^4}{L} \cdot \theta_{max.}$ \qquad ...(2)

where, θ_{max} = Maximum allowable angle of twist (rad)

The strength of the shaft is the maximum value of T_{max} as obtained in equations (1) and (2).

(b) Hollow shaft

The maximum torque which can be transmitted by a hollow shaft,

$$T_{max} = Zp\, \tau_{max} = \frac{\pi}{16}\left(\frac{D^4 - d^4}{D}\right)\tau_{max} \qquad ...(1)$$

where, D = External shaft diameter

d = Internal shaft diameter

τ_{max} = Maximum allowable shear stress

Also, $T_{max} = \dfrac{G\theta_{max}}{L}$ $J = \dfrac{G\theta_{max}}{L} \cdot \dfrac{\pi}{32}(D^4 - d^4)$ \qquad ...(2)

The strength of the shaft is the maximum value of T_{max} as obtained from equations (1) and (2).

10.10.2 Torsional Rigidity

The torsion equation gives

$$\frac{T}{J} = \frac{G\theta}{L}$$

$$\therefore \qquad \theta = \frac{TL}{GJ}$$

The values of G, J and L are constant for a given shaft.

$\therefore$ Angle of twist θ is directly proportional to the torque T.

Now
$$\frac{\theta}{L} = \frac{T}{GJ}$$

or
$$\frac{T}{\theta/L} = GJ. = \text{torsional rigidity.}$$

The torque divided by the angle of twist per unit length of shaft is called *torsional rigidity.*

The parameter GJ is analogous to the quantity EI. The quantity EI is called flexural rigidity in case of beam subjected to bending.

10.10.3 Torsional Stiffness

The torque per unit angle of twist is called torsional stiffness of the shaft.

$$\text{Torsional stiffness} = \frac{T}{\theta} \text{ (N-m/rad).}$$

10.11 STRAIN ENERGY DUE TO TORSION

Energy is stored in the shaft due to angular distortion θ. This is also called *torsional resilience* or torsional energy.

$$\text{Torsional energy} = \text{Work done by the torque}$$
$$= \text{Average torque} \times \text{Angular twist}$$

$$U = \frac{1}{2} T\theta$$

From torsion equation,

$$\frac{T}{J} = \frac{\tau}{R} = \frac{2\tau}{D}$$

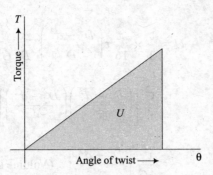

Fig. 10.14 Torsional energy

$\therefore$ $$T = \frac{\tau J}{R}.$$

Also $$\frac{T}{J} = \frac{G\theta}{L} = \frac{\tau}{R}$$

$\therefore$ $$\theta = \frac{\tau L}{GR}$$

$\therefore$ $$U = \frac{1}{2}\left(\frac{\tau J}{R}\right)\left(\frac{\tau L}{GR}\right)$$

$$= \frac{1}{2}\frac{\tau^2}{G} \cdot \frac{\pi R^4}{2R^2} \cdot L$$

$$= \frac{1}{4} \times \frac{\tau^2}{G}(\pi R^2 L) = \frac{1}{4}\frac{\tau^2}{G} \text{ (Volume)}$$

Modulus of resilience $= \dfrac{U}{V} = \dfrac{1}{4}\dfrac{\tau^2}{G}$ for solid shaft.

For a hollow shaft,

$$T = J.\frac{2\tau}{D} = \frac{\pi}{32}(D^2 - d^2)\frac{2\tau}{D}$$

$$= \frac{\pi}{16}\tau\left(\frac{D^4 - d^4}{D}\right)$$

$\therefore$ $$\theta = \frac{2\tau}{D} \cdot \frac{L}{G}$$

$$\therefore \qquad U = \frac{1}{2} \times \frac{\pi}{16} \tau \left(\frac{D^4 - d^4}{D} \right) \cdot \frac{2\tau}{D} \frac{L}{G}$$

$$= \frac{\tau^2}{4} \left[\left(\frac{D^2 + d^2}{D^2} \right) \left(\frac{D^2 - d^2}{4} \right) \right] . \pi L$$

$$= \frac{\tau^2}{4G} \left(\frac{D^2 + d^2}{D^2} \right) \qquad \text{(Volume shaft)}$$

Modulus of resilience = Torsional energy for unit volume

$$\frac{U}{V} = \frac{\tau^2}{4G} \cdot \left(\frac{D^2 + d^2}{D^2} \right)$$

Example 10.19: Find the power transmitted by a solid circular shaft of steel of 50 mm diameter at 120 rpm, if the permissible shear stress is 62.5 N/mm^2.

Solution:
$$D = 50 \text{ mm}$$
$$N = 120 \text{ rpm}$$
$$\tau = 62.5 \text{ N/mm}^2$$

Torque,
$$T = \frac{\pi}{16} D^3 \tau$$

$$= \frac{\pi}{16} (50)^3 \times 62.5$$

$$= 1533980 \text{ N-mm}$$

$$= 1533.98 \text{ N-m}$$

Power,
$$P = \frac{2\pi NT}{60 \times 1000} = \frac{2\pi \times 120 \times 1533.98}{60 \times 1000}$$

$$= 19.276 \text{ kW}.$$

Example 10.20: If the maximum torque transmitted by a solid shaft exceeds the mean torque by 30% in each revolution, find a suitable shaft diameter to transmitted 75 kW power at 200 rpm. Take allowable shear stress as 70 N/mm^2.

Solution:
$$T_{\max} = 1.3 \, T_{\text{mean}}.$$
$$P = 75 \text{ kW}$$
$$N = 200 \text{ rpm}$$
$$\tau_{\max} = 70 \text{ N/mm}^2$$

Power,
$$P = \frac{2\pi N \, T_{\text{mean}}}{60 \times 1000}$$

$\therefore$
$$T_{\text{mean}} = \frac{60,000 P}{2\pi N} = \frac{60,000 \times 75}{2\pi \times 200}$$

$$= 3581 \text{ N-m,}$$
$$T_{\text{max}} = 1.3 \times 3581$$
$$= 4655.3 \text{ N-m.}$$

Now
$$T_{\text{max}} = \frac{\pi}{16} D^3 \tau_{\text{max}}$$

$$D = \sqrt[3]{\frac{T_{\text{max}} \times 16}{\pi \times \tau_{\text{max}}}} = \sqrt[3]{\frac{4655.3 \times 16 \times 10^3}{\pi \times 70}}$$

$$= \mathbf{69.7 \text{ mm}} \quad \mathbf{Ans.}$$

Example 10.21: Find the torque which a shaft of 25 cm diameter can safely transmit, if stress is not to exceed 460 N/mm^2.

Solution:
$$D = 25 \text{ cm} = 250 \text{ mm}$$
$$\tau = 460 \text{ N/mm}^2$$

Now torque,

$$T = \frac{\pi}{16} D^3 \tau$$

$$= \frac{\pi}{16} (250)^3 \times 460 \text{ N-mm}$$

$$= \frac{\pi}{16} (250)^3 \times 460 \times 10^{-3} \text{ N-m}$$

$$= \mathbf{1.6 \times 10^6 \text{ N-m}} \quad \mathbf{Ans.}$$

Example 10.22: Find the diameter of a shaft to transmit a torque of 3×10^4 N-m if allowable shear stress is 50 N/mm^2.

Solution:
$$T = 3 \times 10^4 \text{ N-m} = 3 \times 10^4 \times 10^3 \text{ N-mm}$$
$$\tau = 50 \text{ N/mm}^2$$

Now,
$$T = \frac{\pi}{16} D^3 \tau$$

$\therefore$
$$D = \sqrt[3]{\frac{16 T}{\pi \tau}} = \sqrt[3]{\frac{16 \times 10^4 \times 10^3}{\pi \times 50}}$$

$$= \mathbf{145 \text{ mm}} \quad \mathbf{Ans.}$$

Example 10.23: A hollow shaft with outer diameter 25 cm and inner diameter 15 cm has to transmit maximum torque. Find the value of torque if allowable shear stress is 40 N/mm^2.

Solution:

$$D = 25 \text{ cm} = 250 \text{ mm}$$
$$d = 15 \text{ cm} = 150 \text{ mm}$$
$$\tau = 40 \text{ N/mm}^2$$

Now,

$$T = \frac{\pi}{16}\left(\frac{D^4 - d^4}{D}\right)\tau$$

$$= \frac{\pi}{16}\left(\frac{250^4 - 150^4}{150}\right) \times 40$$

$$= 0.1066 \times 10^9 \text{ N-mm}$$

$$= \mathbf{106.6 \text{ KN-m}} \quad \textbf{Ans.}$$

Example 10.24: For one propeller drive shaft compute the torsional shear stress when it is transmitting a torque of 1.76 kN-m. The shaft is a hollow tube having an outside diameter of 60 mm and an inside diameter of 40 mm. Find the stress at both the outer and inner surfaces. [U.P.T.U. II Sem., 2001-02]

Solution:

$$T = 1.76 \text{ kN-m} = 1.76 \times 10^6 \text{ N mm}$$
$$D = 60 \text{ mm}$$
$$d = 40 \text{ mm}$$

$$T = \frac{\pi}{16}\left(\frac{D^4 - d^4}{D}\right)\tau$$

$$\tau = \frac{16DT}{\pi(D^4 - d^4)} \quad \frac{16 \times 60 \times 1.76 \times 10^6}{\pi(60^4 - 40^4)}$$

$$= 51.7 \text{ N/mm}^2 .$$

This is the stress at outer radius.

The stress varies linearly with the radius.

The stress at the inner radius of shaft,

$$\tau_i = \tau_o \frac{d}{D} = 51.7 \times \frac{40}{60}$$

$$= \mathbf{34.48 \text{ N/mm}^2} \quad \textbf{Ans.}$$

Example 10.25: Show that for a given maximum shear stress, the minimum diameter required for solid circular shaft to transmit P(kW) at N(rpm) can be expressed as $D = \text{constant} \times \left(\dfrac{P}{N}\right)^{\frac{1}{3}}$.

Solution:

$$P = \frac{2\pi NT}{60 \times 1000} \text{ [kW]}$$

$$T = \frac{60,000 P \times 10^3}{2\pi N} \text{ N-mm}$$

$$= 3 \times 10^7 \frac{P}{\pi N} \text{ N-mm}$$

But,

$$T = \frac{\pi}{16} D^3 \tau$$

$\therefore$

$$D^3 = \frac{16T}{\pi \tau} = \frac{16 \times 3 \times 10^7 P}{\pi \tau \times \pi N}$$

$\therefore$

$$D = k \left(\frac{P}{N}\right)^{\frac{1}{3}}$$

where,

$$k = \frac{16 \times 3 \times 10^7}{\pi^2 \tau} = \text{constant}$$

Example 10.26: What external and internal diameters are required for a hollow shaft to transmit 50 kW of power at 300 rpm if the shear stress is limited to 100 MN/m^2. Take outside diameter to be twice of inside diameter.

Solution:

$$P = 50 \text{ kW}$$

$$N = 300 \text{ rpm}$$

$$\tau = 100 \text{ MN/m}^2 = 100 \text{ N/mm}^2$$

$$D = 2d.$$

$$P = \frac{2\pi NT}{60 \times 100D}$$

$$\therefore \qquad T = \frac{60,000P}{2\pi N} = \frac{60,000 \times 50}{2\pi \times 300} \text{ N-m}$$

$$= \frac{60,000 \times 50 \times 10^3}{2\pi \times 300} \text{ N-mm.}$$

Also, $\qquad T = \frac{\pi}{16}\left(\frac{D^4 - d^4}{D}\right)\tau$

$$= \frac{\pi}{16}\left[\frac{D^4 - \left(\frac{D}{2}\right)^4}{D}\right]\tau$$

$$= \frac{\pi}{16} \times \frac{15}{16} D^3 \tau$$

$$\therefore \qquad D = \sqrt[3]{\frac{256}{\pi \times 15} \times \frac{T}{\tau}} = \sqrt[3]{\frac{256}{15\pi} \times \frac{60,000 \times 50 \times 10^3}{2\pi \times 300 \times 100}}$$

$$= 44.2 \text{ mm}$$

$$d = \frac{D}{2} = \frac{44.2}{2} = \textbf{22.1 mm} \quad \textbf{Ans.}$$

Example 10.27: Determine the torque required to give a twist of 10 rounds to a steel wire which is of 0.60 mm diameter and 30 m length. $G = 80$ GPa.

Solution: $\qquad \theta = 2\pi \times 10 \text{ rad.}$

$$D = 0.6 \text{ mm}$$

$$L = 30 \text{ m} = 30 \times 10^3 \text{ mm}$$

$$G = 80 \text{ GPa} = 80 \times 10^3 \text{ N/mm}^2.$$

From torsion equation,

$$\frac{T}{J} = \frac{G\theta}{L}$$

$$T = \frac{G\theta J}{L} = \frac{80 \times 10^3 \times 2\pi \times 10}{30 \times 10^3} \times \left(\frac{\pi}{32} \times 0.6^4\right)$$

$$= \textbf{2.1297 N-mm} \quad \textbf{Ans.}$$

Example 10.28: Calculate the diameter of a circular shaft to transmit 75 kW at 200 rpm. Allowable shear stress is restricted to 50 MN/m^2 and twist 1° in 2 m shaft length. $G = 100$ GN/m^2.

Solution:

$$P = 75 \text{ kW}$$

$$N = 200 \text{ rpm}$$

$$\tau = 50 \text{ MN/m}^2 = 50 \text{ N/mm}^2$$

$$\theta = 1° \times \frac{\pi}{180} = \frac{\pi}{180} \text{ rad}$$

$$L = 2 \text{ m} = 2000 \text{ mm}$$

$$G = 100 \text{ GN/m}^2 = 100 \times 10^3 \text{ N/mm}^2$$

$$P = \frac{2\pi NT}{60,000}$$

$$T = \frac{60,000}{2\pi N} = \frac{60,000 \times 75}{2\pi \times 200}$$

$$= 3581 \text{ N-m.} = 3581 \times 10^3 \text{ N-mm.}$$

(a) *Using shear stress as design basis*

$$T = \frac{\pi}{16} D^3 \tau$$

$$D = \sqrt[3]{\frac{16T}{\pi\tau}} = \sqrt[3]{\frac{358 \times 10^3 \times 16}{\pi \times 50}}$$

$$= 71.45 \text{ mm.}$$

(b) *Using twist as design basic*

$$\frac{T}{J} = \frac{G\theta}{L}$$

$$\therefore \quad J = \frac{TL}{G\theta}$$

$$\frac{\pi}{32} D^4 = \frac{TL}{G\theta}$$

$$\therefore \quad D = \sqrt[4]{\frac{32 \times T \times L}{\pi \times G \times \theta}} = \sqrt[4]{\frac{32 \times 3581 \times 10^3 \times 2000}{\pi \times 10^5 \times \dfrac{\pi}{180}}}$$

$$D = 80.4 \text{ mm}$$

Select $\quad D = 80.4$ mm, i.e, the larger value.

Example 10.29: It is required to transmit 112.5 kW at 200 rpm by a solid circular shaft. Select a suitable diameter of the shaft if allowable twist is 1° in a length of 3 m and allowable shear stress is 75 N/mm². $G = 0.082 \times 10^6$ N/mm².

Solution:
$$P = 112.5 \text{ kW}$$
$$N = 200 \text{ rpm}$$

$$\theta = 1° = \frac{\pi}{180} \text{ rad}$$
$$L = 3 \text{ m} = 3000 \text{ mm.}$$
$$\tau = 75 \text{ N/mm}^2$$
$$G = 0.086 \times 10^6 \text{ N/mm}^2$$

$$P = \frac{2\pi NT}{60 \times 1000}$$

$$T = \frac{60,000\ P}{2\pi N} = \frac{60,000 \times 112.5}{2\pi \times 200} = 5.37 \text{ N-m.}$$

$$= 5.37 \times 10^3 \text{ N-mm}$$

(a) *Using twist as design basis*

$$\frac{T}{J} = \frac{G\theta}{L}$$

$$J = \frac{TL}{G\theta}$$

$$\frac{\pi}{32} D^4 = \frac{TL}{G\theta}$$

$$D = \sqrt[4]{\frac{TL}{G\theta} \times \frac{32}{\pi}} = \sqrt[4]{\frac{5.37 \times 10^3 \times 3000 \times 32}{0.08 \times 10^6 \times \dfrac{\pi}{180} \times \pi}}$$

$$= 103.4 \text{ mm}$$

(b) *Using shear stress as design basis*

$$T = \frac{\pi}{16} D^3 \tau$$

$$D = \sqrt[3]{\frac{16T}{\pi\tau}} = \sqrt[3]{\frac{16 \times 5.37 \times 10^3}{\pi \times 75}}$$

$$= 71.47 \text{ mm.}$$

Select $\qquad D = 103.4$ mm, i.e, the larger value.

Example 10.30: A circular steel shaft 25 mm diameter is subjected to a torque of 500 N-m. Determine:

   (a) The maximum shear stress developed in the shaft.

   (b) The shear stress at a distance 10 mm from the centre of the shaft.

   (c) The angular twist over 1 m length of the shaft. Take $G = 80$ GPa.

Solution:
$$D = 25 \text{ mm}$$
$$R = 12.5 \text{ mm}$$
$$T = 500 \text{ N-m} = 500 \times 10^3 \text{ N-mm}$$

   (a) *Maximum shear stress at the surface of shaft*

$$T = \frac{\pi}{16} D^3 \tau$$

$\therefore$
$$\tau = \frac{16T}{\pi D^3} = \frac{16 \times 500 \times 10^3}{\pi \times (25)^3} = 163 \text{ N/mm}^2$$

   (b) *Shear stress at r = 10 mm*

$$\tau_r = \frac{\tau}{R} \times r = \frac{163}{12.5} \times 10 = 130.4 \text{ N/mm}^2$$

   (c)
$$\frac{T}{J} = \frac{G\theta}{L}$$

$$\theta = \frac{TL}{GJ} = \frac{500 \times 10^3 \times 1000}{\left(80 \times 10^3\right) \times \dfrac{\pi}{32}(25)^4} = 0.163 \text{ rad}$$

$$= 0.163 \times \frac{180}{\pi} = 9.344°.$$

Example 10.31: A hollow shaft is to transmit a uniform torque of 30 kN-m. The total angular twist in 2.5 m length of the shaft is not to exceed 2 degrees and the allowable shear stress is 85 MPa. Determine the external and internal diameters if $G = 80$ GPa.

Solution:
$$T = 30 \text{ kN-m} = 30 \times 10^6 \text{ N-mm}$$
$$L = 2.5 \text{ m} = 2500 \text{ mm}$$

$$\theta = 2° = 2 \times \frac{\pi}{180} \text{ rad}$$

$$\tau = 85 \text{ MPa} = 85 \text{ N/mm}^2$$
$$G = 80 \text{ GPa} = 80 \times 10^3 \text{ N/mm.}$$

(a) *Using twist as design basis*

$$\frac{T}{J} = \frac{G\theta}{L} \qquad \text{(Torsion Equation)}$$

$$J = \frac{TL}{G\theta}$$

$$\frac{\pi}{32}(D^4 - d^4) = \frac{TL}{G\theta}.$$

$$(D^4 - d^4) = \frac{32TL}{\pi G\theta} = \frac{32 \times 30 \times 10^6 \times 2500 \times 180}{\pi \times 80 \times 10^3 \times 2\pi}$$

$$= 273.567 \times 10^6 \text{ mm.} \qquad \qquad ...(1)$$

(b) *Using shear stress as design basis*

$$T = \frac{\pi}{16}\left(\frac{D^4 - d^4}{D}\right)\tau$$

$$\therefore \qquad \left(\frac{D^4 - d^4}{D}\right) = \frac{16T}{\pi\tau} = \frac{16 \times 30 \times 10^6}{\pi \times 85} = 1.798 \times 10^6. \qquad ...(2)$$

From eqns. (1) and (2),

$$1.798 \times 10^6\, D = 273.567 \times 10^6$$

$$\therefore \qquad D = \frac{273.567 \times 10^6}{1.798 \times 10^6} = 152.3 \text{ mm}$$

$$(D^4 - d^4) = 273.567 \times 10^6$$

$$\therefore \qquad d^4 = D^4 - 273.567 \times 10^6$$

$$= (152.3)^4 - 273.567 \times 10^6$$

$$= 264.453 \times 10^6$$

$$\therefore \qquad d = 127.5 \text{ mm.}$$

Example 10.32:   The diameter of a shaft is 20 cm. Find the safe maximum torque which can be transmitted by the shaft if the permissible stress in the shaft material be 4000 N/cm^2 and permissible angle of twist is 0.2 degree per meter length. Take $G = 8 \times 10^6$ N/cm^2. If the shaft rotates at 320 rpm, what maximum power can be transmitted by the shaft? (U.P.T.U. I Sem., 2005-06)

Solution: $D = 20$ cm $= 200$ mm

$$\tau = 4000 \text{ N/cm}^2 = 40 \text{ N/mm}^2.$$

$$\theta = 0.2° = 0.2 \times \frac{\pi}{180} \text{ rad.}$$

$$L = 1000 \text{ mm.}$$

$$G = 8 \times 10^6 \text{ N/cm}^2 = 8 \times 10^4 \text{ N/mm}^2$$

(a) *Using shear stress as design basis*

$$T = \frac{\pi}{16} D^3 \tau$$

$$= \frac{\pi}{16} (200)^3 \times 40 = 62831853 \text{ N-mm}$$

$$= 62.83 \text{ kN-m.}$$

(b) *Using twist as design basis*

$$\frac{T}{J} = \frac{G\theta}{L}$$

$$T = \frac{G\theta J}{L} = \frac{8 \times 10^4 \times 0.2}{1000} \times \frac{\pi}{180} \times \frac{\pi(200)^4}{32}$$

$$= 43864908 \text{ N-mm}$$

$$= 43.865 \text{ kN-m}$$

The safe maximum torque is 43.865 kN-m, the lower value.

Power, $\quad P = \dfrac{2\pi NT}{60 \times 1000}$

$$= \frac{2\pi \times 320 \times 43.865}{60}$$

$$= \mathbf{1470 \text{ kW}} \quad \textbf{Ans.}$$

Example 10.33: A solid circular shaft is to transmit 160 kW at 180 rpm. What will be the suitable diameter of this shaft if the permissible stress in the shaft material should not exceed 2×10^6 Pa and twist per unit length should not exceed 2°. Take $G = 200$ GPa. [U.P.T.U. II Sem., 2005-06]

Solution: $\quad P = 160$ kW

$$N = 180 \text{ rpm}$$

$$\tau = 2 \times 10^6 \text{ Pa}$$

$$\theta = 2° = 2 \times \frac{\pi}{180} \text{ rad.}$$

$$L = 1 \text{ m}$$
$$G = 200 \text{ GPa} = 200 \times 10^9 \text{ Pa.}$$

$$P = \frac{2\pi NT}{60 \times 1000}$$

$\therefore$
$$T = \frac{60,000 \, P}{2\pi N} = \frac{60,000 \times 160}{2\pi \times 180}$$

$$= 8488.26 \text{ N-m}$$

(a) *Using shear stress as design basis*

$$T = \frac{\pi}{16} D^3 \tau$$

$\therefore$
$$D = \sqrt[3]{\frac{16T}{\pi\tau}}$$

$$= \sqrt[3]{\frac{16 \times 8488.26}{\pi \times 2 \times 10^6}}$$

$$= 0.27856 \text{ m}$$
$$= 278.56 \text{ mm}$$

(b) *Using twist as design basis*

$$\frac{T}{J} = \frac{G\theta}{L}$$

$$J = \frac{TL}{G\theta}$$

$$\frac{\pi}{32} D^4 = \frac{TL}{G\theta}$$

$$D = \sqrt[4]{\frac{32TL}{\pi G\theta}} = \sqrt[4]{\frac{32 \times 8488.26 \times 1 \times 180}{\pi \times 200 \times 10^9 \times 2 \times \pi}}$$

$$= 0235.25 \text{ m}$$
$$= 235.25 \text{ mm.}$$

Select $\quad\quad D = 235.25$ mm. i.e, the smaller value.

Example 10.34: A hollow shaft of 3 m length is subjected to a torque such that the maximum shear stress produced is 75 MN/m^2. The external and internal diameters of the shaft are 150 mm and 100 mm respectively. Find the shear stress at the inside surface. Take $G = 75$ MN/m^2. [U.P.T.U. II Sem., 2003-04]

Solution:

$$L = 3 \text{ m}$$
$$\tau = 75 \text{ MN/m}^2$$
$$D = 150 \text{ mm} = 0.15 \text{ m}$$
$$d = 100 \text{ mm} = 0.10 \text{ m}$$
$$G = 75 \text{ MN/m}^2.$$

$$T = \frac{\pi}{16} \left(\frac{D^4 - d^4}{D} \right) \tau$$

$$= \frac{\pi}{16} \left(\frac{0.15^4 - 0.1^4}{0.15} \right) \times 75$$

$$= 0.03988 \text{ MN-m.}$$

$$\frac{T}{J} = \frac{G\theta}{L}$$

∴
$$\theta = \frac{TL}{GJ} = \frac{0.03988 \times 3 \times 16 \times 0.15}{75 \times \pi \left(0.15^4 - 0.1^4\right)} = 3 \text{ rad}$$

$$= 3 \times \frac{180}{\pi} = 171.8°$$

The shear stress varies linearly with radius

∴
$$\frac{\tau_i}{\tau_o} = \frac{d}{D}$$

∴
$$\tau_i = \frac{d}{D} \times \tau_o = \frac{100}{150} \times 75 = \textbf{50 MN/m}^2 \quad \textbf{Ans.}$$

Example 10.35: A solid circular shaft is required to transmit 200 kW at 100 rpm. Determine the diameter of the shaft if permissible shear stress is 60 N/mm². Calculate the energy stored per meter length of the shaft. $G = 1 \times 10^5$ N/mm².

Solution:

$$P = 200 \text{ kW}$$
$$N = 100 \text{ rpm}$$
$$\tau = 60 \text{ N/mm}^2$$
$$G = 1 \times 10^5 \text{ N/mm}^2$$

Now,
$$P = \frac{2\pi NT}{60 \times 1000}$$

∴
$$T = \frac{60,000 P}{2\pi N} = \frac{60,000 \times 200}{2\pi \times 100}$$

$$T = 19108 \text{ N-m}$$
$$= 19108 \times 10^3 \text{ N-mm.}$$

$$T = \frac{\pi}{16} D^3 \tau$$

$$\therefore \qquad D = \sqrt[3]{\frac{16T}{\pi\tau}} = \sqrt[3]{\frac{16 \times 19108 \times 10^3}{\pi \times 60}}$$

$$= 116.95 \text{ mm}$$

Strain energy, $\qquad U = \dfrac{T^2}{4G} \times V$

$$\therefore \qquad U = \frac{60^2}{4 \times (1 \times 10^5)} \left[\frac{\pi}{4} (116.95)^2 \times 1000 \right]$$

$$= 96630 \text{ N-mm} = 96.63 \text{ N-m}$$
$$= \mathbf{96.63 \ J. \quad Ans.}$$

Example 10.36: A solid shaft of 200 mm diameter is to be replaced by a hollow shaft of the same material, weight and length. Find the external and internal diameter of the hollow shaft if its strain energy is 1.15 is higher than that of the solid shaft when transmitting torque at the same maximum shear stress.

Solution: Both shafts have same material, same length and same weight.

Weight of solid shaft $\quad = \rho A_s L$

Weight of hollow shaft $= \rho A_h L$

$$\therefore \qquad A_s = A_h$$

$$\frac{\pi}{4} (200)^2 = \frac{\pi}{4} (D^2 - d^2)$$

$$D^2 - d^2 = 40,000$$

Again strain energy stored by solid shaft

$$U_s = \frac{\tau^2}{4G} \text{ Volume} = \frac{\tau^2}{4G} \cdot \frac{\pi}{4} D^2 L$$

Strain energy stored by hollow shaft,

$$U_h = \frac{\tau^2}{4G} \cdot \frac{\pi}{4} (D^2 - d^2) L \times \left(\frac{D^2 + d^2}{D^2} \right)$$

Now, $\qquad U_h = 1.15 U_s$

Now, the two shafts must have same volume.

$\therefore$ $$D^2 + d^2 = 1.15D^2$$
$$D^2 = 6.67\, d^2$$
But, $$5.67\, d^2 = 40{,}000$$

$$d = \sqrt{\frac{40{,}000}{5.67}} = 84 \text{ m}$$

$$D = \sqrt{6.67} \times 84 = 216.94 \text{ mm}$$

Example 10.37:

(a) Show that for a hollow shaft of external and internal radius r_o and r_i respectively, the torsional strain energy is given by

$$\frac{\tau^4}{4G}\left[\frac{r_0^2 + r_i^2}{r_0^2}\right] \times \text{volume of shaft.}$$

(b) A hollow shaft having the external diameter 1.5 times the internal diameter is subjected to pure torque. If it attains a maximum shear stress τ_{max}, show that strain energy per unit volume is $\dfrac{13}{36G}\,\tau_{max}^2$.

Solution: (a) Strain energy,

$$U = \frac{1}{2}\, T\theta$$

From torsion equation,

$$\frac{T}{J} = \frac{\tau}{R} = \frac{G\theta}{L}$$

$$\theta = \frac{\tau L}{GR} = \frac{\tau L}{Gr_0}$$

and $$T = \frac{\tau J}{R} = \frac{\tau J}{r_0}$$

$\therefore$ $$U = \frac{1}{2}\frac{\tau J}{r_0} \times \frac{\tau L}{Gr_0} = \frac{\tau^2}{2G} \times \frac{JL}{r_0^2}$$

$$= \frac{\tau^2}{4G} \cdot \frac{\pi}{2}\frac{(r_0^4 + r_i^4)}{r_0^2} \cdot L$$

$$= \frac{\tau^4}{4G}\frac{(r_0^2 + r_i^2)}{r_0^2} \times \pi(r_0^2 - r_i^2)L$$

$$U = \frac{\tau^4}{4G} \frac{(r_0^2 + r_i^2)}{r_0^2} \times \text{Vol. of shaft.}$$

(b) Strain energy per unit volume,

$$\frac{U}{V} = \frac{\tau_{max}^2}{4G} \cdot \left(\frac{d_0^2 + d_i^2}{d_0^2} \right)$$

$$= \frac{\tau_{max}^2}{4G} \times \frac{d_0^2 + \left(\frac{2}{3} d_0 \right)^2}{d_0^2}$$

$$= \frac{13}{36G} \tau_{max}^2$$

10.12   COMPARISON OF SOLID AND HOLLOW SHAFTS

In industry power is transmitted both by solid as well as hollow shafts. For same power or torque transmission, hollow shafts are lighter in weight due to unfavourable stress distribution across the cross-section of a solid shaft.

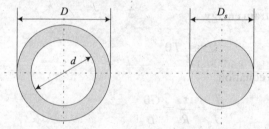

Fig. 10.15   Solid and hollow shafts

1. Geometry. If diameter of solid shaft is D_s and external and internal diameters of hollow shaft are D and d respectively:

Area of cross-section of solid shaft $\qquad = \frac{\pi}{4} D_s^2$

Area of cross-section of hollow shaft $\qquad = \frac{\pi}{4} (D^2 - d^2)$

Polar moment of inertia of solid shaft, $\qquad J = \frac{\pi}{32} D_s^4$

Polar section modulus of solid shaft, $\qquad Z_p = \frac{\pi}{16} D_s^3$

Polar moment of inertia of hollow shaft, $J = \dfrac{\pi}{32}(D^4 - d^4)$

Polar section modulus of hollow shaft, $Zp = \dfrac{\pi}{16}\left(\dfrac{D^4 - d^4}{D}\right)$

2. Weight. The weight of a shaft, $W = \rho g V$ [kN]

where, ρ = Density of material [kg/m³]

g = acceleration due to gravity

$= 9.81$ m/s²

V = volume of the shaft [m³]

For a solid shaft, $V = \dfrac{\pi}{4} D_s^2 L$ [m³]

For a hollow shaft, $V = \dfrac{\pi}{4}(D^2 - d^2)L$ [m³]

For same material and length, the volume of solid and hollow shaft must be same.

$\therefore \qquad D_s^2 = (D^2 - d^2).$

3. Strength. The torque transmission by a shaft

$$T = Zp\tau$$

For a solid shaft,

$$T_s = \dfrac{\pi}{16} D_s^3 \tau$$

For a hollow shaft,

$$T_h = \dfrac{\pi}{16}\left(\dfrac{D^4 - d^4}{D}\right)\tau$$

$\therefore \qquad \dfrac{T_s}{T_h} = \dfrac{D_s^3 \times D}{(D^4 - d^4)},$ if τ is same.

4. Rigidity. Torsional rigidity, $\dfrac{T}{\theta} = \dfrac{GJ}{L}$

$$\dfrac{\text{Torsional rigidity of solid shaft}}{\text{Torsional rigidity of hollow shaft}} = \dfrac{GJ_s/L}{GJ_h/L}$$

For same material and shaft length,

$$\frac{\text{Torsional rigidity of solid shaft}}{\text{Torsional rigidity of hollow shaft}} = \frac{J_s}{J_4}$$

$$= \frac{\dfrac{\pi}{32} \cdot D_s^4}{\dfrac{\pi}{32}(D^4 - d^4)} = \frac{D_s^4}{(D^4 - d^4)}$$

5. Angle of twist. From torsion equation:

$$\frac{T}{J} = \frac{\tau}{R} = \frac{G\theta}{L}$$

∴

$$\theta = \frac{\tau L}{GR}$$

For same material and length,

$$\frac{\theta_s}{\theta_h} = \frac{D_h}{D_s}$$

Example 10.38: Compare the weights of a solid shaft and hollow shaft of same material, same length, same torque and same allowable shear stress. The internal diameter of hollow shaft is 2/3 of its outer diameter.   [U.P.T.U. II Sem., 2003-04]

Solution:

(a) *Solid shaft*

Torque,
$$T = \frac{\pi}{16} D_s^3 \tau$$

Weight,
$$W_s = \frac{\pi}{4} D_s^2 L\rho g$$

(b) *Hollow shaft*

Torque,
$$T = \frac{\pi}{16}\left(\frac{D^4 - d^4}{D}\right)\tau$$

$$= \frac{\pi}{16}\left[\frac{D^4 - \left(\dfrac{2D}{3}\right)^4}{D}\right]\tau$$

$$= \frac{\pi}{16} \times 0.8 D^3 \tau$$

Weight, $\qquad W_h = \dfrac{\pi}{4}(D^2 - d^2)L\rho g$

$$= \dfrac{\pi}{4}\left[D^2 - \left(\dfrac{2D}{3}\right)^2\right]L\rho g$$

$$= \dfrac{\pi}{4}\times 0.55\, D^2 L\rho g$$

(c) *Comparison*

For same torque,

$$\dfrac{\pi}{16}\,D_s^3\tau \;=\; \dfrac{\pi}{16}\times 0.8\, D^3\tau$$

$\therefore \qquad\qquad \dfrac{D_s}{D} = 0.928.$

$$\dfrac{\text{Weight of solid shaft}}{\text{Weight of hollow shaft}} = \dfrac{Ws}{W_h} = \dfrac{\pi/4 D_s^2 L\rho g}{\dfrac{\pi}{4}\times 0.55 D^2 L\rho g}$$

$$= \dfrac{(0.928D)}{0.55D} = \mathbf{1.566} \quad \textbf{Ans.}$$

Example 10.39: A solid round shaft is replaced by a hollow shaft, the external diameter of which is $1\dfrac{1}{4}$ times the internal diameter. Allowing the same intensity of torsional stress in each, compare the weight of the solid shaft with that of hollow shaft.

Solution: The torque transmitted by solid shaft,

$$T = \dfrac{\pi}{16}\,D_s^3\tau$$

The torque transmitted by hollow shaft,

$$T = \dfrac{\pi}{16}\dfrac{(D^4 - d^4)}{D}\tau.$$

For same torque and allowable shear stress

$$\dfrac{\pi}{16}D_s^3\tau \;=\; \dfrac{\pi}{16}\dfrac{(D^4 - d^4)}{D}\tau.$$

$$D_s^3 = \frac{\left[D^4 - \left(\dfrac{D}{1.25}\right)^4\right]}{D}$$

$$\therefore \quad \frac{D_s}{D} = 0.84$$

$$\frac{\text{Weight of solid shaft}}{\text{Weight of hollow shaft}} = \frac{\dfrac{\pi}{4} D_s^2 \rho g L}{\dfrac{\pi}{4}(D^2 - d^2)\rho g L}$$

$$= \frac{\dfrac{\pi}{4}(0.84D)^2}{\dfrac{\pi}{4}\left[D^2 - \left(\dfrac{D}{1.25}\right)^2\right]} = 1.96 \quad \textbf{Ans.}$$

Example 10.40: Two shafts of the same material and same length are subjected to same torque. The first shaft is of a solid circular section and the second shaft is of a hollow circular section with internal diameter 0.6 times the external diameter. Compare the weight of the two shafts if the maximum shear stress developed in each of them is the same.

Solution:

Torque, $\quad T = \dfrac{\pi}{16} D^3 \tau$

For same strength,

$$Z_s = Z_h$$

$$\frac{\pi}{16} D_s^3 = \frac{\pi}{16}\left(\frac{D^4 - d^4}{D}\right) = \frac{\pi}{16}\frac{[D^4 - (0.6D)^4]}{D} = \frac{\pi}{16} 0.87 D^3$$

$$\therefore \quad \frac{D_s}{D} = \sqrt[3]{0.87} = 0.955$$

The ratio of weights,

$$\frac{Wh}{W_s} = \frac{\dfrac{\pi}{4}[D^2 - (0.6D)^2]\rho g L}{\dfrac{\pi}{4}D_s^2 \rho g L} = \frac{D^2 - (0.6D)^2}{(0.87D)^2}$$

$$= 0.475 \quad \textbf{Ans.}$$

10.13   COMPARISON OF BENDING AND TORSION

The various parameters of bending of beams and torsion of shafts are compared in Table 10.1.

Table 10.1   Comparison of bending and torsion.

Sl. No.	Parameter	Bending	Torsion
1.	Loading	Bending of beam under bending moment	Torsion of shafts under torque
2.	Governing equation	Bending equation $$\frac{M}{I} = \frac{\sigma}{Y} = \frac{E}{R}$$	Torsion equation $$\frac{T}{J} = \frac{\tau}{R} = \frac{G\theta}{L}$$
3.	Stress	$\sigma = \dfrac{M}{Z}$. Z = section modulus	$\tau = \dfrac{T}{Z_p}$, Z_p = polar section modulus.
4.	Section modulus	$Z = \dfrac{I}{y}$ Rectangle, $Z = \dfrac{bd^2}{6}$ circle, $Z = \dfrac{\pi d^3}{32}$	$Z_p = \dfrac{J}{R}$ Solid shaft $Z_p = \dfrac{\pi D^3}{16}$ Hollow shaft, $Z_p = \dfrac{\pi(D^4 - d^4)}{16D}$
5.	Rigidity	Flexural Rigidity, $EI = MR$	Tosional Rigidity, $GJ = \dfrac{T}{\theta/L}$
6.	Combined Loading	Bending + direct load $$\sigma = \frac{P}{A} \pm \frac{My}{I}$$ $$= \frac{P}{A} \pm \frac{Pey}{I}$$	Torsion + Bending $$M_e = \frac{1}{2}\left[M + \sqrt{M^2 + T^2}\right]$$ $$T_e = \left[\sqrt{M^2 + T^2}\right]$$
7.	Strain Energy	$U = \dfrac{M^2 L}{2EI}$	$\dfrac{U}{v} = \dfrac{\tau^2}{4G}$, $U = \dfrac{\tau^2 L}{2GJ}$

QUESTION BANK NO. 10

1. What is pure bending? Derive the relation between bending moment and radius of curvature of the beam under pure bending.

2. What are the assumptions made in the theory of pure bending?

3. Define and explain the following:

     (i) Moment of resistance in bending.

    (ii) Section modulus

   (iii) Flexural rigidity

4. Explain the following terms:

    (i) Pure torsion

    (ii) Polar section modulus

   (iii) Torsional rigidity

5. What do you understand by pure torsion? What are its assumptions? Define

   torsion equation, $\dfrac{T}{J} = \dfrac{\tau}{R} = \dfrac{G\theta}{L}$ where symbols are having usual meanings.

6. What is shear stress distribution for a shaft under pure torsion if the shaft cross-section is:

    (i) Solid

    (ii) Hollow.

7. Show that the strain energy stored in a solid shaft for unit volume is $0.25\ \tau^2/G$. Where τ is the shear stress at the outer radius of the shaft and G is the modulus of rigidity.

ANNEXURES

ANNEXURES

ANNEXURE I: INTRODUCTION TO SI UNITS

ANNEXURE II: STEAM TABLES IN SI UNITS

ANNEXURE III: SECTION MODULUS

ANNEXURE IV: QUESTION PAPER UPTU 2006-2007

ANNEXURE I: INTRODUCTION TO SI UNITS

International system of Units (SI units) have seven fundamental units and two supplementary units, which cover the entire field of science and engineering. These basic units are given in Table 1.1.

Table I.1 SI System: Basic Units

S. No.	Physical Quantity	Unit	Symbol
	Fundamental Units		
1.	Length (L)	Metre	m
2.	Mass (M)	Kilogramme	kg
3.	Amount of substance (n)	Mole	Mol
4.	Time (t)	Second	s
5.	Temperature (T)	Kelvin	K
6.	Electric Current (I)	Ampere	A
7.	Luminous intensity (Iv)	Candela	cd
	Supplementary Units		
1.	Plane angle ($\alpha, \beta, \theta, \phi$)	Radian	rad
2.	Solid angle (Ω)	Steradian	sr

The dimensions of all other quantities are derived from the basic units. Some of the derived units are given in Table 1.2.

Table I.2 SI System: Derived Units

S. No.	Quantity	Unit	Symbol	Alternative unit	In basic units
1	2	3	4	5	6
1.	Force, Weight (F, W)	newton	N	–	$kg\,m/s^2$
2.	Work, Energy, Enthalpy (W, E, H)	joule	J	N-m	$kg\,m^2/s^2$
3.	Power (P)	watt	W	J/s	$kg\,m^2/s^2$
4.	Pressure (p)	pascal	Pa	N/m^2	kg/m^2
5.	Gas Constant (R)	-	J/kg K	-	m^2/Ks^2
6.	Molar mass or Molecular mass (M)	-	kg/mol	-	kg/mol
7.	Specific heat (c)	-	J/kg K	-	m/s^2K
8.	Thermal Conductivity (k)	-	w/mK	-	$kg\,m/s^2K$
9.	Mass density (ρ)	-	kg/m^3	-	kg/m^3
10.	Linear velocity (V)	-	m/s	-	m/s
11.	Linear acceleration (a)	-	m/s^2	-	m/s^2
12.	Angular velocity (ω)	-	rad/s	-	rad/s
13.	Angular acceleration (α)	-	rad/s^2	-	rad/s^2
14.	Dynamic viscosity (μ)	-	$N\text{-}s/m^2$	-	kg/sm
15.	Kinematic viscosity (μ)	-	m^2/s	-	m^2/s
16.	Frequency (f)	herz	Hz	-	s^{-1}

It is often convenient and desirable to use multiples of various units, the standard list of which is given in Table 1.3.

Table I.3 SI System: Standard Multipliers

Factor	Prefix	Abbreviation
10^{12}	tera	T
10^{9}	giga	G
10^{6}	mega	M
10^{3}	kilo	k
10^{-3}	milli	m
10^{-6}	micro	m
10^{-9}	nano	n
10^{-12}	pico	P

Rules for S.I. Units

There are certain styles and usage of S.I. units.

1. For large numerals use groups of three without a coma. Grouping should start both sides from decimal. No grouping is done if figures are only 4.

Correct	Incorrect
234 869.739 45	23 4869.7 394 5
40 000	40000
3456.72	3456.7 2
0.123 457	0.123457

2. A dash should be used to separate units that are multiplied together.

Correct	Incorrect
N-m	Nm or mN which stand for millinewton

3. Plurals are never used with symbols or full names. For example, metre or metres are written as *m*.

4. Full names of units or prefixes shall be written in upright roman, lower case letters. No capitals are used even if unit is named after a person.

Correct	Incorrect
kelvin	Kelvin
newton	Newton
joule	Joule
watt	Watt
volt	Volt

5. All symbols are written in small letters except those derived from the proper names.

Correct	Incorrect
m	M
s	S
kg	Kg
W	w
J	j
N	n
K	k

6. No abbreviation should be used for units. Write either as a symbol or full name.

Correct	Incorrect
N or newton	newt
kg	kgram

7. No full stop or a punctuation mark be used.

Correct	Incorrect
3 m	3m,
100.25 kg	100.25 kg.

8. If a prefix is used with a symbol no space be left in between.

Correct	Incorrect
MJ	M J

9. Not more than one prefix be used. For 10^{-9} m.

Correct	Incorrect
nm	m µ m

10. Power to base unit applies to whole and not to base alone.

Correct	Incorrect
$km^2 = (km)^2 = 10^6 m^2$	$km^2 \neq k(m)^2 = 10^3 m^2$

11. Leave a space between numeral and its unit.

Correct	Incorrect
64 kW	64kW

12. A solidus or oblique may be used or a negative power may be used.

Correct

$\frac{m}{s}$, m/s or ms^{-1}

ANNEXURE II: STEAM TABLES IN SI UNITS

Table II.1 Saturated water and steam (temperatures) tables

Temperature in °C	Absolute pressure in bar	Specific volume in m²/kg		Specific enthalpy in kJ/kg			Specific entropy in kJ/kg K			Temperature in °C
		Water	Steam	Water	Evaporation	Steam	Water	Evaporation	Steam	
(t)	(p)	(v_f)	(v_g)	(h_f)	(h_{fg})	(h_g)	(s_f)	(s_{fg})	(s_g)	(t)
0	0.006 11	0.001 000	206.31	0.0	2 501.6	2 501.6	0.000	9.158	9.158	0
1	0.006 57	0.001 000	192.61	4.2	2 499.2	2 503.4	0.015	9.116	9.131	1
2	0.007 06	0.001 000	179.92	8.4	2 496.8	2 505.2	0.031	9.074	9.105	2
3	0.007 58	0.001 000	168.17	12.6	2 494.5	2 507.1	0.046	9.033	9.079	3
4	0.008 13	0.001 000	157.27	16.8	2 492.1	2 508.9	0.061	8.992	9.053	4
5	0.008 72	0.001 000	147.16	21.0	2 489.7	2 510.7	0.076	8.951	9.027	5
6	0.009 35	0.001 000	137.78	25.2	2 487.4	2 512.6	0.091	8.911	9.002	6
7	0.010 01	0.001 000	129.06	29.4	2 485.0	2 514.4	0.106	8.870	8.976	7
8	0.010 72	0.001 000	120.97	33.6	2 482.6	2 516.2	0.121	8.830	8.951	8
9	0.011 47	0.001 000	113.44	37.8	2 480.3	2 518.1	0.136	8.791	8.927	9
10	0.012 27	0.001 000	106.43	42.0	2 477.9	2 519.9	0.151	8.751	8.902	10
11	0.013 12	0.001 000	99.909	46.2	2 475.5	2 521.7	0.166	8.712	8.878	11
12	0.014 01	0.001 000	93.835	50.4	2 473.2	2 523.6	0.181	8.673	8.854	12
13	0.014 97	0.001 001	88.176	54.6	2 470.8	2 525.4	0.195	8.635	8.830	13
14	0.015 97	0.001 001	82.900	58.7	2 468.5	2 527.2	0.210	8.597	8.806	14
15	0.017 04	0.001 001	77.978	62.9	2 466.1	2 529.1	0.224	8.559	8.783	15
16	0.018 17	0.001 001	73.384	67.1	2 463.8	2 530.9	0.239	8.520	8.759	16
17	0.019 36	0.001 001	69.095	71.3	2 461.4	2 532.7	0.253	8.483	8.736	17

(p)	(t)	(v_f)	(v_g)	(h_f)	(h_fg)	(h_g)	(s_f)	(s_fg)	(s_g)	(p)
18	0.020 62	0.001 001	65.087	75.5	2 459.0	2 534.5	0.268	8.446	8.714	**18**
19	0.021 96	0.001 002	61.341	79.7	2 456.7	2 536.4	0.282	8.409	8.691	**19**
20	0.023 37	0.001 002	57.838	83.9	2 454.3	2 538.2	0.296	8.372	8.668	**20**
21	0.024 85	0.001 002	54.561	88.0	2 452.0	2 540.0	0.310	8.336	8.646	**21**
22	0.026 42	0.001 002	51.492	92.2	2449.6	2 541.8	0.325	8.299	8.624	**22**
23	0.28 08	0.001 002	48.619	96.4	2 447.2	2 543.6	0.339	8.263	8.602	**23**
24	0.029 82	0.001 002	45.926	100.6	2 444.9	2 545.5	0.353	8.228	8.581	**24**
25	0.031 66	0.001 003	43.402	104.8	2 442.5	2 547.3	0.367	8.192	8.559	**25**
26	0.033 60	0.001 003	41.034	108.9	2 440.2	2 549.1	0.381	8.157	8.538	**26**
27	0.035 64	0.001 003	38.813	113.1	2 437.8	2 550.9	0.395	8.122	8.517	**27**
28	0.037 78	0.001 004	36.728	117.3	2 435.4	2 552.7	0.409	8.087	8.496	**28**
29	0.040 04	0.001 004	34.769	121.5	2 433.1	2 554.5	0.423	8.052	8.475	**29**
30	0.042 42	0.001 004	32.929	125.7	2 430.7	2 556.4	0.437	8.018	8.455	**30**
31	0.044 91	0.001 005	31.199	129.8	2.428.3	2 558.2	0.450	8.984	8.434	**31**
32	0.047 53	0.001 005	29.572	134.0	2 425.9	2 560.0	0.464	7.950	8.414	**32**
33	0.050 29	0.001 005	28.042	138.2	2 423.6	2 561.8	0.478	7.916	8.394	**33**
34	0.053 18	0.001 006	26.601	142.4	2 421.2	2 563.6	0.491	7.883	8.374	**34**
35	0.056 22	0.001 006	25.245	146.6	2 418.8	2 565.4	0.505	7.849	8.354	**35**
36	0.059 40	0.001 006	23.967	150.7	2 416.4	2 567.2	0.518	7.817	8.335	**36**
37	0.062 74	0.001 007	22.763	154.9	2 414.1	2 569.0	0.532	7.783	8.315	**37**
38	0.066 24	0.001 007	21.627	159.1	2 411.7	2 570.8	0.545	7.751	8.296	**38**
39	0.069 91	0.001 007	20.557	163.3	2 409.3	2 572.6	0.559	7.718	8.277	**39**
40	0.073 75	0.001 008	19.546	167.5	2 406.9	2 574.4	0.572	7.686	8.258	**40**
41	0.077 77	0.001 008	18.592	171.6	2 404.5	2 576.2	0.585	7.654	8.239	**41**
42	0.081 99	0.001 009	17.692	175.8	2 402.1	2 577.9	0.599	7.622	8.221	**42**
43	0.086 39	0.001 009	16.841	180.0	2 399.7	2.579.7	0.612	7.591	8.203	**43**
44	0.091 00	0.001 009	16.036	184.2	2 397.3	2.581.5	0.625	7.559	8.184	**44**

(p)	(t)	(v_f)	(v_g)	(h_f)	(h_fg)	(h_g)	(s_f)	(s_fg)	(s_g)	(p)
45	0.095 82	0.001 010	15.276	188.4	2 394.9	2 583.3	0.638	7.528	8.166	45
46	0.100 80	0.001 010	14.557	192.5	2 392.5	2 585.1	0.651	7.497	8.148	46
47	0.106 12	0.001 011	13.877	196.7	2 390.1	2 586.9	0.664	7.466	8.130	47
48	0.111 62	0.001 011	13.233	200.9	2 387.7	2 588.6	0.678	7.435	8.113	48
49	0.117 36	0.001 012	12.623	205.1	2 385.3	2 590.4	0.691	7.404	8.095	49
50	0.123 35	0.001 012	12.046	209.3	2382.9	2 592.2	0.704	7.374	8.078	50
51	0.129 61	0.001 013	11.499	213.4	2380.5	2 593.9	0.716	7.344	8.060	51
52	0.136 11	0.001 013	10.980	217.6	2378.1	2 595.7	0.729	7.314	8.043	52
53	0.142 93	0.001 014	10.488	221.8	2375.7	2 597.5	0.742	7.284	8.026	53
54	0.150 02	0.001 014	10.022	226.0	2373.2	2 599.2	0.755	7.254	8.009	54
55	0.157 41	0.001 015	9.5789	230.2	2 370.8	2 601.0	2.768	7.225	7.993	55
56	0.165 11	0.001 015	9.1587	234.3	2368.4	2602.7	0.780	7.196	7.976	56
57	0.173 13	0.001 016	8.759 8	238.5	2366.0	2 604.5	0.793	7.166	7.959	57
58	0.181 47	0.001 016	8.380 8	242.7	2 363.5	2 606.2	0.806	7.137	7.943	58
59	0.190 16	0.001 017	8.020 8	246.9	2 361.1	2 608.0	0.818	7.109	7.927	59
60	0.199 20	0.001 017	7.678 5	251.1	2 358.6	2 609.7	0.831	7.080	7.911	60
61	0.208 61	0.001 018	7.353 2	255.3	2 356.1	2 611.4	0.844	7.051	7.895	61
62	0.218 38	0.001 018	7.043 7	259.5	2 353.7	2 613.2	0.856	7.023	7.879	62
63	0.228 55	0.001 019	6.749 3	263.6	2 351.3	2 614.9	0.868	6.995	7.863	63
64	0.239 12	0.001 019	6.469 0	267.8	2 348.8	2 616.6	0.881	6.967	7.848	64
65	0.250 09	0.001 020	6.202 3	272.0	2 346.4	2 618.4	0.893	6.939	7.832	65
66	0.261 50	0.001 020	5.948 2	276.2	2 343.9	2 620.1	0.906	6.911	7.817	66
67	0.273 34	0.001 021	5.706 2	280.4	2 341.4	2 621.8	0.918	6.884	7.802	67
68	0.285 63	0.001 022	5.475 6	284.6	2 338.9	2 623.5	0.930	6.856	7.786	68
69	0.298 38	0.001 022	5.255 8	288.8	2 336.4	2 625.2	0.943	6.828	7.771	69
70	0.311 62	0.001 023	5.046 3	293.0	2 333.9	2 626.9	0.955	6.802	7.757	70
71	0.325 35	0.001 024	4.846 4	297.2	2 331.4	2 628.6	0.967	6.775	7.742	71

(p)	(t)	(v_f)	(v_g)	(h_f)	(h_{fg})	(h_g)	(s_f)	(s_{fg})	(s_g)	(p)
72	0.339 58	0.001 024	4.655 7	301.3	2 329.0	2 630.3	0.979	6.748	7.727	72
73	0.354 34	0.001 025	4.473 7	305.5	2 326.5	2 632.0	0.991	6.721	7.712	73
74	0.369 64	0.001 025	4.300 0	309.7	2 324.0	2 633.7	1.003	6.695	7.698	74
75	0.385 49	0.001 026	4.134 1	313.9	2 321.5	2 635.4	1.015	6.668	7.683	75
76	0.401 91	0.001 027	3.975 7	318.1	2 318.9	2 637.0	1.027	6.642	7.669	76
77	0.418 91	0.001 027	3.824 3	322.3	2 316.4	2 638.7	1.039	6.616	7.655	77
78	0.436 52	0.001 028	3.679 6	326.5	2 313.9	2 640.4	1.051	6.590	7.641	78
79	0.454 74	0.001 029	3.541 3	330.7	2 311.4	2 642.1	1.063	6.564	7.627	79
80	0.473 60	0.001 029	3.409 1	334.9	2 308.9	2 643.8	1.075	6.538	7.613	80
81	0.493 60	0.001 030	3.282 6	339.1	2 306.3	2 645.4	1.087	6.512	7.599	81
82	0.513 29	0.001031	3.161 6	343.3	2 303.8	2 647.1	1.099	6.487	7.586	82
83	0.534 16	0.001 031	3.045 8	347.5	2 301.2	2 648.7	1.111	6.461	7.572	83
84	0.555 73	0.001 032	2.935 0	351.7	2 298.7	2 650.4	1.123	6.436	7.559	84
85	0.578 03	0.001 033	2.828 8	355.9	2 296.1	2 652.0	1.134	6.411	7.545	85
86	0.601 08	0.001 033	2.727 2	360.1	2 293.5	2 653.6	1.146	6.386	7.532	86
87	0.624 89	0.001 034	2.629 8	364.3	2 291.0	2 655.3	1.158	6.361	7.519	87
88	0.649 48	0.001 035	2.536 5	368.5	2 288.4	2 656.9	1.169	6.337	7.506	88
89	0.674 87	0.001 035	2.447 0	372.7	2 285.8	2 658.5	1.181	6.312	7.493	89
90	0.701 09	0.001 036	2.361 3	376.9	2 283.2	2 660.1	1.193	6.287	7.480	90
91	0.728 15	0.001 037	2.279 1	381.1	2 280.6	2 661.7	1.204	6.263	7.467	91
92	0.756 06	0.001 038	2.200 2	385.4	2 278.0	2 663.4	1.216	6.238	7.454	92
93	0.784 89	0.001 038	2.124 5	389.6	2 275.4	2 665.0	1.227	6.215	7.442	93
94	0.814 61	0.001 039	2.051 9	393.8	2 272.8	2 666.6	1.239	6.190	7.429	94
95	0.845 26	0.001 040	1.982 2	398.0	2 270.1	2 668.1	1.250	6.167	7.417	95
96	0.876 86	0.001 041	1.915 2	402.2	2 267.5	2 669.7	1.261	6.143	7.404	96
97	0.909 44	0.001 041	1.851 0	406.4	2 264.9	2 671.3	1.273	6.119	7.392	97

(p)	(t)	(v_f)	(v_g)	(h_f)	(h_fg)	(h_g)	(s_f)	(s_fg)	(s_g)	(p)
98	0.943 01	0.001 042	1.789 2	410.6	2 262.3	2 672.9	1.284	6.096	7.380	98
99	0.977 61	0.001 043	1.730 0	414.8	2 259.6	2 674.4	1.296	6.072	7.368	99
100	1.013 3	0.001 044	1.673 0	419.1	2 256.9	2 676.0	1.307	6.048	7.355	100
102	1.087 6	0.001 045	1.565 5	427.5	2 251.6	2 679.1	1.329	6.002	7.331	102
104	1.166 8	0.001 047	1.466 2	335.9	2 246.3	2 682.2	1.352	5.956	7.308	104
106	1.250 4	0.001 048	1.347 2	444.4	2 240.9	2 685.3	1.374	5.910	7.284	106
108	1.339 0	0.001 050	1.288 9	452.9	2 235.4	2 688.3	1.396	5.865	7.261	108
110	1.432 7	0.001 052	1.209 9	461.3	2 230.0	2 691.3	1.418	5.821	7.239	110
112	1.531 6	0.001 054	1.136 6	469.8	2 224.5	2 694.3	1.440	5.776	7.216	112
114	1.636 2	0.001 055	1.068 5	478.3	2 218.9	2 697.2	1.462	5.732	7.194	114
116	1.746 5	0.001 057	1.005 2	486.7	2 213.5	2 700.2	1.484	5.688	7.172	116
118	1.862 8	0.001 059	0.946 34	495.2	2 207.9	2 703.1	1.506	5.645	7.151	118
120	1.985 4	0.001 061	0.891 52	503.7	2 202.3	2 706.0	1.528	5.601	7.129	120
122	2.114 5	0.001 063	0.840 45	512.2	2 196.6	2 708.8	1.549	5.559	7.108	122
124	2.250 4	0.001 064	0.792 83	520.7	2 190.9	2 711.6	1.570	5.517	7.087	124
126	2.393 3	0.001 066	0.748 40	529.2	2 185.2	2 714.4	1.592	5.475	7.067	126
128	2.543 5	0.001 068	0.706 91	537.8	2 179.4	2 717.2	1.613	5.433	7.046	128
130	2.701 3	0.001 070	0.668 14	546.3	2 173.6	2 719.9	1.634	5.392	7.026	130
132	2.867 0	0.001 072	0.631 88	554.8	2 167.8	2 722.6	1.655	5.351	7.006	132
134	3.040 7	0.001 074	0.597 95	563.4	2 161.9	2 725.3	1.676	5.310	6.986	134
136	3.222 9	0.001 076	0.566 18	572.0	2 155.9	2 727.9	1.697	5.270	6.967	136
138	3.413 8	0.001 078	0.536 41	580.5	2 150.0	2 730.5	1.718	5.229	6.947	138
140	3.613 9	0.001 080	0.508 49	589.1	2 144.0	2 733.1	1.739	5.189	6.928	140
142	3.823 1	0.001 082	0.482 30	597.7	2 137.9	2 735.6	1.760	5.150	6.910	142
144	4.042 0	0.001 084	0.457 71	606.3	2 131.8	2 738.1	1.780	5.111	6.891	144
146	4.270 9	0.001 086	0.434 60	614.9	2 125.7	2 740.6	1.801	5.071	6.872	146
148	4.510 1	0.001 089	0.412 88	623.5	2 119.5	2 743.0	1.821	5.033	6.854	148

(p)	(t)	(v_f)	(v_g)	(h_f)	(h_{fg})	(h_g)	(s_f)	(s_{fg})	(s_g)	(p)
150	4.760 0	0.001 091	0.392 45	632.2	2 113.2	2 745.4	1.842	4.994	6.836	150
155	5.433 3	0.001 096	0.346 44	653.8	2 097.4	2 751.2	1.892	4.899	6.791	155
160	6.180 6	0.001 102	0.306 76	675.5	2 081.2	2 756.7	1.943	4.805	6.748	160
165	7.007 7	0.001 108	0.272 40	697.2	2 064.8	2 762.0	1.992	4.713	6.705	165
170	7.920 2	0.001 114	0.242 55	719.1	2 048.0	2 767.1	2.042	4.621	6.663	170
175	8.924 4	0.001 121	0.216 54	741.1	2 030.7	2 771.8	2.091	4.531	6.622	175
180	10.027	0.001 128	0.193 80	763.1	2 013.2	2 776.3	2.139	4.443	6.582	180
185	11.233	0.001 135	0.173 86	785.3	1 995.1	2 780.4	2.187	4.355	6.542	185
190	12.551	0.001 142	0.156 32	807.5	1 976.8	2 784.3	2.236	4.268	6.504	190
195	13.987	0.001 149	0.140.84	829.9	1 957.9	2 287.8	2.283	4.182	6.465	195
200	15.549	0.001 156	0.127 16	852.4	1 938.5	2 790.9	2.331	4.097	6.428	200
205	17.243	0.001 164	0.115 03	875.0	1 918.8	2 793.8	2.378	4.013	6.391	205
210	19.077	0.001 172	0.104 24	897.7	1 898.5	2 796.2	2.425	3.929	6.354	210
215	21.060	0.001 181	0.094 625	920.6	1 877.7	2 798.3	2.471	3.846	6.317	215
220	23.198	0.001 190	0.086 038	943.7	1 856.2	2 799.9	2.518	3.764	6.282	220
225	25.501	0.001 199	0.078 349	966.9	1 834.3	2 801.2	2.564	3.682	6.246	225
230	27.976	0.001 209	0.071 450	990.3	1 811.7	2 802.0	2.610	3.601	6.211	230
235	30.632	0.001 219	0.065 245	1 013.8	1788.5	2 802.3	2.656	3.519	6.175	235
240	33.478	0.001 229	0.059 654	1 037.6	1 764.6	2 802.2	2.702	3.439	6.141	240
245	36.523	0.001 240	0.054 606	1 061.6	1 740.0	2 801.6	2.748	3.358	6.106	245
250	39.776	0.001 251	0.050 037	1 085.8	1 714.6	2 800.4	2.794	3.277	6.071	250
255	43.246	0.001 263	0.045 896	1 110.2	1 688.5	2 798.7	2.839	3.197	6.036	255
260	46.943	.001 276	0.042 134	1 134.9	1 661.5	2 796.4	2.885	3.116	6.001	260
265	50.877	0.001 289	0.038 710	1 .159.9	1 633.6	2 793.5	2.931	3.035	5.966	265
270	55.058	0.001 303	0.035 588	1 185.2	1604.7	2 789.9	2.976	2.954	5.930	270

(p)	(s_g)	(s_{fg})	(s_f)	(h_g)	(h_{fg})	(h_f)	(v_g)	(v_f)	(t)	(p)
275	5.895	2.873	3.022	2 785.5	1 574.7	1 210.8	0.032 746	0.001 317	59.496	**275**
280	5.858	2.790	3.068	2 780.4	1 543.6	1 236.8	0.030 126	0.001 332	64.202	**280**
285	5. 822	2.707	3.115	2 774.5	1 511.3	1 263.2	0.027 733	0.001 349	69.186	285
290	5.785	2.624	3.161	2 767.6	1 477.6	1 290.0	0.025 535	0.001 366	74.461	290
295	5.447	2.539	3.208	2 759.8	1 442.5	1 317.3	0.023 513	0.001 384	80.037	295
300	5.708	2.453	3.255	2 751.0	1 406.0	1 345.0	0.021 649	0.001 404	85.927	**300**
305	5.669	2.366	3.303	2 741.1	1 367.7	1 373.4	0.019 927	0.001 425	92.144	**305**
310	5.628	2.277	3.351	2 730.0	1 327.6	1 402.4	0.018 334	0.001 448	98.700	**310**
315	5.586	2.186	3.400	2 717.6	1 285.5	1 432.1	0.016 856	0.001 473	105.61	**315**
320	5.542	2.092	3.450	2 703.7	1 241.1	1 462.6	0.015 480	0.001 500	112.89	**320**
325	5.497	1.996	3.501	2 688.0	1 194.0	1 494.0	0.014 195	0.001 529	120.56	**325**
330	5.449	1.896	3.553	2 670.2	1 143.7	1 526.5	0.012 989	0.001 562	128.63	**330**
335	5.398	1.792	3.606	2 649.7	1 089.5	1 560.2	0.011 854	0.001 598	137.12	**335**
340	5.343	1.681	3.662	2 626.2	1 030.7	1 595.5	0.010 780	0.001 639	146.05	**340**
345	5.283	1.564	3.719	2 598.9	966.4	1 632.5	0.097 631	0.001 686	155.45	**345**
350	5.218	1.438	3.780	2 567.7	895.8	1 671.9	0.087 991	0.001 741	165.35	**350**
355	5.144	1.295	3.849	2 530.4	813.8	1 716.6	0.078 592	0.001 809	175.77	**355**
360	5.060	1.139	3.921	2 485.4	721.2	1 764.2	0.069 398	0.001 896	186.75	**360**
365	4.958	0.956	4.002	2 428.0	610.0	1 818.0	0.060 116	0.002 016	198.33	**365**
370	4.814	0.703	4.111	2 342.8	452.6	1 890.2	0.049 728	0.002 214	210.54	370
374.15	4.443	0.000	4.443	2 107.4	0.0	2 107.4	0.003 170	0.003 170	221.20	**374.15**

Table II.2 Saturated water and steam (pressure) tables.

Absolute pressure in bar (p)	Temperature in °C (t)	Specific volume in m³/kg		Specific enthalpy in kJ/kg			Specific entropy in kJ/kg K			Absolute pressure in bar (p)
		Water (v_f)	Steam (v_g)	Water (h_f)	Evaporation (h_{fg})	Steam (h_g)	Water (s_f)	Evaporation (s_{fg})	Steam (s_g)	
0.0061	0.000	0.001 000	206.31	0.0	2 501.6	2 501.6	0.000	9.158	9.158	**0.006 1**
0.010	6.983	0.001 000	129.21	29.3	2 485.1	2 514.4	0.106	8.871	8.977	**0.010**
0.015	13.04	0.001 001	87.982	54.7	2 470.8	2 525.5	0.196	8.624	8.830	**0.015**
0.020	17.51	0.001 001	67.006	73.5	2 460.1	2 533.6	0.261	8.464	8.725	**0.020**
0.025	21.10	0.001 002	54.256	88.4	2 451.8	2 540.2	0.312	8.333	8.645	**0.025**
0.030	24.10	0.001 003	45.667	101.0	2 444.6	2 545.6	0.354	8.224	8.578	**0.030**
0.035	26.69	0.001 003	39.479	111.8	2 438.6	2 550.4	0.391	8.132	8.523	**0.035**
0.040	28.98	0.001 004	34.802	121.4	2 433.1	2 554.5	0.423	8.053	8.476	**0.040**
0.045	31.03	0.001 005	31.141	130.0	2 428.2	2 558.2	0.451	7.983	8.434	**0.045**
0.050	32.90	0.001 005	28.194	137.8	2 423.8	2 561.6	0.476	7.920	8.396	**0.050**
0.060	36.18	0.001 006	23.741	151.5	2 416.0	2 567.5	0.521	7.810	8.331	**0.06**
0.070	39.03	0.001 007	20.531	163.4	2 409.2	2 572.6	0.559	7.718	8.277	**0.07**
0.080	41.53	0.001 008	18.105	173.9	2 403.2	2 577.1	0.593	7.637	8.230	**0.08**
0.090	43.79	0.001 009	16.204	183.3	2 397.8	2 581.1	0.622	7.566	8.188	**0.09**
0.100	45.83	0.001 010	14.675	191.8	2 392.9	2 584.7	0.649	7.502	8.151	**0.10**
0.11	47.71	0.001 011	13.416	199.7	2 388.4	2 588.1	0.674	7.444	8.118	**0.11**
0.12	49.45	0.001 012	12.362	206.9	2 384.3	2 591.2	0.696	7.391	8.087	**0.12**
0.13	51.06	0.001 013	11.466	213.7	2 380.3	2 594.0	0.717	7.342	8.059	**0.13**

(p)	(t)	(v_f)	(v_g)	(h_f)	(h_fg)	(h_g)	(s_f)	(s_fg)	(s_g)	(p)
0.14	52.57	0.001 013	10.694	220.0	2 376.7	2 596.7	0.737	7.296	8.033	0.14
0.15	54.00	0.001 014	10.023	226.0	2 373.2	2 599.2	0.755	7.254	8.009	0.15
0.16	55.34	0.001 015	9.433 1	231.6	2 370.0	2 601.6	0.772	7.215	7.987	0.16
0.17	56.62	0.001 015	8.911 1	236.9	2 366.9	2 603.8	0.788	7.178	7.996	0.17
0.18	57.83	0.001 016	8.445 2	242.0	2 363.9	2 605.9	0.804	7.142	7.946	0.18
0.19	58.98	0.001 017	8.027 2	246.8	2 361.1	2 607.9	0.818	7.109	7.927	0.19
0.20	60.09	0.001 017	8.649 8	251.5	2 358.4	2 609.9	0.832	7.077	7.909	0.20
0.21	61.15	0.001 018	7.367 3	255.9	2 355.8	2 611.7	0.845	7.047	7.892	0.21
0.22	62.16	0.001 018	6.995 1	260.1	2 353.4	2 613.5	0.858	7.018	7.876	0.22
0.23	63.14	0.001 019	6.709 3	264.2	2 351.0	2 615.2	0.870	6.991	7.861	0.23
0.24	64.08	0.001 019	6.446 7	268.2	2 348.6	2 616.8	0.882	6.964	7.846	0.24
0.25	64.99	0.001 020	6.204 5	272.0	2 346.3	2 618.3	0.893	6.939	7.832	0.25
0.26	65.87	0.001 020	5.980 3	275.7	2 344.2	2 619.9	0.904	6.915	7.819	0.26
0.27	66.72	0.001 021	5.772 4	279.2	2 342.1	2 621.3	0.915	6.891	7.806	0.27
0.28	67.55	0.001 021	5.577 8	282.7	2 340.0	2 622.7	0.925	6.868	7.793	0.28
0.29	68.35	0.001 022	5.398 2	286.0	2 338.1	2 624.1	0.935	6.847	7.781	0.29
0.30	69.12	0.001 022	5.229 3	289.3	2 336.1	2 625.4	0.944	6.825	7.769	0.30
0.32	70.62	0.001 023	4.922 0	295.6	2 332.4	2 628.0	0.962	6.785	7.747	0.32
0.34	72.03	0.001 024	4.650 4	301.5	2 328.9	2 630.4	0.980	6.747	7.727	0.34
0.36	73.37	0.001 025	4.407 6	307.1	2 325.5	2 632.6	0.996	6.711	7.707	0.36
0.38	74.66	0.001 026	4.190 0	312.5	2 322.3	2 634.8	1.011	6.677	7.688	0.38
0.40	75.89	0.001 027	3.993 4	317.7	2 319.2	2 636.9	1.026	6.645	7.671	0.40
0.42	77.06	0.001 027	3.814 8	322.6	2 316.3	2 638.9	1.040	6.614	7.654	0.42
0.44	78.19	0.001 028	3.652 2	327.3	2 313.4	2 640.7	1.054	6.584	7.638	0.44
0.46	79.28	0.001 029	3.503 2	331.9	2 310.7	2 642.6	1.067	6.556	7.623	0.46
0.48	80.33	0.001 029	3.366 3	336.3	2 308.0	2 644.3	1.079	6.530	7.609	0.48
0.50	81.35	0.001 030	3.240 1	340.6	2 305.4	2 646.0	1.091	6.504	7.595	0.50

(p)	(t)	(v_f)	(v_g)	(h_f)	(h_fg)	(h_g)	(s_f)	(s_fg)	(s_g)	(p)
0.52	82.33	0.001 031	3.123 3	344.7	2 302.9	2 647.6	1.103	6.478	7.581	0.52
0.54	83.28	0.001 031	3.014 8	348.7	2 300.5	2 649.2	1.114	6.455	7.569	0.54
0.56	84.19	0.001 032	3.913 9	352.5	2 298.2	2 650.7	1.125	6.431	7.556	0.56
0.58	85.09	0.001 033	3.819 7	356.3	2 295.8	2 652.1	1.135	6.409	7.544	0.58
0.60	85.95	0.001 033	3.731 7	359.9	2 293.7	2 653.6	1.145	6.388	7.533	0.60
0.62	86.80	0.001 034	2.649 1	363.5	2 291.4	2 654.9	1.155	6.367	7.552	0.62
0.64	87.62	0.001 034	2.571 5	366.9	2 289.4	2 656.3	1.165	6.346	7.511	0.64
0.66	88.42	0.001 035	2.498 5	370.3	2 287.3	2 657.6	1.174	6.326	7.500	0.66
0.68	89.20	0.001 036	2.429 7	373.6	2 285.2	2 658.8	1.183	6.307	7.490	0.68
0.70	89.96	0.001 036	2.364 7	376.8	2 283.3	2 660.1	1.192	6.288	7.480	0.70
0.72	90.70	0.001 037	2.303 1	379.9	2 281.4	2 661.3	1.201	6.270	7.471	0.72
0.74	91.43	0.001 037	2.244 8	382.9	2 279.5	2 662.4	1.209	6.253	7.462	0.74
0.76	92.14	0.001 038	2.189 5	385.9	2 277.7	2 663.4	1.217	6.235	7.452	0.76
0.78	92.83	0.001 038	2.136 9	388.9	2 275.8	2 664.7	1.225	6.219	7.444	0.78
0.80	93.51	0.001 039	2.086 9	391.7	2 274.1	2 665.8	1.235	6.202	7.435	0.80
0.85	95.15	0.001 040	1.972 1	398.6	2 269.8	2 668.4	1.252	6.163	7.415	0.85
0.90	96.71	0.001 041	1.869 1	405.2	2 265.7	2 670.9	1.270	6.125	7.395	0.90
0.95	98.20	0.001 042	1.777 1	411.5	2 261.7	2 673.2	1.287	6.091	7.378	0.95
1.00	99.63	0.001 043	1.693 8	417.5	2 257.9	2 675.4	1.303	6.057	7.360	1.00
1.013 25	100.00	0.001 044	1.673 0	419.1	2 256.9	2 676.0	1.307	6.048	7.355	1.013 25
1.05	101.0	0.001 045	1.618 1	423.3	2 254.3	2 670.6	1.318	6.025	7.343	1.05
1.10	102.3	0.001 046	1.549 2	428.8	2 250.8	2 679.6	1.333	5.995	7.328	1.10
1.15	103.6	0.001 047	1.486 1	434.2	2 247.4	2 681.6	1.347	5.966	7.313	1.15
1.20	104.8	0.001 048	1.428 1	439.3	2 244.1	2 683.4	1.361	5.937	7.298	1.20
1.25	106.0	0.001 049	1.374 6	444.4	2 240.8	2 685.2	1.374	5.911	7.285	1.25
1.30	107.1	0.001 050	1.325 0	449.2	2 237.8	2 687.0	1.387	5.885	7.272	1.30
1.35	108.2	0.001 050	1.279 1	453.4	2 234.8	2 688.7	1.399	5.860	7.259	1.35

(p)	(t)	(v_f)	(v_g)	(h_f)	(h_fg)	(h_g)	(s_f)	(s_fg)	(s_g)	(p)
1.40	109.3	0.001 051	1.236 3	458.4	2 231.9	2 690.3	1.411	5.836	7.247	1.40
1.45	110.4	0.001 052	1.196 3	462.8	2 229.0	2 691.8	1.423	5.812	7.235	1.45
1.50	111.4	0.001 053	1.159 0	467.1	2 226.3	2 693.4	1.433	5.790	7.223	1.50
1.60	113.3	0.001 055	1.091 1	475.4	2 220.8	2 696.2	1.455	5.747	7.202	1.60
1.70	115.2	0.001 056	1.030 9	483.2	2 215.8	2 699.0	1.475	5.706	7.181	1.70
1.80	116.9	0.001 058	0.977 18	490.7	2 210.8	2 701.5	1.494	5.668	7.162	1.80
1.90	118.6	0.001 059	0.928 95	497.9	2 206.1	2 704.0	1.513	5.631	7.144	1.90
2.00	120.2	0.001 061	0.885 40	504.7	2 201.6	2 706.3	1.530	5.597	7.127	2.00
2.1	121.8	0.001 062	0.845 86	511.3	2 197.2	2 708.5	1.547	5.564	7.111	2.1
2.2	123.3	0.001 064	0.809 80	517.6	2 193.0	2 710.6	1.563	5.532	7.095	2.2
2.3	124.7	0.001 065	0.776 77	523.7	2 188.9	2 712.6	1.578	5.502	7.080	2.3
2.4	126.1	0.001 066	0.746 41	529.6	2 184.9	2 714.5	1.593	5.473	7.066	2.4
2.5	127.4	0.001 068	0.718 40	535.3	2 181.1	2 716.4	1.607	5.445	7.052	2.5
2.6	128.7	0.001 069	0.692 47	540.9	2 177.3	2 718.2	1.621	5.418	7.039	2.6
2.7	130.0	0.001 070	0.668 40	546.2	2 173.7	2 719.9	1.634	5.392	7.026	2.7
2.8	131.2	0.001 071	0.646 00	551.4	2 170.1	2 721.5	1.647	5.367	7.014	2.8
2.9	132.4	0.001 072	0.625 09	556.5	2 166.6	2 723.1	1.660	5.342	7.002	2.9
3.0	133.5	0.001 074	0.605 53	561.5	2 163.2	2 724.7	1.672	5.319	6.991	3.0
3.1	134.7	0.001 075	0.587 18	566.2	2 159.9	2 726.1	1.683	5.297	6.980	3.1
3.2	135.8	0.001 076	0.569 95	570.9	2 156.7	2 727.6	1.695	5.274	6.969	3.2
3.3	136.8	0.001 077	0.553 73	575.5	2 153.5	2 729.0	1.706	5.253	6.959	3.3
3.4	137.9	0.001 078	0.538 43	579.9	2 150.4	2 730.3	1.717	5.232	6.949	3.4
3.5	138.9	0.001 079	0.523 97	584.3	2 147.3	2 731.6	1.727	5.212	6.939	3.5
3.6	139.9	0.001 080	0.510 29	588.5	2 144.4	2 732.9	1.738	5.192	6.930	3.6
3.7	140.8	0.001 081	0.497 33	592.7	2 141.4	2 734.1	1.748	5.173	6.921	3.7
3.8	141.8	0.001 082	0.485 02	596.7	2 138.6	2 735.3	1.758	5.154	9.912	3.8
3.9	142.7	0.001 083	0.473 33	600.8	2 135.7	2 736.5	1.767	5.136	9.903	3.9

(p)	(t)	(v_f)	(v_g)	(h_f)	(h_fg)	(h_g)	(s_f)	(s_fg)	(s_g)	(p)
4.0	143.6	0.001 084	0.462 20	604.7	2 132.9	2 737.6	1.776	5.118	6.894	4.0
4.1	144.5	0.001 085	0.451 59	608.5	2 130.2	2 738.7	1.786	5.100	6.886	4.1
4.2	145.4	0.001 086	0.441 47	612.3	2 127.5	2 739.8	1.795	5.083	6.878	4.2
4.3	146.3	0.001 087	0.431 81	616.0	2 124.9	2 740.9	1.803	5.067	6.870	4.3
4.4	147.1	0.001 088	0.422 57	619.6	2 122.3	2 741.9	1.812	5.050	6.862	4.4
4.5	147.9	0.001 089	0.413 73	623.2	2 119.7	2 742.9	1.820	5.035	6.855	4.5
4.6	148.7	0.001 090	0.405 26	626.7	2 117.2	2 743.9	1.829	5.018	6.847	4.6
4.7	149.5	0.001 090	0.397 14	630.1	2 114.7	2 744.8	1.837	5.003	6.840	4.7
4.8	150.3	0.001 091	0.389 34	633.5	2 112.2	2 745.7	1.845	4.988	6.833	4.8
4.9	151.1	0.001 092	0.381 86	636.8	2 109.8	2 746.6	1.853	4.973	6.826	4.9
5.0	151.8	0.001 093	0.374 66	640.1	2 107.4	2 747.5	1.860	4.959	8.819	5.0
5.2	153.3	0.001 095	0.361 06	646.5	2 102.7	2 749.2	1.875	4.931	6.806	5.2
5.4	154.8	0.001 096	0.348 44	652.8	2 098.1	2 750.9	1.890	4.903	6.793	5.4
5.6	156.2	0.001 098	0.336 69	658.8	2 093.7	2 752.5	1.904	4.877	6.781	5.6
5.8	157.5	0.001 099	0.325 72	664.7	2 089.3	2 754.0	1.918	4.851	6.769	5.8
6.0	158.8	0.001 101	0.315 46	670.4	2 085.1	2 755.5	1.931	4.827	6.758	6.0
6.2	160.1	0.001 102	0.305 84	676.1	2 080.8	2 756.9	1.944	4.803	6.747	6.2
6.4	161.4	0.001 104	0.296 80	681.5	2 076.7	2 758.2	1.956	4.780	6.736	6.4
6.6	162.6	0.001 105	0.288 29	686.8	2 072.7	2 759.5	1.968	4.757	6.725	6.6
6.8	163.8	0.001 107	0.280 26	692.0	2 068.8	2 760.8	1.980	4.735	6.715	6.8
7.0	165.0	0.001 108	0.272 68	697.1	2 064.9	2 762.0	1.992	4.713	6.705	7.0
7.2	166.1	0.001 110	0.265 50	702.0	2 061.2	2 763.2	2.003	4.693	6.696	7.2
7.4	167.2	0.001 111	0.258 70	706.9	2 057.4	2 764.3	2.014	4.672	6.686	7.4
7.6	168.3	0.001 112	0.252 24	711.7	2 053.7	2 765.4	2.025	4.652	6.677	7.6
7.8	169.4	0.001 114	0.246 10	716.3	2 050.1	2 766.4	2.035	4.633	6.668	7.8
8.0	170.4	0.001 115	0.240 26	720.9	2 046.5	2 767.4	2.046	4.614	6.660	8.0

(p)	(t)	(v_f)	(v_g)	(h_f)	(h_fg)	(h_g)	(s_f)	(s_fg)	(s_g)	(p)
8.2	171.4	0.001 116	0.234 69	725.4	2 043.0	2 768.4	2.056	4.595	6.651	8.2
8.4	172.4	0.001 118	0.229 38	729.9	2 039.6	2 769.4	2.066	4.577	6.643	8.4
8.6	173.4	0.001 119	0.224 31	734.2	2 036.2	2 770.4	2.075	4.560	6.635	8.6
8.8	174.4	0.001 120	0.219 46	738.5	2 032.8	2 771.3	2.085	4.542	6.627	8.8
9.0	175.4	0.001 121	0.214 82	742.6	2 029.5	2 772.1	2.094	4.525	6.619	9.0
9.2	176.3	0.001 123	0.210 37	746.8	2 026.2	2 773.0	2.103	4.509	6.612	9.2
9.4	177.2	0.001 124	0.206 10	750.8	2 023.0	2 773.8	2.112	4.492	6.604	9.4
9.6	178.1	0.001 125	0.202 01	754.8	2 019.8	2 774.6	2.121	4.476	6.597	9.6
9.8	179.0	0.001 126	0.198 08	758.7	2 016.7	2 775.4	2.130	4.460	6.590	9.8
10.0	179.9	0.001 127	0.194 30	762.6	2 013.6	2 776.2	2.138	4.445	6.583	10.0
10.5	182.0	0.001 130	0.185 48	772.0	2 006.0	2 778.0	2.159	4.407	6.566	10.5
11.0	184.1	0.001 133	0.177 39	781.1	1 998.6	2 779.7	2.179	4.371	6.550	11.0
11.5	186.0	0.001 136	0.170 02	789.9	1 991.4	2 781.3	2.198	4.336	6.534	11.5
12.0	188.0	0.001 139	0.163 21	798.4	1 984.3	2 782.7	2.216	4.303	6.519	12.0
12.5	189.8	0.001 141	0.156 96	806.7	1 977.5	2 784.2	2.234	4.271	6.505	12.5
13.0	191.6	0.001 144	0.151 14	814.7	1 970.7	2 785.4	2.251	4.240	6.491	13.0
13.5	193.3	0.001 146	0.145 76	822.5	1 964.2	2 786.7	2.267	4.211	6.478	13.5
14.0	195.0	0.001 149	0.140 73	830.1	1 957.7	2 787.8	2.284	4.181	6.465	14.0
14.5	196.7	0.001 151	0.136 06	837.5	1 951.2	2 788.9	2.299	4.154	6.453	14.5
15.0	198.3	0.001 154	0.131 67	844.6	1 945.3	2 789.9	2.314	4.127	6.441	15.0
15.5	199.8	0.001 156	0.127 56	851.6	1 939.2	2 790.8	2.329	4.100	6.429	15.5
16.0	201.4	0.001 159	0.123 70	858.5	1 933.2	2 791.7	2.344	4.074	6.418	16.0
16.5	202.9	0.001 161	0.120 06	865.3	1 952.3	2 792.6	2.358	4.049	6.407	16.5
17.0	204.3	0.001 163	0.116 64	871.8	1 921.6	2 793.4	2.371	4.025	6.396	17.0
17.5	205.7	0.001 166	0.113 40	878.2	1 915.9	2 794.1	2.384	4.001	6.385	17.5
18.0	207.1	0.001 168	0.110 33	884.5	1 910.3	2 794.8	2.398	3.977	6.375	18.0
18.5	208.5	0.001 170	0.107 42	890.7	1 904.8	2 795.5	2.410	3.955	6.365	18.5

(p)	(t)	(v_f)	(v_g)	(h_f)	(h_fg)	(h_g)	(s_f)	(s_fg)	(s_g)	(p)
19.0	209.8	0.001 172	0.104 67	896.8	1 899.3	2 796.1	2.423	3.933	6.356	19.0
19.5	211.1	0.001 174	0.102 04	902.7	1 894.0	2 796.7	2.435	3.911	6.346	19.5
20.0	212.4	0.001 177	0.099 55	908.5	1 888.7	2 797.2	2.447	3.890	6.337	20.0
21.0	214.8	0.001 181	0.094 902	919.9	1 878.3	2 798.2	2.470	3.849	6.319	21.0
22.0	217.2	0.001 185	0.090 663	930.9	1 868.1	2 799.1	2.492	3.809	6.301	22.0
23.0	219.6	0.001 189	0.086 780	941.6	1 858.2	2 799.8	2.514	3.771	6.285	23.0
24.0	221.8	0.001 193	0.083 209	951.9	1 848.5	2 800.4	2.534	3.735	6.269	24.0
25.0	223.9	0.001 197	0.079 915	961.9	1 839.1	2 801.0	2.554	3.699	6.253	25.0
26.0	226.0	0.001 201	0.076 865	971.7	1 829.7	2 801.4	2.574	3.665	6.239	26.0
27.0	228.1	0.001 205	0.074 033	981.2	1 820.5	2 801.7	2.592	3.632	6.224	27.0
28.0	230.0	0.001 209	0.071 396	990.5	1 811.5	2 802.0	2.611	3.600	6.211	28.0
29.0	232.0	0.001 213	0.068 935	999.5	1 802.7	2 802.2	2.628	3.569	6.197	29.0
30.0	233.8	0.001 216	0.066 632	1008.3	1 794.0	2 802.3	2.646	3.538	6.184	30.0
31.0	235.7	0.001 220	0.064 473	1017.1	1 785.4	2 802.3	2.662	3.509	6.171	31.0
32.0	237.4	0.001 224	0.062 443	1025.4	1 776.9	2 802.3	2.679	3.480	6.159	32.0
33.0	239.2	0.001 227	0.060 533	1033.7	1 768.6	2 802.3	2.694	3.452	6.146	33.0
34.0	240.9	0.001 231	0.058 731	1041.8	1 760.3	2 802.1	2.710	3.424	6.134	34.0
35.0	242.5	0.001 235	0.057 028	1049.7	1 752.3	2 802.0	2.725	3.398	6.123	35.0
36.0	244.2	0.001 238	0.055 417	1057.5	1 744.2	2 801.7	2.740	3.371	6.111	36.0
37.0	245.8	0.001 242	0.053 889	1065.2	1 736.2	2 801.4	2.755	3.345	6.100	37.0
38.0	247.3	0.001 245	0.052 439	1072.7	1 728.4	2 801.1	2.769	3.321	6.090	38.0
39.0	248.8	0.001 249	0.051 061	1080.1	1 720.7	2 800.8	2.783	3.296	6.079	39.0
40.0	250.3	0.001 252	0.049 749	1087.4	1 712.9	2 800.3	2.797	3.272	6.069	40.0
42.0	253.2	0.001 259	0.047 306	1101.6	1 697.8	2 799.4	2.823	3.225	6.048	42.0
44.0	256.1	0.001 266	0.045 078	1115.4	1 682.9	2 798.3	2.849	3.180	6.029	44.0
46.0	258.8	0.001 273	0.043 036	1128.8	1 668.2	2 797.0	2.874	3.136	6.010	46.0
48.0	261.4	0.001 279	0.041 158	1141.8	1 653.9	2 795.7	2.897	3.094	5.991	48.0
50.0	263.9	0.001 286	0.039 425	1154.5	1 639.7	2 794.2	2.921	3.053	5.974	50.0

(p)	(t)	(v_f)	(v_g)	(h_f)	(h_{fg})	(h_g)	(s_f)	(s_{fg})	(s_g)	(p)
52.0	266.4	0.001 293	0.037 820	1166.9	1 625.7	2 792.6	2.943	3.013	5.956	52.0
54.0	268.8	0.001 299	0.036 330	1179.0	1 611.8	2 790.8	2.965	2.974	5.939	54.0
56.0	271.1	0.001 306	0.034 942	1190.8	1 598.2	2 789.0	2.986	2.937	5.923	56.0
58.0	273.4	0.001 312	0.033 646	1202.4	1 584.6	2 787.0	3.007	2.899	5.906	58.0
60.0	275.5	0.001 319	0.032 433	1213.7	1 571.3	2 785.0	3.027	2.863	5.890	60.0
62.0	277.7	0.001 325	0.031 295	1224.9	1 558.0	2 782.9	3.047	2.828	5.875	62.0
64.0	279.8	0.001 332	0.030 225	1235.8	1 544.8	2 780.6	3.066	2.794	5.860	64.0
66.0	281.9	0.001 338	0.029 218	1246.5	1 531.8	2 778.3	3.085	2.760	5.845	66.0
68.0	283.9	0.001 345	0.028 267	1257.1	1 518.8	2 775.9	3.104	2.727.	5.831	68.0
70.0	285.8	0.001 351	0.027 368	1267.4	1 506.0	2 773.4	3.122	2.694	5.816	70.0
72.0	287.7	0.001 358	0.026 517	1277.7	1 493.2	2 770.9	3.140	2.662	5.802	72.0
74.0	289.6	0.001 365	0.025 711	1287.8	1 480.4	2 768.2	3.157	2.631	5.788	74.0
76.0	291.4	0.001 371	0.024 944	1297.8	1 467.8	2 765.5	3.174	2.600	5.774	76.0
78.0	293.2	0.001 378	0.024 215	1307.5	1 455.4	2 762.7	3.191	2.569	5.760	78.0
80.0	295.0	0.001 384	0.023 521	1317.5	1 442.7	2 759.9	3.208	2.539	5.747	80.0
82.0	296.7	0.001 391	0.022 860	1326.7	1 430.3	2 757.0	3.224	2.510	5.734	82.0
84.0	298.4	0.001 398	0.022 228	1336.2	1 417.8	2 754.0	3.240	2.481	5.721	84.0
86.0	300.1	0.001 404	0.021 624	1345.4	1 405.5	2 750.9	3.256	2.452	5.708	86.0
88.0	301.7	0.001 411	0.021 046	1354.7	1 393.1	2 747.8	3.271	2.424	5.695	88.0
90.0	303.3	0.001 418	0.020 493	1363.8	1 380.8	2 744.6	3.287	2.395	5.682	90.0
92	304.9	0.001 425	0.019 962	1372.8	1 368.5	2 741.3	3.302	2.367	5.669	92
94	306.5	0.001 432	0.019 453	1381.7	1 356.3	2 738.0	3.317	2.340	5.657	94
96	308.0	0.001 439	0.018 964	1390.6	1 344.1	2 734.7	3.332	2.313	5.644	96
98	309.5	0.001 446	0.018 493	1396.4	1 331.9	2 731.2	3.346	2.286	5.632	98
100	311.0	0.001 453	0.018 041	1408.0	1 319.7	2 727.7	3.361	2.259	5.620	100
105	314.6	0.001 470	0.016 981	1429.5	1 289.9	2 718.7	3.396	2.194	5.590	105
110	318.0	0.001 489	0.016 007	1450.5	1 258.8	2 709.3	3.430	2.129	5.560	110

(p)	(t)	(v_f)	(v_g)	(h_f)	(h_fg)	(h_g)	(s_f)	(s_fg)	(s_g)	(p)
115	321.4	0.001 508	0.015 114	1471.3	1 228.2	2 699.5	3.464	2.066	5.530	115
120	324.6	0.001 527	0.014 285	1491.7	1 197.5	2 698.2	3.497	2.003	5.500	120
125	327.8	0.001 547	0.013 518	1511.9	1 166.5	2 678.4	3.530	1.941	5.471	125
130	330.8	0.001 567	0.012 800	1531.9	1 135.1	2 667.0	3.561	1.880	5.441	130
135	333.8	0.001 588	0.012 130	1551.8	1 103.3	2 655.1	3.593	1.818	5.411	135
140	336.6	0.001 611	0.011 498	1571.5	1 070.9	2 642.4	3.624	1.756	5.380	140
145	339.4	0.001 634	0.010 905	1591.3	1 037.9	2 629.2	3.655	1.694	5.349	145
150	342.1	0.001 658	0.010 343	1610.9	1 004.2	2 615.1	3.686	1.632	5.318	150
155	344.8	0.001 683	0.009 813	1630.7	969.7	2 600.4	3.716	1.570	5.286	155
160	347.3	0.001 710	0.009 310	1650.4	934.5	2 584.9	3.747	1.506	5.253	160
165	349.7	0.001 739	0.008 833	1670.4	898.5	2 568.9	3.778	1.442	5.220	165
170	352.3	0.001 770	0.008 372	1691.6	860.0	2 551.6	3.811	1.375	5.186	170
175	354.6	0.001 803	0.007 927	1713.3	820.0	2 533.3	3.844	1.306	5.150	175
180	357.0	0.001 840	0.007 497	1734.8	779.1	2 513.9	3.877	1.236	5.113	180
185	359.0	0.001 881	0.007 082	1756.5	736.5	2 493.0	3.910	1.164	5.074	185
190	361.4	0.001 926	0.006 676	1778.7	691.8	2 470.5	3.943	1.090	5.033	190
195	363.6	0.001 978	0.006 276	1801.9	643.9	2 445.8	3.978	1.011	4.989	195
200	365.7	0.002 037	0.005 875	1826.6	591.6	2 418.2	4.015	0.926	4.941	200
205	367.8	0.002 110	0.005 462	1854.2	532.0	2 386.2	4.056	0.830	4.886	205
210	369.8	0.002 202	0.005 023	1886.3	461.2	2 347.5	4.105	0.717	4.822	210
215	371.8	0.002 342	0.004 509	1929.4	365.2	2 294.6	4.170	0.566	4.736	215
220	373.7	0.002 668	0.003 735	2010.3	186.3	2 196.6	4.293	0.288	4.581	220
221.2	374.15	0.003 170	0.003 170	2107.4	000.0	2 107.4	4.443	0.000	4.443	221.2

Table II.3 Specific volume of superheated steam

Absolute pressure in bar (p)	Saturation Temperature in °C (t_s)	Specific volume (v) in m³/kg at various temperature in °C										
		100	150	200	250	300	350	400	500	600	700	800
0.02	17.5	86.08	97.63	109.2	120.7	132.2	143.8	155.3	178.4	201.5	224.6	247.6
0.04	29.0	43.03	48.81	54.58	60.35	66.12	71.89	77.66	89.20	100.7	112.3	123.8
0.06	36.2	28.68	32.53	36.38	40.23	44.08	47.93	51.77	59.47	67.16	74.85	82.54
0.08	41.5	21.50	24.40	27.28	30.17	33.06	35.94	38.83	44.60	50.37	56.14	61.91
0.10	45.8	17.20	19.51	21.83	24.14	26.45	28.75	31.06	35.68	40.30	44.91	49.53
0.15	54.0	11.51	13.06	14.61	16.16	17.71	19.25	20.80	23.89	26.98	30.07	33.16
0.20	60.1	8.585	9.748	10.91	12.07	13.22	14.37	15.53	17.84	20.15	22.45	24.76
0.25	65.0	6.874	7.808	8.737	9.665	10.59	11.52	12.44	14.29	16.14	17.99	19.84
0.30	69.1	5.714	6.493	7.268	8.040	8.811	9.581	10.35	11.89	13.43	14.70	16.51
0.35	72.7	4.898	5.568	6.233	6.896	7.557	8.218	8.879	10.20	11.52	12.84	14.16
0.40	75.9	4.279	4.866	5.448	6.028	6.607	7.185	7.763	8.918	10.07	11.23	12.38
0.45	78.7	3.803	4.325	4.844	5.360	5.875	6.389	6.903	7.930	8.957	9.984	10.99
0.50	81.3	3.418	3.889	4.356	4.821	5.284	5.747	6.209	7.134	8.057	8.981	9.904
0.60	86.0	2.844	3.238	3.628	4.016	4.402	4.788	5.174	5.944	6.714	7.484	8.254
0.70	90.0	2.434	2.773	3.108	3.441	3.772	4.103	4.434	5.095	5.755	6.415	7.074
0.80	93.5	2.126	2.425	2.718	3.010	3.300	3.590	3.879	4.457	5.035	5.613	6.190
0.90	96.7	1.887	2.153	2.415	2.674	2.933	3.190	3.448	3.962	4.475	4.989	5.502
1.00	99.6	1.696	1.936	2.172	2.406	2.639	2.871	3.103	3.565	4.028	4.490	4.952
1.50	111.4	...	1.285	1.444	1.601	1.757	1.912	2.067	2.376	2.685	2.993	3.301
2.00	120.2	...	0.9595	1.080	1.199	1.316	1.433	1.549	1.781	2.013	2.244	2.475
2.5	127.4	...	0.7641	0.8620	0.9574	1.052	1.145	1.239	1.424	1.610	1.795	1.980
3.0	133.5	...	0.6337	0.7164	0.7964	0.8753	0.9535	1.031	1.187	1.341	1.496	1.650
3.5	138.9	...	0.5406	0.6123	0.6814	0.7493	0.8166	0.8835	1.017	1.149	1.282	1.414

(p)	(t_s)	100	150	200	250	300	350	400	500	600	700	800
4.0	143.6	..	0.4707	0.5343	0.5952	0.6549	0.7139	0.7725	0.8892	1.005	1.121	1.237
4.5	147.9	..	0.4165	0.4738	0.5284	0.5817	0.6343	0.6865	0.7905	0.8939	0.9971	1.100
5.0	151.8	..	..	0.4250	0.4744	0.5226	0.5701	0.6172	0.7108	0.8040	0.8969	0.9896
6.0	158.8	..	..	0.3520	0.3939	0.4344	0.4742	0.5136	0.5918	0.6696	0.7471	0.8245
7.0	165.0	..	..	0.2999	0.3364	0.3714	0.4057	0.4396	0.5069	0.5737	0.6402	0.7066
8.0	170.4	..	..	0.2608	0.2932	0.3241	0.3543	0.3842	0.4432	0.5017	0.5600	0.6181
9.0	175.4	..	..	0.2303	0.2596	0.2874	0.3144	0.3410	0.3936	0.4458	0.4976	0.5493
10.0	179.9	..	..	0.2059	0.2328	0.2580	0.2824	0.3065	0.3540	0.4010	0.4477	0.4943
11.0	184.1	..	..	0.1859	0.2108	0.2339	0.2563	0.2782	0.3215	0.3644	0.4069	0.4492
12.0	188.0	..	..	0.1692	0.1924	0.2139	0.2345	0.2547	0.2945	0.3338	0.3729	0.4118
13.0	191.6	..	..	0.1551	0.1769	0.1969	0.2161	0.2348	0.2716	0.3080	0.3441	0.3800
14.0	195.0	..	..	0.1429	0.1636	0.1823	0.2002	0.2177	0.2520	0.2859	0.3194	0.3528
15.0	198.3	..	..	0.1324	0.1520	0.1697	0.1865	0.2029	0.2350	0.2667	0.2980	0.3292
16.0	201.4	..	..	..	0.1419	0.1587	0.1745	0.1900	0.2202	0.2499	0.2793	0.3086
17.0	204.3	..	..	..	0.1329	0.1489	0.1640	0.1786	0.2070	0.2351	0.2628	0.2904
18.0	207.1	..	..	..	0.1250	0.1402	0.1546	0.1684	0.1954	0.2219	0.2481	0.2742
19.0	209.8	..	..	..	0.1179	0.1325	0.1461	0.1593	0.1849	0.2101	0.2350	0.2597
20.0	212.4	..	..	..	0.1115	0.1255	0.1386	0.1511	0.1756	0.1995	0.2232	0.2467
22.0	217.2	..	..	..	0.1004	0.1134	0.1255	0.1370	0.1593	0.1812	0.2028	0.2242
24.0	221.8	..	..	..	0.09108	0.1034	0.1146	0.1252	0.1458	0.1659	0.1858	0.2054
26.0	226.0	..	..	..	0.08321	0.09483	0.1053	0.1153	0.1344	0.1530	0.1714	0.1895
28.0	230.0	..	..	..	0.07644	0.08751	0.09740	0.1067	0.1246	0.1419	0.1590	0.1759
30.0	233.8	..	..	..	0.07055	0.08116	0.09053	0.09931	0.1161	0.1323	0.1483	0.1641
32.0	237.4	..	..	..	0.06538	0.07559	0.08451	0.09283	0.1087	0.1239	0.1390	0.1538
34.0	240.9	..	..	..	0.06080	0.07068	0.07920	0.08711	0.1021	0.1165	0.1307	0.1447
36.0	244.2	..	..	..	00.5670	0.06630	0.07448	0.08202	0.09626	0.1100	0.1234	0.1366
38.0	247.3	..	..	..	0.05302	0.06237	0.07025	0.07747	0.09104	0.1041	0.1168	0.1294
40.0	250.3	..	..	..	..	0.05883	0.06645	0.07338	0.08634	0.09876	0.1109	0.1229

(p)	(t_s)	100	150	200	250	300	350	400	500	600	700	800
42.0	253.2	⋮	⋮	⋮	⋮	0.055 63	0.063 00	0.069 67	0.082 09	0.093 97	0.1056	0.117 0
44.0	256.0	⋮	⋮	⋮	⋮	0.052 70	0.059 86	0.066 30	0.078 23	0.089 61	0.1007	0.1116
46.0	258.8	⋮	⋮	⋮	⋮	3.050 03	0.056 99	0.063 22	0.074 70	0.085 62	0.096 26	0.1067
48.0	261.4	⋮	⋮	⋮	⋮	0.047 57	0.054 36	0.060 39	0.071 47	0.081 97	0.092 19	0.1022
50.0	263.9	⋮	⋮	⋮	⋮	0.045 30	0.051 94	0.057 79	0.068 49	0.078 62	0.088 45	0.098 09
55.0	269.9	⋮	⋮	⋮	⋮	0.043 43	0.046 66	0.052 13	0.062 02	0.071 31	0.080 31	0.089 12
60.0	275.6	⋮	⋮	⋮	⋮	0.036 15	0.042 22	0.047 38	0.056 59	0.065 18	0.073 48	0.081 59
65.0	280.8	⋮	⋮	⋮	⋮	0.032 58	0.038 48	0.043 38	0.052 03	0.060 03	0.067 74	0.075 26
70.0	285.8	⋮	⋮	⋮	⋮	0.029 46	0.035 23	0.039 92	0.048 09	0.055 59	0.062 79	0.069 80
75.0	290.5	⋮	⋮	⋮	⋮	0.026 72	0.032 43	0.036 94	0.044 69	0.051 76	0.058 52	0.065 09
80.0	295.0	⋮	⋮	⋮	⋮	0.024 26	0.029 95	0.034 31	0.041 70	0.048 39	0.054 77	0.060 96
85.0	299.2	⋮	⋮	⋮	⋮	0.021 91	0.027 76	0.032 00	0.039 08	0.045 44	0.051 48	0.057 32
90.0	303.3	⋮	⋮	⋮	⋮	⋮	0.025 79	0.029 93	0.036 74	0.042 80	0.048 53	0.054 08
95.0	307.2	⋮	⋮	⋮	⋮	⋮	0.024 03	0.028 08	0.034 65	0.040 45	0.045 91	0.051 19
100.0	311.0	⋮	⋮	⋮	⋮	⋮	0.022 42	0.026 41	0.032 76	0.038 32	0.043 55	0.048 58
110.0	318.0	⋮	⋮	⋮	⋮	⋮	0.019 61	0.023 51	0.029 50	0.034 66	0.039 47	0.044 08
120.0	324.6	⋮	⋮	⋮	⋮	⋮	0.017 21	0.021 08	0.026 79	0.031 60	0.036 07	0.040 33
130.0	330.8	⋮	⋮	⋮	⋮	⋮	0.015 10	0.019 02	0.024 49	0.029 02	0.033 19	0.037 16
140.0	336.6	⋮	⋮	⋮	⋮	⋮	0.013 21	0.017 23	0.022 51	0.026 80	0.030 72	0.034 44
150.0	342.1	⋮	⋮	⋮	⋮	⋮	0.011 46	0.015 66	0.020 80	0.024 88	0.028 59	0.032 09
160.0	347.3	⋮	⋮	⋮	⋮	⋮	0.009 76	0.014 28	0.019 29	0.023 20	0.026 72	0.030 03
170.0	352.3	⋮	⋮	⋮	⋮	⋮		0.013 03	0.017 97	0.021 72	0.025 07	0.028 21
180.0	357.0	⋮	⋮	⋮	⋮	⋮		0.011 91	0.016 79	0.020 40	0.023 00	0.026 59
190.0	361.4	⋮	⋮	⋮	⋮	⋮		0.010 89	0.015 73	0.019 22	0.022 29	0.025 15
200.0	365.7	⋮	⋮	⋮	⋮	⋮		0.009 95	0.014 77	0.018 16	0.021 11	0.023 85
210.0	369.8	⋮	⋮	⋮	⋮	⋮		0.009 07	0.013 91	0.017 20	0.020 04	0.022 67
220.0	373.7	⋮	⋮	⋮	⋮	⋮		0.008 25	0.013 12	0.016 33	0.019 07	0.021 60
221.2	374.15	⋮	⋮	⋮	⋮	⋮		0.008 16	0.013 03	0.016 22	0.018 95	0.021 35

Table II.4 Specific enthalpy of superheated steam

Absolute pressure in bar	Saturation Temperature in °C	Specific enthalpy (h) in kJ/kg at various temperature in °C										
(p)	(t_s)	100	150	200	250	300	350	400	500	600	700	800
0.02	17.5	2688.5	2783.7	2880.0	2977.7	3076.8	3177.5	3279.7	3489.2	3705.6	3928.8	4158.7
0.04	29.0	2688.3	2783.5	2879.9	2977.6	3076.8	3177.4	3279.7	3489.2	3705.6	3928.8	4158.7
0.06	36.2	2688.0	2783.4	2879.8	2977.6	3076.7	3177.4	3279.6	3489.2	3705.6	3928.8	4158.7
0.08	41.5	2687.8	2783.2	2879.7	2977.5	3076.7	3177.3	3279.6	3489.1	3705.5	3928.8	4158.7
0.10	45.8	2687.5	2783.1	2879.6	2977.4	3076.6	3177.3	3279.6	3489.1	3705.5	3928.8	4158.7
0.15	54.0	2686.9	2782.4	2879.5	2977.3	3076.5	3177.7	3279.5	3489.1	3705.5	3928.7	4158.7
0.20	60.1	2686.3	2782.3	2879.2	2977.1	3076.4	3177.1	3279.4	3489.0	3705.4	3928.7	4158.7
0.25	65.0	2685.7	2782.0	2879.0	2977.0	3076.3	3177.0	3279.3	3489.0	3705.4	3928.7	4158.6
0.30	69.1	2685.1	2781.6	2878.7	2976.8	3076.1	3176.9	3279.3	3488.9	3705.4	3928.7	4158.6
0.35	72.7	2684.5	2781.2	2878.5	2976.7	3076.0	3176.8	3279.2	3488.9	3705.3	3928.7	4158.6
0.40	75.9	2683.8	2780.9	2878.2	2976.5	3075.9	3176.8	3279.1	3488.8	3705.3	3928.6	4158.6
0.45	78.7	2683.2	2780.5	2878.0	2976.3	3075.8	3176.7	3279.1	3488.8	3705.2	3928.6	4158.5
0.50	81.3	2682.6	2780.1	2877.7	2976.1	3075.7	3176.6	3279.0	3488.7	3705.2	3928.6	4158.5
0.60	86.0	2681.3	2779.4	2877.3	2975.8	3075.4	3176.4	3278.8	3488.6	3705.1	3928.5	4158.5
0.70	90.0	2680.0	2778.6	2876.8	2975.5	3075.2	3176.2	3278.7	3488.5	3705.0	3928.4	4158.4
0.80	93.5	2678.8	2777.8	2876.3	2975.2	3075.0	3176.0	3278.5	3488.4	3705.0	3928.4	4158.4
0.90	96.7	2677.5	2777.1	2875.8	2974.8	3074.7	3175.8	3278.4	3488.3	3704.9	3928.3	4158.3
1.00	99.6	2676.2	2776.3	2875.4	2974.5	3074.5	3175.6	3278.2	3488.1	3704.8	3928.2	4158.3
1.50	111.4	...	2772.5	2872.9	2972.9	3073.3	3174.7	3277.5	3487.6	3704.4	3927.9	4158.0
2.00	120.2	...	2768.5	2870.5	2971.2	3072.1	3173.8	3276.7	3487.0	3704.0	3927.6	4157.8
2.5	127.4	...	2764.5	2868.0	2969.6	3070.9	3172.8	3275.9	3486.5	3703.6	3927.3	4157.6
3.0	133.5	...	2760.4	2865.5	2967.9	3069.7	3171.9	3275.2	3486.0	3703.2	3927.0	4157.3
3.5	138.9	...	2756.3	2863.0	2966.2	3068.4	3170.9	3274.4	3485.4	3702.7	3926.7	4157.1

(p)	(t_s)	100	150	200	250	300	350	400	500	600	700	800
4.0	143.6	..	2752.0	2860.4	2964.5	3067.2	3170.0	3273.6	3484.9	3702.3	3926.4	4156.9
4.5	147.9	..	2746.7	2857.8	2962.8	3066.0	3169.1	3272.9	3484.3	3701.9	3926.1	4156.7
5.0	151.8	..	..	2855.1	2961.1	3064.8	3168.1	3272.1	3483.8	3701.5	3925.8	4156.4
6.0	158.8	..	..	2849.7	2957.6	3062.3	3166.2	3270.6	3482.7	3700.7	3925.1	4155.9
7.0	165.0	..	..	2844.2	2954.0	3059.8	3164.3	3269.0	3481.6	3699.9	3924.5	4155.5
8.0	170.4	..	..	2838.6	2950.4	3057.3	3162.4	3267.5	3480.5	3699.1	3923.9	4155.0
9.0	175.4	..	..	2832.7	2946.8	3054.7	3160.5	3266.0	3479.4	3698.2	3923.3	4154.5
10.0	179.9	..	..	2826.8	2943.0	3052.1	3158.5	3264.4	3478.3	3697.4	3922.7	4154.1
11.0	184.1	..	..	2820.7	2939.3	3049.6	3156.6	3262.9	3477.2	3696.6	3922.0	4153.6
12.0	188.0	..	..	2814.4	2935.4	3046.9	3154.6	3261.3	3476.1	3695.8	3921.4	4153.1
13.0	191.6	..	..	2808.0	2931.5	3044.3	3152.7	3259.7	3475.0	3695.0	3920.8	4152.7
14.0	195.0	..	..	2801.4	2927.6	3041.6	3150.7	3258.2	3473.9	3694.1	3920.2	4152.2
15.0	198.3	..	..	2794.7	2923.5	3038.9	3148.7	3256.6	3472.8	3693.3	3919.6	4151.7
16.0	201.4	..	..	..	2919.4	3036.2	3146.7	3255.0	3471.7	3692.5	3918.9	4151.3
17.0	204.3	..	..	..	2915.3	3033.5	3144.7	3253.5	3470.6	3691.7	3918.3	4150.8
18.0	207.1	..	..	..	2911.0	3030.7	3142.7	3251.9	3469.5	3690.9	3917.7	4150.3
19.0	209.8	..	..	..	2906.7	3027.9	3140.7	3250.3	3468.4	3690.0	3917.1	4149.8
20.0	212.4	..	..	..	2902.4	3025.0	3138.6	3248.7	3467.3	3689.2	3916.5	4149.4
22.0	217.2	..	..	..	2893.4	3019.3	3134.5	3245.5	3465.1	3687.6	3915.2	4148.4
24.0	221.8	..	..	..	2884.2	3013.4	3130.4	3242.3	3462.9	3685.9	3914.0	4147.5
26.0	226.0	..	..	..	2874.7	3007.4	3126.1	3239.0	3460.6	3684.3	3912.7	4146.6
28.0	230.0	..	..	..	2864.9	3001.3	3121.9	3235.8	3458.4	3682.6	3911.5	4145.6
30.0	233.8	..	..	..	2854.8	2995.1	3117.5	3232.5	3456.2	3681.0	3910.3	4144.7
32.0	237.4	..	..	..	2844.4	2988.7	3113.2	3229.2	3454.0	3679.3	3909.0	4143.8
34.0	240.9	..	..	..	2833.6	2982.2	3108.7	3225.9	3451.7	3677.7	3907.8	4142.8
36.0	244.2	..	..	..	2822.5	2975.6	3104.2	3222.5	3449.5	3676.1	3906.5	4141.9
38.0	247.3	..	..	..	2811.0	2968.9	3099.7	3219.1	3447.2	3674.4	3905.3	4141.0
40.0	250.3	..	..	..	..	2962.0	3095.1	3215.7	3445.0	3672.8	3904.1	4140.0

(p)	(t_s)	100	150	200	250	300	350	400	500	600	700	800
42.0	253.2	...	...	...	...	2 955.0	3 090.4	3 212.3	3 442.7	3 671.1	3 902.8	4 139.1
44.0	256.0	...	...	...	...	2 947.8	3 085.7	3 208.8	3 440.5	3 669.5	3 901.6	4 138.2
46.0	258.8	...	...	...	...	2 940.8	3 080.9	3 205.3	3 438.2	3 667.8	3 900.3	4 137.2
48.0	261.4	...	...	...	...	2 933.1	3 076.1	3 201.8	3 435.9	3 666.2	3 899.1	4 136.3
50.0	263.9	...	...	...	...	2 925.5	3 071.2	3 198.3	3 433.7	3 664.5	3 897.9	4 135.3
55.0	269.9	...	...	...	...	2 905.6	3 058.7	3 189.3	3 427.9	3 660.4	3 894.8	4 133.0
60.0	275.6	...	...	...	...	2 885.0	3 045.8	3 180.1	3 422.2	3 656.2	3 891.7	4 130.7
65.0	280.8	...	...	...	...	2 863.0	3 032.4	3 170.8	3 416.4	3 652.1	3 888.6	4 128.8
70.0	285.8	...	...	...	...	2 839.0	3 018.7	3 161.2	3 410.6	3 647.9	3 885.4	4 126.0
75.0	290.5	...	...	...	...	2 814.1	3 004.5	3 151.6	3 404.7	3 643.7	3 882.4	4 123.7
80.0	295.0	...	...	...	...	2 786.6	2 989.9	3 141.6	3 398.8	3 639.5	3 879.2	4 121.3
85.0	299.2	...	...	...	...	2 757.1	2 974.7	3 131.5	3 392.8	3 635.4	3 876.1	4 119.0
90.0	303.3	...	...	...	...	...	2 959.0	3 121.2	3 386.8	3 631.1	3 873.0	4 116.7
95.0	307.2	...	...	...	...	...	2 942.7	3 110.7	3 380.7	3 627.0	3 869.9	4 114.4
100.0	311.0	...	...	...	...	...	2 925.8	3 099.9	3 374.6	3 622.7	3 866.8	4 112.0
110.0	318.0	...	...	...	...	...	2 889.6	3 077.8	3 362.2	3 614.2	3 860.5	4 107.0
120.0	324.6	...	...	...	...	...	2 849.7	3 054.8	3 349.6	3 605.7	3 854.3	4 102.7
130.0	330.8	...	...	...	...	...	2 805.0	3 030.7	3 336.8	3 597.1	3 848.0	4 098.0
140.0	336.6	...	...	...	...	...	2 754.2	3 005.6	3 323.8	3 588.5	3 841.7	4 093.0
150.0	342.1	...	...	...	...	...	2 694.8	2 979.1	3 310.6	3 579.8	3 835.4	4 088.6
160.0	347.3	...	...	...	...	...	2 620.8	2 951.3	3 297.1	3 571.0	3 829.1	4 084.0
170.0	352.3	...	...	...	...	...	...	2 921.7	3 283.5	3 562.2	3 822.8	4 079.3
180.0	357.0	...	...	...	...	...	...	2 890.3	3 269.6	3 553.4	3 816.5	4 074.0
190.0	361.4	...	...	...	...	...	...	2 856.7	3 255.4	3 544.5	3 810.2	4 070.0
200.0	365.7	...	...	...	...	...	...	2 820.5	3 241.1	3 535.5	3 803.8	4 065.3
210.0	369.8	...	...	...	...	...	...	2 781.3	3 226.5	3 526.5	3 797.5	4 060.6
220.0	373.7	...	...	...	...	...	...	2 738.8	3 211.7	3 517.4	3 791.1	4 055.9
221.2	374.15	...	...	...	...	...	...	2 734.5	32 10.7	3 516.4	3 789.1	4 054.7

Table II.5 Specific entropy of superheated steam

Absolute Pressure in bar	Saturation Temperature in °C	Specific entropy (s) in kJ/kg-K at various temperature in °C										
(p)	(t_s)	100	150	200	250	300	350	400	500	600	700	800
0.02	17.5	9.193	9.433	9.648	9.844	10.025	10.193	10.351	10.641	10.904	11.146	11.371
0.04	29.0	8.873	9.113	9.328	9.524	9.705	9.874	10.031	10.321	10.585	10.827	11.051
0.06	36.2	8.685	8.925	9.141	9.337	9.518	9.686	9.844	10.134	10.397	10.639	10.864
0.08	41.5	8.552	8.792	9.008	9.204	9.385	9.554	9.711	10.001	10.265	10.507	10.731
0.10	45.8	8.449	8.689	8.905	9.101	9.282	9.450	9.608	9.898	10.162	10.404	10.628
0.15	54.0	8.261	8.502	8.718	8.915	9.096	9.264	9.422	9.712	9.975	10.217	10.442
0.20	60.1	8.126	8.368	8.584	8.781	8.962	9.130	9.288	9.578	9.842	10.084	10.309
0.25	65.0	8.022	8.264	8.481	8.678	8.859	9.028	9.186	9.476	9.739	9.981	10.206
0.30	69.1	7.936	8.179	8.396	8.593	8.774	8.943	9.101	9.391	9.654	9.897	10.121
0.35	72.7	7.864	8.107	8.325	8.522	8.703	8.872	9.030	9.320	9.583	9.826	10.050
0.40	75.9	7.801	8.045	8.263	8.460	8.641	8.810	8.968	9.258	9.522	9.764	9.989
0.45	78.7	7.745	7.990	8.208	8.405	8.587	8.755	8.914	9.204	9.467	9.709	9.934
0.50	81.3	7.695	7.941	8.159	8.356	8.538	8.707	8.865	9.155	9.419	9.661	9.886
0.60	86.0	7.609	7.855	8.074	8.272	8.454	8.622	8.781	9.071	9.334	9.576	9.801
0.70	90.0	7.535	7.783	8.002	8.200	8.382	8.551	8.709	9.000	9.263	9.505	9.730
0.80	93.5	7.470	7.720	7.940	8.138	8.320	8.489	8.648	8.938	9.214	9.444	9.669
0.90	96.7	7.413	7.664	7.884	8.083	8.266	8.435	8.593	8.884	9.147	9.389	9.614
1.00	99.6	7.362	7.614	7.835	8.034	8.217	80386	8.544	8.835	9.098	9.341	9.565
1.50	111.4	..	7.419	7.644	7.845	8.028	8.198	8.356	8.647	8.911	9.153	9.378
2.00	120.2	..	7.279	7.507	7.710	7.894	8.064	8.223	8.514	8.778	9.020	9.245
2.5	127.4	..	7.169	7.400	7.604	7.789	7.960	8.119	8.410	8.674	8.917	9.142
3.0	133.5	..	7.077	7.312	7.518	7.703	7.874	8.034	8.326	8.590	8.833	9.058
3.5	138.5	..	6.998	7.237	7.444	7.631	7.802	7.962	8.254	8.518	8.761	8.986

(p)	(t_s)	100	150	200	250	300	350	400	500	600	700	800
4.0	143.6	..	6.929	7.171	7.380	7.568	7.740	7.899	8.192	8.456	8.699	8.925
4.5	147.9	..	6.866	7.112	7.323	7.512	7.684	1.844	8.137	8.402	8.645	8.870
5.0	151.8	..	:	7.059	7.272	7.461	7.634	7.795	8.088	8.353	8.596	8.821
6.0	158.8	:	:	6.966	7.183	7.374	7.548	7.709	8.003	8.268	8.511	8.737
7.0	165.0	:	:	6.886	7.107	7.300	7.475	7.636	7.931	8.196	8.440	8.665
8.0	170.4	:	:	6.815	7.040	17.235	7.411	7.573	7.868	8.134	8.377	8.603
9.0	175.4	:	:	6.751	6.980	7.177	7.354	7.517	7.812	8.079	8.323	8.549
10.0	179.9	:	:	6.692	6.926	7.125	7.303	7.467	7.763	8.029	8.273	8.500
11.0	184.1	:	:	6.638	6.876	7.078	7.257	7.42]	7.718	7.985	8.229	8.455
12.0	188.0	:	:	6.587	6.831	7.034	7.214	7.379	7.677	7.944	8.188	8.415
13.0	191.6	:	:	6.539	6.788	6.994	7.175	7.340	7.639	7.906	8.151	8.378
14.0	195.0	:	:	6.494	6.748	6.956	7.139	7.305	7.603	7.871	8.116	8.343
15.0	198.3	:	:	6.451	6.710	6.921	7.104	7.271	7.570	7.839	8.084	8.311
16.0	201.4	:	:	:	6.674	6.887	7.072	7.239	7.540	7.808	8.054	8.281
17.0	204.3	:	:	:	6.640	6.856	7.042	7.210	7.511	7.779	8.025	8.252
18.0	207.1	:	:	:	6.607	6.826	7.013	7.182	7.483	7.752	7.998	8.226
19.0	209.8	:	:	:	6.576	6.797	6.986	7.155	7.457	7.727	7.973	8.200
20.0	212.4	:	:	:	6.545	6.770	6.960	7.130	7.432	7.702	7.949	8.176
22.0	217.2	:	:	:	6.488	6.718	6.911	7.082	7.386	7.657	7.904	8.132
24.0	221.8	:	:	:	6.434	6.670	6.866	7.038	7.344	7.615	7.862	8.091
26.0	226.0	:	:	:	6.382	6.625	6.824	6.998	7.305	7.577	7.825	8.053
28.0	230.0	:	:	:	6.333	6.582	6.784	6.960	7.269	7.541	7.789	8.018
30.0	233.8	:	:	:	6.286	6.542	6.747	6.925	7.235	7.508	7.756	7.986
32.0	237.4	:	:	:	6.240	6.504	6.712	6.891	7.203	7.477	7.726	7.955
34.0	240.9	:	:	:	6.195	6.467	6.679	6.860	7.172	7.447	7.697	7.927
36.0	244.2	:	:	:	6.151	6.432	6.647	6.829	7.144	7.420	7.669	7.900
38.0	247.3	:	:	:	6.109	6.397	6.616	6.801	7.117	7.393	7.643	7.874
40.0	250.3	:	:	:	..	6.364	6.587	6.773	7.091	7.368	7.619	7.850

(p)	(t_s)	100	150	200	250	300	350	400	500	600	700	800
42.0	253.2	:	:	:	:	6.332	6.559	6.747	7.066	7.344	7.595	7.826
44.0	256.0	:	:	:	:	6.301	6.532	6.722	7.043	7.321	7.573	7.804
46.0	258.8	:	:	:	:	6.270	6.505	6.697	7.020	7.299	7.551	7.783
48.0	261.4	:	:	:	:	6.240	6.479	6.674	6.998	7.278	7.531	7.763
50.0	263.9	:	:	:	:	6.211	6.455	6.651	6.977	7.258	7.511	7.743
55.0	269.9	:	:	:	:	6.139	6.395	6.597	6.928	7.210	7.464	7.697
60.0	275.6	:	:	:	:	6.069	6.339	6.546	6.882	7.166	7.422	7.655
65.0	280.8	:	:	:	:	6.001	6.285	6.499	6.839	7.126	7.382	7.617
70.0	285.8	:	:	:	:	5.933	6.233	6.454	6.799	7.088	7.346	7.581
75.0	290.5	:	:	:	:	5.864	6.184	6.411	6.762	7.053	7.311	7.547
80.0	295.0	:	:	:	:	5.794	6.135	6.369	6.726	7.019	7.279	7.516
85.0	299.2	:	:	:	:	5.744	6.088	6.330	6.692	6.987	7.249	7.486
90.0	303.3	:	:	:	:	:	6.041	6.292	6.660	6.957	7.220	7.458
95.0	307.2	:	:	:	:	:	5.995	6.254	6.629	6.929	7.192	7.431
100.0	311.0	:	:	:	:	:	5.949	6.218	6.599	6.901	7.166	7.406
110.0	318.0	:	:	:	:	:	5.857	6.148	6.543	6.850	7.117	7.358
120.0	324.6	:	:	:	:	:	5.764	6.081	6.491	6.802	7.072	1.315
130.0	330.8	:	:	:	:	:	5.666	6.016	6.441	6.758	7.030	7.274
140.0	336.6	:	:	:	:	:	5.562	5.951	6.394	6.716	6.991	7.237
150.0	342.1	:	:	:	:	:	5.447	5.888	6.349	6.676	6.954	7.201
160.0	347.3	:	:	:	:	:	5.311	5.824	6.305	6.639	6.919	7.168
170.0	352.3	:	:	:	:	:	:	5.760	6.264	6.603	6.886	7.137
180.0	357.0	:	:	:	:	:	:	5.695	6.223	6.569	6.854	7.107
190.0	361.4	:	:	:	:	:	:	5.628	6.184	6.536	6.824	7.078
200.0	365.7	:	:	:	:	:	:	5.559	6.146	6.504	6.795	7.051
210.0	369.8	:	:	:	:	:	:	5.486	6.108	6.474	6.768	7.025
220.0	373.7	:	:	:	:	:	:	5.410	6.072	6.444	6.741	7.000
221.2	374.15	:	:	:	:	:	:	5.399	6.068	6.441	6.738	6.994

ANNEXURE III: SECTION MODULUS

Section modulus is the ratio of moment of inertia of a beam section about neutral axis (passing through the centre of gravity of the section) to the distance of extreme fibre from neutral axis.

Mathematically, section modulus,

$$Z = \frac{I}{y_{max}}$$

where, I = M.O.I. about neutral axis

y_{max} = Distance of the outermost layer of beam form neutral axis

The bending equation for a beam is

$$\frac{M}{I} = \frac{\sigma}{y} = \frac{E}{R}$$

The stress will be maximum where y is maximum.

$\therefore$
$$\frac{M}{I} = \frac{\sigma_{max}}{y_{max}}$$

$\therefore$
$$M = \sigma_{max} \cdot \frac{I}{y_{max}} = \sigma_{max} \cdot Z$$

where,
$$Z = \frac{I}{y_{max}}$$

The moment of resistance of a section depends upon section modulus. Therefore, section modulus represents the strength of the section against bending moment.

I. SECTION MODULUS FOR COMMON BEAM SECTIONS

1. Rectangular Section

If b is the width and d the depth of a rectangular section, the M.O.I. about neutral axis passing through $c.g.$ is given by

$$I = \frac{bd^3}{12}$$

$$y_{max} = \frac{d}{2}$$

$\therefore$ Section modulus,

$$Z = \frac{I}{y_{max}} = \frac{bd^3}{12 \times \left(\frac{d}{2}\right)} = \frac{bd^2}{6}.$$

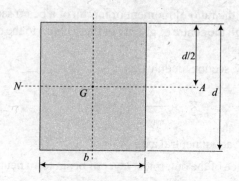

Fig. III.1 Rectangular section

2. Hollow Rectangular Section

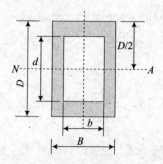

Fig. III.2 Hollow rectangular section

$$\text{M.O.I.} = \frac{BD^3}{12} - \frac{bd^3}{12}$$

$$= \frac{BD^3 - bd^3}{12}$$

$$y_{max} = \frac{D}{2}$$

$$\therefore \quad Z = \frac{I}{y_{max}}$$

$$= \frac{[BD^3 - bd^3]}{12 \times \dfrac{D}{2}}$$

$$= \left[\frac{BD^3 - bd^3}{6D}\right]$$

3. Circular Section

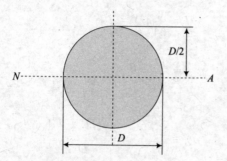

Fig. III.3   Circular section

M.O.I. of a circular section of diameter D.

$$I = \frac{\pi}{64} D^4$$

$$y_{max} = \frac{D}{2}$$

$\therefore$
$$Z = \frac{I}{y_{max}} = \frac{\frac{\pi}{64} D^4}{\frac{D}{2}}$$

$$= \frac{\pi}{32} D^3.$$

4. Hollow Circular Section

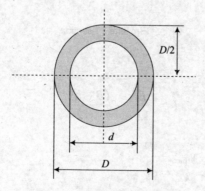

Fig. III.4   Hollow circular section

M.O.I. of a hollow circular section,

$$I = \frac{\pi}{64}[D^4 - d^4]$$

$$y_{max} = \frac{D}{2}$$

$\therefore$

$$Z = \frac{I}{y_{max}} = \frac{\frac{\pi}{64}[D^4 - d^4]}{\frac{D}{2}}$$

$$= \frac{\pi}{32}\left[\frac{D^4 - d^4}{D}\right]$$

II. SECTION MODULUS FOR SYMMETRICAL SECTIONS

5. I-Section

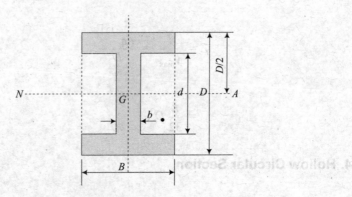

Fig. III.5   I-section

M.O.I. of I-section

$$I = \frac{BD^3}{12} - \frac{(B-b)d^3}{12}$$

$$y_{max} = \frac{D}{2}$$

$$Z = \frac{[BD^3 - (B-b)d^3]}{12 \times \dfrac{D}{2}}$$

$$= \frac{BD^3 - (B-b)d^3}{6D}$$

III. SECTION MODULUS FOR UNSYMMETRICAL SECTIONS

6. Unsymmetrical I-Section

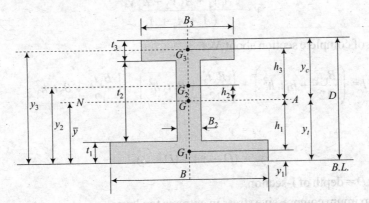

Fig. III.6 Unsymmetrical I-section

The dimensions of unsymmetrical I-section are as follows:

Bottom flange

Width = B_1

Thickness = t_1

Distance between G_1 (*c.g.* of flange section) and G (*c.g.* of total section) = h_1

Distance between G_1 and bottom layer = y_1

Area $\qquad A_1 = B_1 t_1$

Web

Width = B_2

Depth = t_2

Distance between G_2 (*c.g.* of web section) and G (*N-A*) = h_2

Distance between G_2 and bottom layer of section = y_2

Area of web, $\qquad A_2 = B_2 t_2$.

Top flange

    Width $= B_3$

    Thickness $= t_3$

Distance between G_3 (*c.g.* of top flange section) and G (*N-A*) $= h_3$

Distance between G_3 and bottom layer of section $= y_3$

    Area $= A_3 = B_3 t_3$

    $y_t =$ Distance of extreme bottom layer to *N-A*

    $y_c =$ Distance of extreme top layer to *N-A*

    $\bar{y} =$ Distance of G from bottom layer

$$\bar{y} = \frac{A_1 y_1 + A_2 y_2 + A_3 y_3}{(A_1 + A_2 + A_3)}$$

M.O.I. of complete section about *N-A*,

$$I = \left(\frac{B_1 t_1^3}{12} + B_1 t_1 h_1^2 \right) + \left(\frac{B_2 t_2^3}{12} + B_2 t_2 h_2^2 \right) + \left(\frac{B_3 t_3^3}{12} + B_3 t_3 h_3^2 \right)$$

$$y_t = \bar{y}$$

$$y_c = (D - \bar{y}) = (D - y_t)$$

where, $D =$ depth of I-section.

    Maximum compressive stress in extreme top layer,

$$\sigma_c = \frac{M y_c}{I}$$

    Maximum tensile stress in extreme bottom layer,

$$\sigma_t = \frac{M y_t}{I}$$

7. T-section

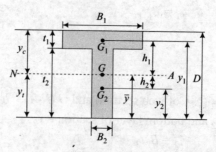

Fig. III.7 *T-section*

The dimensions of *T*-section are:

Top flange

Width = B_1

Thickness = t_1

Area, $\qquad\qquad A_1 = B_1 t_1$

Distance of G_1 (*c.g.* of flange) form bottom line = y_1

Distance of G_1 form *N-A* = h_1

Web

Width = B_2

Thickness = t_2

Area $A_2 = B_2 t_2$

Distance of G_2 (*c.g.* of web) from bottom line = y_2

Distance of G_2 from *N-A* = h_2

Section

Total depth = D

Distance of G (*N-A*) from bottom line = $\bar{y}$

Distance of top layer form *N-A* = y_c

Distance of bottom layer form *N-A* = y_t

$$\bar{y} = \frac{A_1 y_1 + A_2 y_2}{(A_1 + A_2)}$$

$$I = \left(\frac{B_1 t_1^3}{12} + B_1 t_1 h_1^2 \right) + \frac{B_2 t_2^3}{12} + (B_2 t_2 h_2^2)$$

$$y_t = \bar{y}$$

$$y_c = (D - \bar{y})$$

Maximum compressive stress in extreme top layer,

$$\sigma_c = \frac{M y_c}{I}$$

Maximum tensile stress in extreme bottom layer,

$$\sigma_t = \frac{M y_t}{I}$$

IV. POLAR SECTION MODULUS

Polar section modulus or polar modulus is the ratio of polar moment of inertia of a shaft section and its outer radius. Polar section modulus represents the strength of a section against twisting or turning moment.

The torsion formula gives

$$\frac{T}{J} = \frac{\tau}{R} = \frac{G\theta}{L}$$

where, J = Polar moment of Inertia,

$\therefore$

$$T = \frac{\tau J}{R} = \tau Z_p$$

where, $Z_p = \dfrac{J}{R}$ = Polar section modulus.

V. POLAR SECTION MODULUS FOR COMMON SHAFT SECTIONS

1. Solid Shaft

Polar moment of inertia,

$$J = \frac{\pi}{32} D^4$$

Polar section modulus,

$$Z_p = \frac{J}{R} = \frac{\dfrac{\pi}{32} D^4}{\dfrac{D}{2}}$$

$$= \frac{\pi}{16} D^3$$

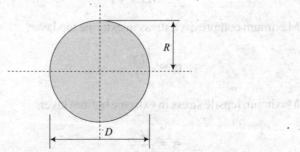

Fig. III.8 Solid circular shaft

2. Hollow Shaft

Polar moment of inertia,

$$J = \frac{\pi}{32} (D^4 - d^4)$$

Polar section modulus,

$$Z_p = \frac{J}{R} = \frac{\frac{\pi}{32}(D^4 - d^4)}{\frac{D}{2}}$$

$$= \frac{\pi}{16}\left[\frac{D^4 - d^4}{D}\right].$$

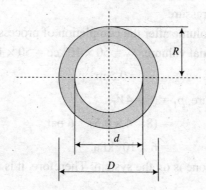

Fig. III.9 Hollow circular shaft

ANNEXURE IV: A SAMPLE PAPER WITH SOLUTION
U.P. Technical University, Lucknow.
B.Tech
(Sem. II) Examination, 2006-2007
MECHANICAL ENGINEERING

[Time: 3 Hours] *[Total Marks: 100]*

1 Attempt any **four** parts of the following: $4 \times 5 = 20$

(a) Explain the concept of continuum, with suitable examples.

Solution: See para 1.5.2

(b) A closed system whose initial volume is 50×10^4 **cc** undergoes a non-flow reversible process for which pressure and volume correlation is given by:

$p = (8 - 4\ V)$ where p is in bar and V in m^3. If **200 kJ** of work is supplied to the system, determine:

(i) final pressure

(ii) final volume after the completion of process.

Solution: Initial volume, $V_1 = 50 \times 10^4$ cc $= 50 \times 10^4 \times 10^{-6}\,m^3$

$$= 0.5\ m^3$$

Initial pressure, $p_1 = (8 - 4V_1)$

$$= (8 - 4 \times 0.5) = 6\ bar$$

$$= 6 \times 10^2\ kPa.$$

The work done is on the system: Therefore, it is negative

$$-W_{1-2} = \int_{V_1}^{V_2} pdV = \int_{V_1}^{V_2} (8 - 4V) \times 10^2 dV\ (kJ)$$

$$\therefore \qquad = \left| 8V - \frac{4V^2}{2} \right|_{0.5}^{V_2} \times 10^2$$

$$-200 = (8V_2 - 2V_2^2) - (8 \times 0.5 - 2 \times 0.5^2)$$

$$\therefore \qquad 2V_2^2 - 8V_2 + 1.5 = 0$$

or $$V_2^2 - 4V_2 + 0.75 = 0$$

$$V_2 = \frac{4 \pm \sqrt{4^2 - 4 \times 0.75}}{2} = 3.9365\ m^3\ \text{or}\ 0.0635\ m^3$$

If V_2 is selected as $3.9365\ m^3$, p_2 is negative.

(i) Final volume, $V_2 = \textbf{0.0635}\ \textbf{m}^3$ **Ans.**

(ii) Final pressure,

$$p_2 = (8 - 4V_2)$$

$$= (8 - 4 \times 0.0635)$$

$$= 7.746 \text{ bar} \quad \textbf{Ans.}$$

(c) Steam enters into a steam turbine with a velocity of **30 m/s** and enthalpy of **2610 kJ/kg** and leaves with a velocity of **10 m/s** and enthalpy of **2050 kJ/kg**. Heat is lost to the surrounding due to temperature difference is **280 kJ/min** and steam consumption rate of the turbine is **6000 kg/hr**. Stating your assumptions, calculate the power developed by the steam turbine.

Solution:

1. *Inlet data*

$$\dot{m} = 6000 \text{ kg/hr}$$

$$= \frac{6000}{3600} = \frac{5}{3} \text{ kg/sec.}$$

$$V_1 = 30 \text{ m/sec.}$$

$$h_1 = 2610 \text{ kJ/kg.}$$

2. *Outlet data*

$$V_2 = 10 \text{ m/sec.}$$

$$h_2 = 2050 \text{ kJ/sec.}$$

3. *Steady flow energy equation*

$$Q_{1-2} - W_{1-2} = \dot{m} \left[(h_2 - h_1) + \frac{V_2^2 - V_1^2}{2 \times 10^3} + \frac{g(Z_2 - Z_1)}{10^3} \right]$$

$$Q_{1-2} = -280 \text{ kJ/min} = -\frac{280}{60} = -4.667 \text{ kJ/sec.}$$

Assume $Z_2 = Z_1$

$$\therefore \quad -4.667 - W_{1-2} = \frac{5}{3} \left[(2050 - 2610) + \frac{(10)^2 - (30)^2}{2 \times 10^3} \right]$$

$$= \frac{5}{3} [-560 - 0.4] = -934$$

$$\therefore \quad W_{1-2} = \textbf{929.33 kW.} \quad \textbf{Ans.}$$

(d) Which is more effective way to increase the efficiency of a reversible heat engine (i) to increase the source temperature T_1 while sink temperature T_2 kept constant or (ii) to decrease the sink temperature by the same amount while source temperature is constant.

Solution: See example 2.27

(e) What is entropy? When entropy is defined only in terms of reversible process, how can then it be evaluated for an irreversible process?

Solution: See para 2.15, 2.15.8 and 2.15.9

For a reversible process,

$$ds = \frac{\delta Q_{rev}}{T}$$

For any irreversible process,

$$\therefore \qquad ds > \frac{\delta Q}{T}$$

$$\therefore \qquad ds \geq \frac{\delta Q}{T}$$

or

$$s_2 - s_1 \geq \int_1^2 \frac{\delta Q}{T}$$

The equality sign holds good for a reversible process and the inequality sign for an irressible process. δQ may be for any reversible or irreversible process.

(f) A metal block of **5 kg** and **200°C** is cooled in a surrounding of air which is at **30°C**. If specific heat of metal is **0.4 kJ/kgK,** calculate the following:

(i) entropy change of block

(ii) entropy change of surrounding & universe.

Solution: Mass of metal block,

$$m = 5 \text{ kg.}$$

$$T_1 = 300°C + 273 = 573 \text{ K.}$$

$$C = 0.4 \text{ kJ/kg K.}$$

$$T_2 = 30°C + 273 = 303 \text{ K.}$$

(i) Heat lost by metal block,

$$Q_{1-2} = mC(T_1 - T_2)$$

$$= 5 \times 0.4 \,(573 - 303)$$

$$= 540 \text{ kJ.}$$

Change of entropy,

$$\Delta s_{\text{system}} = mC \ln \frac{T_2}{T_1} = 5 \times 0.4 \ln \frac{303}{573} = -1.2743 \text{ kJ.}$$

(ii) Heat gained by surrounding = 540 kJ.

Change of entropy of surrounding,

$$\Delta s_{\text{surrounding}} = \frac{Q_{1-2}}{T_2} = \frac{540}{303} = \textbf{1.7822 kJ.}$$

Change of entropy of universe,

$$\Delta s_{\text{universe}} = \Delta s_{\text{system}} + \Delta s_{\text{surrounding}}$$

$$= -1.2743 + 1.7822$$

$$= \textbf{+0.5079 kJ increase} \quad \textbf{Ans.}$$

2 Attempt any **two** parts: $\qquad$ **10 × 2 = 20**

(a) (i) With the help of neat sketches explain the working of a **4** stroke **SI** engine.

Solution: See para 4.5

(ii) With the help of $T - s$ diagrams, explain as to how the Rankine cycle overcomes the limitations of carnot vapour cycle for steam turbine power plant.

Solution: See para 3.9.2 and Table 3.3

(b) For a steam power plant following observation was made:

Supply condition of steam: **60 bar, 450°C**

Condenser pressure: **0.10 bar**

Steam flow rate: **5000 kg/hr.**

Calculate the following:

(i) Turbine work

(ii) % of pump work compared to turbine work

(iii) Heat addition in boiler

(iv) Heat rejection in condenser

(v) Thermal efficiency

Solution:

(i) *Steam turbine*

Using superheated steam tables, at

$$p_1 = 60 \text{ bar and} \quad t_1 = 450°C$$

$$h_1 = \frac{3180.1 + 3422.2}{2} = 3301.15 \text{ kJ/kg.}$$

$$s_1 = \frac{6.546^2 + 6.882}{2} = 6.714 \text{ kJ/kg-K.}$$

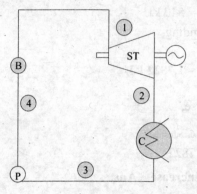

Power plant

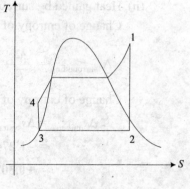

T-s diagram

Process 1-2 is assumed reversible and isentropic

$\therefore$ $\qquad$ $s_1 = s_2$.

At turbine exit, using saturated steam tables,

$$p_2 = 0.10 \text{ bar}$$

$$s_{f_2} = 0.649 \text{ kJ/kgK}$$

$$s_{fg_2} = 7.502 \text{ kJ/kgK}$$

$$h_{f_2} = 191.8 \text{ kJ/kg}.$$

$$h_{fg_2} = 2392.9 \text{ kJ/kg}.$$

$$s_1 = s_2 = s_{f_2} + x_2 s_{fg_2}.$$

$\therefore$ $\qquad$ $x_2 = \dfrac{s_1 - s_{f_2}}{s_{fg_2}} = \dfrac{6.714 - 0.649}{7.502} = 0.80845$

$\therefore$ $\qquad$ $h_2 = h_{f_2} + x_2 h_{fg_2}$

$$= 191.8 + (0.80845)(2392.9) = 2126.34 \text{ kJ/kg}.$$

Turbine work,

$$W_T = \dot{m}(h_1 - h_2) = (3301.15 - 2126.34) \times \frac{5000}{3600}$$

$$= \mathbf{1631.68 \text{ kW}}. \quad \textbf{Ans.}$$

(ii) *Condenser*

At inlet,

$$h_2 = 2126.34 \text{ kJ/kg}.$$

At outlet,

$$h_3 = h_{f_3} = 191.8 \text{ kJ/kg}$$

Heat rejected

$$Q_{2-3} = \dot{m}(h_2 - h_3)$$

$$= \frac{5000}{3600}(2126.34 - 191.8)$$

$$= \textbf{2686.86 kJ. \quad Ans.}$$

(iii) *Pump*

At pump inlet,

$$h_3 = 191.8 \text{ kJ/kg.}$$

At pump outlet,

$$h_4 = h_3 + w_p$$

Assuming water as incompressible,

Pump work,

$$W_p = \dot{m} \, v_{f_3} \, (p_4 - p_3) \times 10^2$$

$$= \frac{5000}{3600} \times 0.001010(60 - 0.1) \times 10^2$$

$$= 8.4 \text{ kW}$$

Pump work as % age of turbine work

$$= \frac{Wp}{WT} \times 100 = \frac{8.4}{1631.68} \times 100$$

$$= \textbf{0.515\% \quad Ans.}$$

(iv) *Boiler*

At inlet of boiler,

$$h_4 = h_3 + w_p = 191.8 + \frac{8.4 \times 3600}{5000}$$

$$= 197.848 \text{ kJ/kg.}$$

At boiler outlet,

$$h_1 = 3301.15 \text{ kJ/kg.}$$

Heat addition in boiler,

$$Q_{4-1} = \dot{m}(h_1 - h_4)$$

$$= \frac{5000}{3600}(3301.15 - 197.848) = \textbf{4310.14 kJ \quad Ans.}$$

(v) *Steam power plant*

Thermal efficiency,

$$\eta = \frac{\text{Network output}}{\text{Heat supplied}}$$

$$= \frac{W_T - W_p}{Q_{4-1}} = \left(\frac{1631.68 - 8.4}{4310.14}\right) \times 100$$

$$= 37.66\% \quad \textbf{Ans.}$$

(c) An engine working on diesel cycle has air intake condition of **1 bar** and **310° K** and compression ratio is **17**. Heat added at constant pressure is **1250 kJ/kg**. Make calculations for the maximum temperature of the cycle, net power output and thermal efficiency of the cycle.

Solution: The p-V diagram of diesel cycle is shown

$p_1 = 1$ bar

$T_1 = 310$ K.

(i) For compression process 1-2,

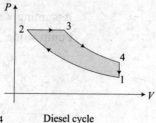

Diesel cycle

$$\frac{T_2}{T_1} = \left(\frac{V_1}{V_2}\right)^{\gamma-1}$$

$$T_2 = T_1(r_c)^{\gamma-1} = 310(17)^{0.4}$$

$$= 962.8 \text{ K.}$$

(ii) Heat supplied per kg of air,

$$Q_{2-3} = m\, Cp(T_3 - T_2)$$

$$1250 = 1 \times 1.005(T_3 - 962.8)$$

∴ $$T_3 = \frac{1250}{1.005} + 962.8 = \textbf{2206.58 K.} \quad \textbf{Ans.}$$

(iii) $$\rho = \frac{V_3}{V_2} = \frac{T_3}{T_2} = \frac{2206.5}{962.8} = 2.29$$

Cycle efficiency,

$$\eta = 1 - \frac{1}{(r_c)^{\gamma-1}}\left[\frac{\rho^{\gamma}-1}{\gamma(\rho-1)}\right]$$

$$= 1 - \frac{1}{(17)^{0.4}}\left[\frac{17^{1.4}-1}{1.4(2.29-1)}\right] = \textbf{80\%} \quad \textbf{Ans.}$$

(iv) Net power output

$$W = \eta(Q_{2-3}) = 0.8 \times 1250 = \textbf{1000 kJ/kg.} \quad \textbf{Ans.}$$

3 Attempt any **two** parts of the following: $\qquad$ **10 × 2 = 20**

(a) Explain the followings:

(i) Necessary and sufficient conditions of equilibrium of a system of coplanar concurrent forces.

Solution: See para 5.7.2

(ii) Concept of free body diagram with the help of suitable examples.

Solution: See para 5.10

(iii) Angle of repose and its applications

Solution: See para 6.3.3(C)

(iv) Belt friction and its applications.

Solution: See para 6.4.5

(b) A plate measuring (4×4) **m²** is acted upon by **5** forces in its plane as shown. Determine the magnitude and direction of the resultant force.

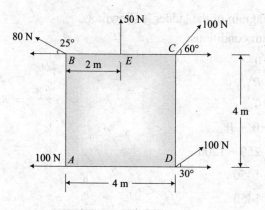 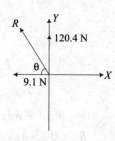

Solution:

1. Resolve the forces along x-axis and y-axis.

$$\Sigma F_x = 100 - 80 \cos 25° + 100 \cos 60° - 100 \cos 30°$$

$$= 100 - 72.5 + 50 - 86.6$$

$$= -9.1 \text{ N.}$$

$$\Sigma F_y = 80 \sin 25° + 50 + 100 \sin 60° - 100 \sin 30°$$

$$= 33.8 + 50 + 86.6 - 50$$

$$= 120.4 \text{ N.}$$

2. The resultant,

$$R = \sqrt{(\Sigma F_x)^2 + (\Sigma F_y)^2}$$

$$= \sqrt{(-9.1)^2 + (120.4)^2}$$

$$= \textbf{120.74 N} \quad \textbf{Ans.}$$

3. The direction of resultant with x-axis.

$$\tan \theta = \frac{\Sigma F_y}{\Sigma F_x} = \frac{120.4}{-9.1}$$

$$\theta = \textbf{--85.677° with --ve } x\textbf{-axis} \quad \textbf{Ans.}$$

(c) A ladder **3 m** long and weighing **250 N** is placed against a wall with end **B** at floor level and **A** on the wall. In addition to self weight, the ladder supports a man weighing **1200 N** at **2.5 m** from **B** on the ladder. If co-efficient of friction at wall is **0.25** and at floor is **0.35** and if ladder makes an angle **60°** with the floor, find the minimum horizontal force which if applied at **B** will prevent the slipping of the ladder.

Solution:

(i) The free body diagram of the ladder is shown.

(ii) Apply equilibrium conditions:

(a) $\qquad \Sigma F_x = 0$

$\qquad f_B + P = R_A.$

(b) $\qquad \Sigma F_y = 0$

$$R_B + f_A = W + W_m$$

$$R_B + \mu_A R_A = 250 + 1200$$

$$R_B + \mu_A(f_B + P) = 1450$$

$$R_B + 0.25(\mu_B R_B + P) = 1450$$

$$R_B + 0.25(0.35 R_B + P) = 1450$$

$$1.0875 R_B + 0.25 P = 1450$$

(c) $\qquad \Sigma M_O = R_B = 1333.33 - 0.23 P \qquad \qquad$ ...(1)

Distance $\qquad OB = l \cos \theta = 3 \cos 60° = 1.5$ m

$$OC = OB - l/2 \cos 60° = 1.5 - 1.5 \cos 60°$$

$$= 0.75 \text{ m}$$

$$OD = OB - 2.5 \cos 60° = 0.25 \text{ m.}$$

$$R_B \times 1.5 = W \times OC + W_m \times OD + R_A(OA)$$

$$= 250 \times 0.75 + 1200 \times 0.25 + R_A (3 \sin 60°)$$

$$= 187.5 + 300 + 2.598R_A.$$

$$R_B = 325 + 1.732R_A$$

$$= 325 + 1.732 (0.25R_B + P)$$

$$= 325 + 0.433R_B + 1.732P$$

$$0.567R_B = 325 + 1.732P$$

$$R_B = 573.19 + 3.0546P \qquad\qquad ...(2)$$

From (1) and (2)

$$1333.33 - 0.23P = 573.19 + 3.0546P$$

$$\therefore \quad 3.28467P = 760.74$$

$$P = \textbf{231.42N} \quad \textbf{Ans.}$$

4 Attempt any **two** parts: $\qquad\qquad$ **10 × 2 = 20**

(a) (i) Define and differentiate between a perfect, deficient and redundant truss.

Solution: See para 8.2.2

(ii) Derive the relationship between shear force, bending moment and the loading for a beam. What are the assumptions required for this derivation?

Solution: See para 7.5.3

(b) Determine the magnitude and nature of forces in the members of truss shown.

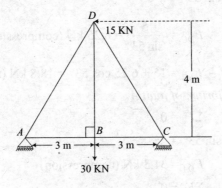

Solution:

(i) *Support reactions*

The truss is supported on rollers at A.

∴ Horizontal reaction at $A = 0$

Horizontal reaction at $C = 15$ kN

Apply equilibrium condition

$$\Sigma M_C = 0$$

$$R_A \times 6 - 15 \times 4 - 30 \times 3 = 0.$$

∴ $$R_A = 25 \text{ kN.}$$

$$\Sigma F_y = 0$$

$$R_A + R_{V_C} = 30$$

∴ $$R_{V_C} = 30 - 25 = 5 \text{ kN.}$$

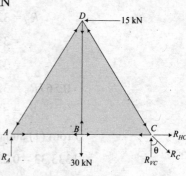

$$R_C = \sqrt{R_{V_C}^2 + R_{H_c}^2} = \sqrt{(5)^2 + (15)^2}$$

$$= 15.8 \text{ kN}$$

$$\tan \theta = \frac{R_{H_C}}{R_{V_C}} = \frac{15}{5} = 3$$

∴ $\theta = 71.56°$ with the vertical

(ii) *Equilibrium of joint C*

$$\Sigma F_x = 0.$$

$$F_{BC} - F_{DC} \cos 53° = 15$$

$$\Sigma F_y = 0.$$

$$F_{DC} \sin 53° = 5$$

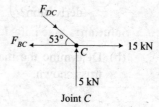

Joint C

∴ $$F_{DC} = \frac{5}{\sin 53°} = 6.25 \text{ kN (compression)}$$

∴ $$F_{BC} = 15 + 6.25 \cos 53° = 18.8 \text{ kN (tension)}$$

(iii) *Equilibrium of joint A*

$$\Sigma F_y = 0$$

$$F_{DA} \sin 53° = 25$$

∴ $$F_{DA} = 31.3 \text{ kN (compression)}$$

$$\Sigma F_x = 0$$

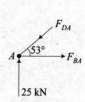

Joint A

$$F_{BA} = F_{DA} \cos 53° = 31.3 \cos 53°$$
$$= 18.8 \text{ kN (tensile)}$$

(iv) *Equilibrium of joint B*

$$F_{DB} = 30 \text{ kN (tension)}$$

$$F_{AB} = F_{CB} = 18.8 \text{ kN (tension)}$$

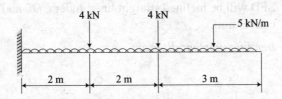

Joint *B*

(v) *Draw force table and diagram*

Member	Force (kN)	Nature
AB	18.8	T
BC	18.8	T
AD	31.3	C
DB	30.0	T
CD	6.25	C

(c) Draw the SF and B.M Diagram for the beam shown.

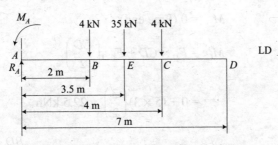

Solution:

1. Draw equivalent *loading diagram*

A load of $5 \times 7 = 35$ kN will be acting at 3.5 m from *A*.

2. *Support reactions*

Apply equilibrium conditions

(i) $\quad \Sigma Fy = 0$

∴ $\quad R_A = 4 + 35 + 4 = 43$ kN.

(ii) $\Sigma M_A = 0$

$$M_A = 4(2) + 35(3.5) + 4(4)$$
$$= 646.5 \text{ kN-m.}$$

3. *Shear force diagram*

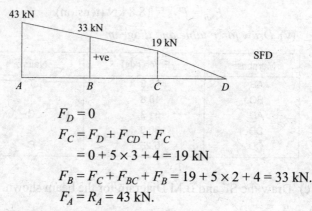

$$F_D = 0$$
$$F_C = F_D + F_{CD} + F_C$$
$$= 0 + 5 \times 3 + 4 = 19 \text{ kN}$$
$$F_B = F_C + F_{BC} + F_B = 19 + 5 \times 2 + 4 = 33 \text{ kN.}$$
$$F_A = R_A = 43 \text{ kN.}$$

SFD will be inclined straight lines under *UDL* and +ve as per sign convention.

4. *Bending moment diagram*

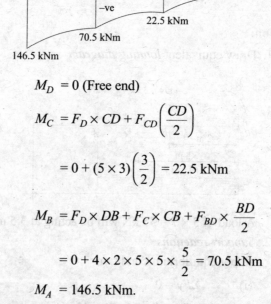

$$M_D = 0 \text{ (Free end)}$$

$$M_C = F_D \times CD + F_{CD}\left(\frac{CD}{2}\right)$$

$$= 0 + (5 \times 3)\left(\frac{3}{2}\right) = 22.5 \text{ kNm}$$

$$M_B = F_D \times DB + F_C \times CB + F_{BD} \times \frac{BD}{2}$$

$$= 0 + 4 \times 2 \times 5 \times 5 \times \frac{5}{2} = 70.5 \text{ kNm}$$

$$M_A = 146.5 \text{ kNm.}$$

BMD will be parabolas under *UDL* and will be –ve for hogging beam.

5 Attempt any **four** parts of the following: **5 × 4 = 20**

(a) Draw stress-strain curve for a ductile and brittle material on a simple diagram. What are the differences between these two curves.

Solution: See para 9.5 and 9.6

(b) Determine the stress in all the three sections and total deformation of the steel rod shown. Cross sectional area = **10 cm²** and $E = 200$ **GN/m²**.

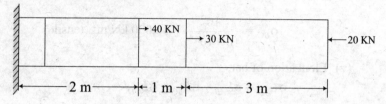

Solution:

(i) Area of cross-section, $A = 10$ cm² $= 10 \times 10^{-4}$ m²

$$E = 200 \text{ GN/m}^2 = 200 \times 10^6 \text{ kN/m}^2.$$

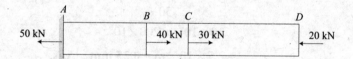

(ii) For equilibrium of the steel rod, the reaction at A will be $(40 + 30 - 20) = 50$ kN left side direction.

(iii) Draw free body diagram of all the sections

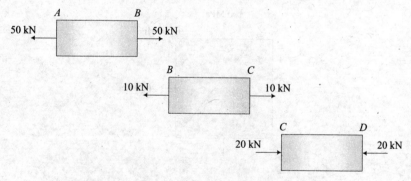

Bar *CD* is under a compression load of 20 kN.

Bar *AB* is under a tensile load of 50 kN.

Bar *BC* is under a tensile load of 10 kN.

(iv) Stress calculation

Stress in section *CD*,

$$\sigma_{CD} = \frac{20}{10 \times 10^{-4}} = 20,000 \text{ kN/m}^2 \text{ (compression)}$$

$$\sigma_{AB} = \frac{50}{10 \times 10^{-4}} = 50,000 \text{ kN/m}^2 \text{ (tensile)}$$

$$\sigma_{BC} = \frac{10}{10 \times 10^{-4}} = 10,000 \text{ kN/m}^2 \text{ (tensile)}$$

(v) Elongation of bar,

$$\sigma L_{AD} = \sigma L_{AB} + \sigma L_{BC} + \sigma L_{CD}$$

$$= \sum \frac{PL}{AE}$$

$$= \frac{1}{AE} [50 \times 2 + 10 \times 1 - 20 \times 3]$$

$$= \frac{50}{10 \times 10^{-4} \times 200 \times 10^6}$$

$$= 2.5 \times 10^{-4} \text{ m} = \textbf{0.25 mm} \quad \textbf{Ans.}$$

(c) Calculate the normal and shear stress on the plane inclined at an angle **60°** for the stress shown. Also calculate the value of principal stress and its location.

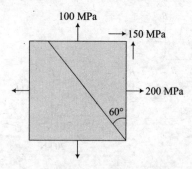

Solution: $\sigma_x = 200$ MPa

$\sigma_y = 100$ MPa

$\tau_{xy} = 150$ MPa

$\alpha = 60°$.

(a) *Normal and shear stress*

Normal stress

$$\sigma_n = \frac{\sigma_x + \sigma_y}{2} + \frac{\sigma_x - \sigma_y}{2} \cos 2\alpha + \tau_{xy} \sin 2\alpha$$

$$= \frac{200 + 100}{2} + \frac{200 - 100}{2} \cos 120° + 150 \sin 120°$$

$$= 150 - 25 + 129.9$$

$$= \textbf{254.9 MPa} \quad \textbf{Ans.}$$

Shear stress,

$$\tau = \frac{\sigma_x - \sigma_y}{2} \sin 2\alpha - \tau_{xy} \cos 2\alpha$$

$$= \frac{200 - 100}{2} \sin 120° - 150 \cos 120°$$

$$= 43.30 + 75$$

$$= \textbf{118.3 MPa} \quad \textbf{Ans.}$$

(b) *Principal stresses*

Major principal stress,

$$\sigma_1 = \frac{\sigma_x + \sigma_y}{2} + \sqrt{\left(\frac{\sigma_x - \sigma_y}{2}\right)^2 + \tau_{xy}^2}$$

$$= \frac{200 + 100}{2} + \sqrt{\left(\frac{200 - 100}{2}\right)^2 + (150)^2}$$

$$= 150 + \sqrt{2500 + 22500}$$

$$= 150 + 158$$

$$= \textbf{308 MPa (Tensile)} \quad \textbf{Ans.}$$

Minor principal stress,

$$\sigma_2 = \frac{\sigma_x + \sigma_y}{2} - \sqrt{\left(\frac{\sigma_x - \sigma_y}{2}\right)^2 + \tau_{xy}^2}$$

$$= 150 - 158$$

$$= \textbf{8 MPa (compressive)} \quad \textbf{Ans.}$$

$$\tan \theta_1 = \frac{\sigma_1 - \sigma_x}{\tau_{xy}} = \frac{308 - 200}{150} = 0.72$$

$\theta_1 = 35.75°$. The angle of major principal plane with plane of σ_x.

$$\tan \theta_2 = \theta_1 + 90° = 35.75 + 90$$

$= 125.75°$. The angle of minor principal plane with plane of σ_x.

(d) Derive the torsion formula.

$$\frac{T}{J} = \frac{\tau}{r} = \frac{G\theta}{L}$$

Enumerate the assumptions that are made in deriving this formula.

Solution: See para 10.7 and 10.8

(e) Determine the dimensions of a simply supported rectangular steel beam **6m** long to carry a brick wall **250 mm** thick and **3 m** high, if the brick work weights **19.2 kN/m³** and maximum permissible bending tress is **800 N/cm²**. The depth of beam is **3/2** times its width.

Solution:

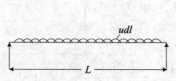

Simply Supported Beam

$$L = 6 \text{ m}$$

$$\frac{d}{b} = \frac{3}{2}$$

Uniformly distributed load.

Total volume of wall $= \dfrac{250}{1000} \times 3 \times 6 = 4.5 \text{ m}^3$

Total weight of wall $= 4.5 \times 19.2 = 86.4$ kN.

$$\text{UdL, } w = \frac{86.4}{6} = 14.4 \text{ kN/m}$$

$$= 14.4 \times 10^3 \text{ N/m}.$$

The maximum bending moment,

$$M = \frac{wL^2}{8}$$

$$= \frac{14.4 \times 10^3 \times 6^2}{8} = 64800 \text{ N-m}$$

The permissible bending stress,

$$\sigma_b = 800 \text{ N/cm}^2 = 800 \times 10^4 \text{ N/m}^2.$$

$$M = \sigma_b Z$$

$$\therefore \quad Z = \frac{M}{\sigma_b} = \frac{64800}{800 \times 10^4} = 8.1 \times 10^{-3} \text{ m}^3$$

$$= 8.1 \times 10^6 \text{ mm}^3$$

$$Z = \frac{bd^2}{6} = \frac{b \times (1.5b)^2}{6} = 0.375b^3$$

$$= 8.1 \times 10^6$$

$$\therefore \quad b = \sqrt[3]{\frac{8.1 \times 10^6}{0.375}} = 278.5 \text{ mm}$$

$$d = 1.5b = 417.7 \text{ mm}$$

The beam dimensions = **278.5 mm × 417.7 m   Ans.**

(f) A solid circular shaft transmits **75 kW** power at **180 rpm**. Calculate the shaft diameter if the twist in the shaft is not to exceed **1 degree** in **2 m** length and shear stress is limited to **50 MN/m²**. Take modulus of rigidity, $G = 100$ **GN/m²**.

Solution:

$$P = 75 \text{ kW}$$

$$N = 180 \text{ rpm.}$$

$$\tau = 50 \text{ MN/m}^2 = 50 \times 10^6 Pa.$$

$$\theta = 1° = \frac{\pi}{180} \text{ rad.}$$

$$L = 2 \text{ m}$$

$$G = 100 \text{ GN/m}^2 = 100 \times 10^9 Pa.$$

$$P = \frac{2\pi NT}{60 \times 1000}$$

$$\therefore \qquad T = \frac{60,000P}{2\pi N} = \frac{60,000 \times 75}{2\pi \times 180} = 3978.87 \text{ N-m.}$$

(a) Using shear stress as design basis

$$T = \frac{\pi}{16} D^3 \tau$$

$$\therefore \qquad D = \sqrt[3]{\frac{16T}{\pi \tau}} = \sqrt[3]{\frac{16 \times 3978.87}{\pi \times 50 \times 10^6}} = 0.074 \text{ m} = 74 \text{ mm}$$

(b) Using twist as design basis

$$\frac{T}{J} = \frac{G\theta}{L}$$

$$\therefore \qquad J = \frac{TL}{G\theta}$$

$$\frac{\pi}{32} D^4 = \frac{TL}{G\theta}$$

$$\therefore \qquad D = \sqrt[4]{\frac{32TL}{\pi G\theta}} = \sqrt[4]{\frac{32 \times 3978.87 \times 2 \times 180}{\pi \times 100 \times 10^9 \times \pi}} = 0.08255 \text{ m}$$

$$= 82.55 \text{ mm.}$$

Select $D = $ **74 mm, i.e., the smaller value.   Ans.**

Index